HUMAN RESOURCE MANAGEMENT IN HEALTH CARE

Edited by

Montague Brown, MBA, DrPH, JD
Editor, *Health Care Management Review*
and
Chairman
Strategic Management Services, Inc.
Washington, D.C. and
Shawnee Mission, Kansas

HEALTH CARE MANAGEMENT REVIEW

AN ASPEN PUBLICATION®
Aspen Publishers, Inc.
Gaithersburg, Maryland
1992

Library of Congress Cataloging-in-Publication Data

Human resource management in health care / edited by Montague Brown.
p. cm.
Collection of articles previously published in Health care management review.
Includes bibliographical references and index.
ISBN: 0-8342-0340-5
1. Medical facilities—Personnel management. I. Brown, Montague. II. Health care management review.
[DNLM: 1. Health Services—organization & administration—collected works. 2. Personnel Management—collected works. WX 159 H9175]
RA971.35.H84 1992 352.1'1'0683—dc20
DNLM/DLC
for Library of Congress
92-7173
CIP

Copyright © 1992 by Aspen Publishers, Inc.
All rights reserved.

Aspen Publishers, Inc., grants permission for photocopying for limited personal or internal use. This consent does not extend to other kinds of copying, such as copying for general distribution, for advertising or promotional purposes, for creating new collective works, or for resale. For information, address Aspen Publishers, Inc., Permissions Department, 200 Orchard Ridge Drive, Suite 200, Gaithersburg, Maryland 20878.

Editorial Services: Ruth Bloom

Library of Congress Catalog Card Number: 92-7173
ISBN: 0-8342-0340-5

Printed in the United States of America

1 2 3 4 5

Contents

vii   **Preface**
       MONTAGUE BROWN

xi   **Acknowledgments**

I   High Performance and Leadership

3   **Improving Health Care Productivity through High-Performance Managerial Development**
JOHN R. SCHERMERHORN, JR.
High–performance managerial development improves productivity by promoting better use of human resources.

11   **Career Development of Women in Health Care Administration: A Preliminary Consideration**
CYNTHIA CARTER HADDOCK AND NANCY ARIES
A report of findings from interviews of women administrators regarding their career development.

19   **High-Performing Managers: Leadership Attributes for the 1990s**
MONTAGUE BROWN AND BARBARA P. McCOOL
Successful leaders work on a variety of challenging projects and are able to play multiple roles. Aspects of effective leadership are explored.

27   **Women Health Care Managers: An Economic Update**
DEBRA L. CAPLAN, LISA LeROY, JACQUELINE M. ROSENTHAL, AND LINDA J. SHYAVITZ
A report is given of a follow–up survey on women health care managers, showing gains in salary and responsibility.

37   **Wage Differences and the Concentration of Women in Hospital Occupations**
ANDREAS MULLER, JAMES J. VITALI, AND DIANE BRANNON
The effect on wage rates of the concentration of women within selected hospital jobs is examined.

47   **Empowering Middle Managers in Hospitals with Team-Based Problem Solving**
ROBERT DAILEY, FREDERICK YOUNG, AND CAMERON BARR
This article describes how a team–oriented problem–solving procedure using management project teams was developed to improve quality of care and productivity in a private, nonprofit hospital.

II Culture, Behavior and Motivation: Tools for Change

59 Cultural Change versus Behavioral Change
RICHARD D. NORDSTROM AND BRUCE H. ALLEN
Health care culture is a powerful force in hospitals and must be taken into consideration in attempting to effect changes in employees' behaviors.

67 Using Managerial Role Motivation Theory To Predict Career Success
MAX G. HOLLAND, CAMERON H. BLACK, AND JOHN B. MINER
Managerial role motivation theory can be used to understand executive performance in health care organizations.

75 Problem Solving by Hospital Managers
TERESA M. STEFFEN AND PAUL C. NYSTROM
Analysis of problem–solving attitudes of over 100 women and men who confront complex problems in hospital settings.

83 Quality Circles: The Myth and Reality of Hospital Management
NAOKI IKEGAMI AND SETH B. GOLDSMITH
Although quality circles have been successfully introduced into U.S. hospitals, they have not fared well in Japan. It is useful to examine the reasons for Japanese reluctance to accept quality circles.

93 Human Resource Indicators for Hospital Managers
JOHN D. ARAM, PAUL F. SALIPANTE, JR., AND JAMES W. KNAUF
The concept of a human resource indicator system is developed, and methods for deriving meaning from the indicators are discussed.

101 An Integrated Approach to Board Development
ROBERT A. McGOWAN
Growing demands on hospital boards call for systematic efforts to increase the effectiveness of individual board members.

III Appraisal, Suggestions and Incentives

109 Performance Appraisal Systems in Rural Western Hospitals
THOMAS C. TIMMRECK
Performance appraisal basics have failed to reach many practicing managers.

123 Performance Appraisal Systems in Health Care Administration
MICHAEL D. WIATROWSKI AND DENNIS S. PALKON
Performance appraisal systems offer the opportunity to reward and improve productivity by identifying performance criteria and relating them to promotion, retention, and evaluation.

133 Assessing the Value of Employee Training
ROBERT BLOMBERG, ELIZABETH LEVY, AND AILENE ANDERSON
It is helpful to review the role evaluation plays in training and apply cost–benefit assessment to management development programming.

141 **Financial Incentives for Middle Managers: Pilot Program in an Inner City, Municipal Teaching Hospital**
LINDA SHYAVITZ, DAVID ROSENBLOOM, AND LYNN CONOVER
The experience of a large public hospital shows incentive programs can work in not-for-profit settings.

149 **An Incentive Program To Increase Revenue in a Public Hospital**
LINDA S. CHAN, PARK W. WAGERS, RAMONA HERNANDEZ, AND SOL BERNSTEIN
Financial reward systems can enhance revenue generation and promote cost savings in public hospitals.

161 **Hospital Cost Savings: Resembling Business**
NANCY L. DAVIS AND THOMAS CHOI
Cost savings can be realized when hospital benefits packages resemble those used in the business community.

171 **Employee Suggestion Programs in Nonprofit Hospitals**
SUSAN G. RICHER AND DAVID M. WEISS
A review of employee suggestion programs is presented, addressing factors that influence hospital employees' attitudes toward participation in such programs and the design and organization of successful programs.

IV Downsizing, Role Stress and Other Issues

181 **The Hospital Merger: Its Effect on Employees**
MARCIA K. PETCHERS, SANDRA SWANKER, AND MARK I. SINGER
The results of a study that assessed the impact of a hospital closing on employees.

187 **Substance Abuse and Mandatory Drug Testing in Health Care Institutions**
JOHN TANNER, JERRY KINARD, SAM CAPPEL, AND PETER WRIGHT
This study of substance abuse in the workplace reveals the attitudes of hospital personnel managers toward drug abuse and mandatory drug testing.

197 **Downsizing: How One Hospital Responded to Decreasing Demand**
ANNE D. MULLANEY
The downsizing experience of one hospital system is described in this case study for the purpose of sharing the lessons learned in the process.

205 **Role Stress in Hospital Executives and Nursing Executives**
GEORGE C. BURKE, III AND CYNTHIA C. SCALZI
Analysis of role stress in hospital and nursing executives reveals similarities and differences in the stress each group faces.

211 **Employee Assistance Programs in the Hospital Industry**
JOHN C. HOWARD AND DAVID SZCZERBACKI
New research shows the degree to which employee assistance programs are used in the hospital industry and ways to determine their success.

219 Commitment and Discipline in Hospitals: Leadership Protocols and Legal Precedents
SANDRA L. GILL, ERIC W. SPRINGER, AND ANDRE L. DELBECQ
Leadership protocols may be used to develop commitment to organizational goals and norms.

227 Negligent Hiring and Retention: Some Evidence of Hospital Vulnerability
JAMES W. FENTON, JR., JERRY L. KINARD, AND FRED R. DAVID
This article describes the results of a survey of hospital human resource managers that was conducted to determine their understanding of negligent hiring employment law and the tools used in employment screening.

237 Index

243 About the Editor

Preface

OVERVIEW

Health Care Management Review publishes articles that bridge the various disciplines of management such as human resources and the field of application: health care. Moreover, *Health Care Management Review* narrows its focus even more with articles that bridge the gap between the management generalist at the CEO level, the clinical specialist-manager, and the management specialist at the general management level.

The articles in this volume will appeal to health care managers throughout the top ranks of management but will be of special interest to those responsible for the human resource function who need ideas, technologies, and results-oriented information to help in their daily work. Those in teaching and learning roles who seek bridging materials that relate theory closely to practice will also want to read this volume and the others in the *Health Care Management Review* series of books of collected works from *Health Care Management Review*.

HIGH PERFORMANCE AND LEADERSHIP

"Improving Health Care Productivity through High-Performance Managerial Development" by Schermerhorn presents a theme likely to be heard throughout the industry for the next decade. The nation clearly needs its health care system to be as productive as possible and that will require the best of the people in the industry. Given the large proportion of women in this sector of our society, to do better undoubtedly means looking hard at career development of women (Haddock and Aries, "Career Development of Women in Health Care Administration: A Preliminary Consideration"), finding ways of helping women and men become high-performing leaders (Brown and McCool, "High-Performing Managers: Leadership Attributes for the 1990s"), and dealing more effectively with economic issues of employees, especially women (Caplan, LeRoy, Rosenthal, and Shyavitz, "Women Health Care Managers: An Economic Update" and Muller, Vitali, and Brannon, "Wage Differences and the Concentration of Women in Hospital Occupations"). Getting the best employees means empowerment and problem solving teams and methods (Dailey, Young, and Barr, "Empowering Middle Managers in Hospitals with Team-Based Problem Solving"). Many other articles published in *Health Care Management Review* and other books in this series deal with issues of this nature and can be found in *Physicians and Management in Health Care* and

Nursing Management: Issues and Ideas published by Aspen Publishers, Inc.

The human resource function in health care should be an exploding function in this decade *if* the work involved in high performance, leadership, quality, and productivity is carried out by the leadership in this area of management responsibility. Full control of these issues is likely to be shared with others. Given the truncated nature of health facilities and the range of disciplines (nursing and medicine to name two) that seek control over their own development, human resource managers will do well to search their own discipline for ideas and methods which can then be shared with others in a team fashion.

CULTURE, BEHAVIOR AND MOTIVATION: TOOLS FOR CHANGE

Human resource managers have many tools to approach change in health care organizations (Nordstrom and Allen, "Cultural Change versus Behavioral Change"; Holland, Black, and Miner, "Using Managerial Role Motivation Theory To Predict Career Success"; Steffen and Nystrom, "Problem Solving by Hospital Managers"; Ikegami and Goldsmith, "Quality Circles: The Myth and Reality of Hospital Management"; Aram, Salipante, and Knauf, "Human Resource Indicators for Hospital Managers"). As U.S. hospitals move to adopt "new" approaches such as total quality management, human resource executives would be well advised to look at an early contribution to this debate by Ikegami and Goldsmith who report their experience in looking at how the Japanese approach quality.

In general the titles of the articles in this section deal more directly with such methods and approaches, the first section and the ones following this draw upon a broad array of social, behavioral, cultural, and economic methods grounded in such theories and crafted for application to health care.

The range and depth of resources and their application to health care problems is further illustrated by McGowan ("An Integrated Approach to Board Development") who brings the work of our field to designing an approach for board member development! Although not employees, these very talented and important people must not, indeed cannot, be assumed to be qualified merely by being named to a board! They, too, need explicit attention; indeed they need the kind of attention that human resource people lavish on those insiders slated for development for major responsibilities.

APPRAISAL, SUGGESTIONS AND INCENTIVES

Human resource departments build the framework and methods for handling one of every employer's major responsibilities, namely, appraising employee performance (Timmreck, "Performance Appraisal Systems in Rural Western Hospitals"; Wiatrowski and Palkon, "Performance Appraisal Systems in Health Care Administration"), their training (Blomberg, Levy, and Anderson, "Assessing the Value of Employee Training"), and looking at incentives (Shyavitz, Rosenbloom, and Conover, "Financial Incentives for Middle Managers: Pilot Program in an Inner City, Municipal Teaching Hospital" and Chan, Wagers, Hernandez, and Bernstein, "An Incentive Program To Increase Revenue in a Public Hospital"), benefit structures (Davis and Choi, "Hospital Cost Savings: Resembling Business") and finding ways to get more ideas through sugestion systems (Richer and Weiss, "Employee Suggestions Programs in Nonprofit Hospitals"). Much more is involved than appears in these articles alone but many of the key issues that bedevil generalist and specialist alike are discussed here.

DOWNSIZING, ROLE STRESS AND OTHER ISSUES

Hospital mergers (Petchers, Swanker, and Singer, "The Hospital Merger: Its Effect on Employees"), substance abuse (Tanner, Kinnard, Cappel, and Wright, "Substance Abuse and Mandatory Drug Testing in Health Care Institutions") and downsizing (Mullaney, "Downsizing: How One Hospital Responded to Decreasing Demand") along with myriad other issues that create nightmare after nightmare for human resource executives. It is no wonder that *Health Care Management Review* includes articles dealing with issues of role stress (Burke and Scalzi, "Role Stress in Hospital Executives and Nursing Executives"), employee assistance programs (Howard and Szczerbacki, "Employee Assistance Programs in the Hospital Industry"), discipline and legal precedents (Gill, Springer, and Delbecq, "Commitment and Discipline in Hospitals: Leadership Protocols and Legal Precedents") and negligent hiring as evidence of hospital vulnerability (Fenton, Kinard, and David, "Negligent Hiring and Retention: Some Evidence of Hospital Vulnerability"). Had an article on early retirement among human resource administrators been available, it might well have been slotted here.

With the flowering of managed care, tougher and tougher competition among providers, hospital stays being shortened, and reduced payment for care rendered, the issues outlined here will seem lightweight as the decade unfolds.

CONCLUSION

Human resource executives in health care have a wide range of tools, theories and programs that can be used to improve the condition of employment for health care workers, their ability to perform at peak and ultimately improve the cost effectiveness and quality of care rendered to patients. The articles in this volume, and indeed throughout *Health Care Management Review*, can assist those seeking such outcomes.

—Montague Brown

Acknowledgments

Editors of volumes such as this get principal billing on the cover although as the readers quickly discover, the real feast is in the articles themselves. The real authors are those whose names appear on each article.

But it takes many people to make such articles possible. Before submission, authors typically have many others read and comment on their writings. When it arrives on the editor's desk, two or more outside reviewers are assigned to read and critique the article using criteria established by *Health Care Management Review*. These anonymous readers spend countless hours to provide insights and suggestions that often greatly strengthen the authors' initial efforts. Frequently, authors revise their articles. Once accepted at this level, Aspen editors copyedit the article to fit *HCMR* style requirements.

Thus, each article here represents the work of the named authors and the work of countless others who make substantial contributions to the final product.

This book is dedicated to the unpaid editorial board members and other reviewers who served *HCMR* during the period when these papers were originally published.

Dedicated to

Associate Editors

Barbara P. McCool, Ph.D.
President
Strategic Management Services, Inc.

B. Jon Jaeger, Ph.D.
Professor
Duke University Medical Center

HCMR Past and Present Board Members

F.K. Ackerman, Jr.
Gary D. Aden
Paul Anderson
Don L. Arnwine
Alexander Balc, Jr.
H. David Banta
Thomas W. Bice
Jan E. Blanpain
Phillip Caper
Raymond G. Davis
Carole E. Esley
A.A. Gavazzi
John R. Griffith
William Hejna

R. Mark Herring
Sagar C. Jain
Donald E.L. Johnson
Everett A. Johnson
John C. Johnson
Arnold D. Kaluzny
Frank Karel
Karl S. Klicka
Richard M. Knapp
Anthony R. Kovner
John Kralewski
Lowell C. Kruse
John J. Laverty
Samuel Levey

Stephen H. Lipson
Wallace Lonergan
James E. Ludlam
Robert O. Lunn
J. Joel May
Carol M. McCarthy
John P. McDaniels
Curtis P. McLaughlin
Matthew F. McNulty, Jr.

Robert A. Milch
William F. Moreland
Anthony T. Mott
Duncan Neuhauser
Harry A. Nurkin
Robert J. O'Brien
Nora O'Malley
Barry A. Passett
David A. Pearson

David Pitts
Dennis D. Pointer
Lawrence D. Prybil
John H. Renner
Carolyn C. Roberts
Gerald Rosenthal
Lou Rossiter
William R. Roy
Stephen M. Shortell

C. Thomas Smith
David B. Starkweather
Donald C. Wegmiller
M. Keith Weikel
Thomas P. Weil
John H. Westerman
Richard D. Wittrup
G. Rodney Wolford
Janice B. Wyatt

In addition, *HCMR* wishes to thank the many ad hoc reviewers for their exemplary service in bringing out the best in the articles presented here.

PART I

HIGH PERFORMANCE AND LEADERSHIP

Improving health care productivity through high-performance managerial development

John R. Schermerhorn, Jr.

High-performance managerial development improves productivity by promoting a better use of human resources. The approach offers well-focused strategies useful to any health care manager.

Hospitals and other health service organizations face continuing pressure to improve productivity in a dynamic and complex operating environment. Indeed, it is hard to find a group, professional, scholarly, consumer, or governmental, who does not echo this theme on a regular basis. The challenge must be successfully met by individual health care institutions and by the industry as a whole. As long as productivity problems remain, health care organizations will be criticized for failing to maintain a positive performance edge and clearly establish excellence as the benchmark of the health care delivery system.

In between the criticisms and calls for action stand practicing health care managers. These are department heads, clinic managers, service managers, unit supervisors, charge nurses, and numerous other people who work in managerial capacities in health service organizations. They are the people who share the common dilemma of having a higher-level manager who holds them accountable for task accomplishments, while being dependent on subordinates, some of whom are also supervisors, to get these tasks done.

It is in the context of this manager's challenge—being accountable for work mostly done by others—that health care managers are expected to answer the calls for productivity improvement and excellent performance. Their responsibilities for doing so are real. After all, the foundations for high productivity lie with the accomplishments of people working at all levels and in all capacities throughout an organization. Through good management, these human resources can be mobilized for productivity gains through high performance; under bad management, they are neglected and left to suffer productivity losses through low performance.

Many popular books on management emphasize the lessons executives, organizations, and industries can learn from the successes of others. The available sources include *In Search of Excellence, Iacocca, Managing, High Output Management*, and many other best sellers. Unfortunately, much of this advice for achiev-

John R. Schermerhorn, Jr., *Ph.D., is Professor of Management in the Department of Management, Southern Illinois University at Carbondale. He also serves as a management consultant for a variety of health service organizations. Dr. Schermerhorn is past chairperson of the Management Education and Development Division of the Academy of Management and a previous contributor to HCMR.*

ing productivity improvement is pitched toward senior executives and top managers. It is not so well applied to the needs of managers working at the essential middle and lower ranks.

This neglect creates a call for action. Health care managers as a whole, not just executives, deserve guidance on developing realistic and practical productivity strategies given their job responsibilities. This means that the strategies must be usable and doable at the same time that daily work requirements, things that always must be done, are still being met.

What follows is an approach to productivity improvement that begins where much of the action really is in any health care organization—with the managers. Called high-performance managerial development, this approach directs the attention of managers toward actions capable of having a high impact on the day-to-day accomplishments of people as the essential resources of any organization. The approach can be used by managers interested in their own development. It can also be easily transferred throughout an institution by proper modeling and team building to achieve a total managerial climate fostering excellent performance.

ELEMENTS OF HIGH-PERFORMANCE MANAGEMENT

Managers, in general, can benefit by developing a well-focused and systematic perspective on high performance in the work place. The following scene took place at a retreat for the management team of a medium-sized medical center. The author was a consultant to the group. He was leading a session on the productivity implications of good management and chose individual performance as a theme, focusing on the question, what can managers do on a day-to-day basis to facilitate high levels of work performance by their subordinates?

The 15 managers listened intently as the consultant asked, "Which of the following factors is the most frequent cause of poor performance by your subordinates: (1) lack of ability, (2) lack of support, (3) lack of effort?" They responded quickly and diligently. But then came the second question. It caught them off guard and caused consternation at first. Some laughed, some looked nervously at others, and some just looked down at their papers. The consultant had asked, "What is the most frequent cause of poor performance by yourself?" He offered the same choice of responses.

Once the data were in and the results tabulated, the session became very serious. The participants were surprised by patterns in the data (see Table 1) and wanted to know more. These data helped the group think more seriously and realistically about the questions.

The data in the first column confirm findings from an area of management research known as attribution theory.[1] When supervisors are asked to identify or attribute the causes of poor performance by subordinates, they more frequently focus on internal deficiencies of the individuals, such as lack of effort or ability, than on external deficiencies in the situation, such as lack of support.

The data show that the managers chose lack of effort as the most frequent cause of poor performance by subordinates. When performance is low or suffering, it seems that someone or something must be blamed. Frequently, the individual worker and his or her lack of effort are singled out as the culprits. Attribution theory predicts that managers who feel this way are likely to spend considerable time and energy trying to correct the perceived problem by motivating subordinates to work harder.

It is interesting that when the consultant's question shifted to the reasons for managers' performance problems, these managers overwhelmingly chose lack of support as the primary cause. By implication they are expressing desire for better support from

TABLE 1

MANAGER'S ATTRIBUTIONS OF CAUSES OF POOR PERFORMANCE

Causes	Frequency of cause in subordinates	Frequency of cause in themselves
Lack of ability	7	1
Lack of support	5	23
Lack of effort	12	1
Total	24*	25

* Totals vary because of nonrespondents.

their supervisors to improve their work. Their motivation to exert the necessary effort is not an issue.

The data in the two columns of the table contrast remarkably with one another. In the meeting at the retreat, in fact, these data initially startled the participating managers and piqued their interest in the training session. The unspoken question at the time, and one later raised directly by the consultant, was, why, if the managers only needed better support to improve their work performance, could the same not be said for the persons they supervised?

THE HIGH-PERFORMANCE EQUATION

The lesson behind the illustration is that when it comes to high performance, the manager must be concerned with much more than the basic motivation of people to work hard. One way to broaden managerial viewpoints is to examine what can be called the high-performance equation[2]: performance = ability × support × effort. The logic of this equation is straightforward. For high levels of individual work performance to be achieved, people must have the right abilities and support, and must be motivated to exert the necessary work effort. Without these factors, performance losses occur. By the same token, these three factors can and should be maximized in any situation for high performance to be consistently achieved. When managers recognize the importance of all three elements in the high-performance equation, the stage is well set for productivity improvements through better human resource management. The elements in the equation suggests some useful points for managerial development.

Performance begins with ability

For people to achieve high performance in their work, they must first and foremost have the requisite abilities. These establish the capacity to perform, providing the basic foundation of aptitudes and skills from which a person's performance potential can be developed and maintained.

Ability should be the first individual performance factor of concern to the manager. It is directly addressed through good employee selection, and good training and development. To achieve a true match between job demands and individual talents, a manager must choose well when hiring and retaining employees, and must provide appropriate training and development opportunities to maintain their abilities over time. An effective manager creates the capacity for high performance by staffing the jobs that need to be done with capable people.

Performance requires support

Even the most capable employee will not achieve high performance unless proper support for the required work activities is available. This is an opportunity issue. It involves creating a work environment that allows a person's abilities to be fully used.

Even the most capable employee will not achieve high performance unless proper support for the required work activities is available.

Like ability, support is a tangible and straightforward performance factor. To do their jobs well, people need good tools, equipment, and facilities; sufficient resources; proper goals and direction; freedom from unnecessary work constraints and performance obstacles; and market-competitive wages or salaries. Providing this support is a major part of any manager's job. Doing it right requires a willingness to get to know the jobs to be done, to identify what is needed by the people charged with doing the work, and to help these people obtain the best available support. An effective manager does all these things well, with the result that capable people have maximum opportunity to achieve high performance.

Performance involves effort

Effort represents the willingness to perform. It is the motivation factor that so predominates the concerns of managers and those consultants and writers who advise them about management practices. People, even capable and well-supported ones, will not achieve high performance unless they exert the necessary effort in their work, and the decision to do so rests squarely with the individual. All any manager can do is create conditions under which the answer to the question, "Shall I work hard today?" is more likely to be yes than no.

In response to this challenge, managers find a variety of well-intentioned advice on how to motivate people through proper reinforcement techniques (the more positive the better), reward-incentive schemes (the more performance contingent the better), and successful leadership styles (the more transformational the better). This advice is useful, but there is another side to motivation that can and should be tapped for additional support. The issue is competency—something that all too often is neglected as a motivational resource.

A sense of competency has motivational properties. Psychologists call this the effectance motive and relate it to the gratification of achieving a sense of mastery over one's environment.[3] Simply put, when people feel competent in their work, they can be expected to work harder at it. Competency, in this sense, is a natural turn on. It is an internal self-renewing force that motivates, stimulates, and enthuses a person without outside help.

To the extent that a manager can assist others to be or feel more competent, they can make an important contribution to their subordinates' performance potential. Interestingly enough, one way of doing so is to ensure that people have both the necessary abilities and support for the jobs they need to do. Competency comes from ability; it is a feeling that one's skills and aptitudes match the tasks at hand. Competency also comes from support; it is a feeling that the work environment helps, rather than hinders, the successful completion of assigned tasks.

Andrew Grove, President of Intel and author of *High Output Management*, calls this concept leverage.[4] By broadening attention to include actions that maximize ability and support in the work setting, the more tangible of the performance factors, the high-performance equation shows managers how to gain a substantial spillover impact on effort, the least tangible factor. When people have the necessary abilities and the proper support to do their jobs, it is likely that feelings of personal competency will motivate them to work hard and do these jobs well.

Synthesis of the elements

The health care managers at the retreat spent the rest of the evening discussing the high-performance equation and its work applications. Everyone seemed more confident in having and sharing a systematic framework for performance improvement through better human resource utilization in the medical center. They were also pleased with the group consensus that a lot of leverage on this goal could be attained if everyone worked hard to improve the levels of ability and support available in their respective areas of managerial responsibility. "After all," the consultant said in closing, "where else in management can one address such concrete factors and achieve such impact on the desired high-performance edge?"

HIGH-PERFORMANCE MANAGERIAL DEVELOPMENT

The basis for high-performance management at any level of supervisory responsibilities lies in the ability of managers to apply elements of the equation to day-to-day problems and opportunities. Health care managers can benefit from planned managerial development activities of three complementary types: (1) developing high-performance action agendas, (2) developing high-performance management teams, and (3) developing a high-performance management culture.

Developing high-performance action agendas

An important first step in the continued development of any manager is to learn how to directly influence each of the three high-performance factors. The following three action agendas offer starting points from which concerned health care managers can develop realistic, people-oriented productivity strategies. The agendas identify the fundamentals required to ensure proper coverage of each performance factor in any situation. As the foundations for managerial action, these agendas can and should be supplemented by additional strategies that take into account unique features of a given work setting.

Productivity agenda no. 1: Focusing on ability

Action objective: Staff the jobs that need to be done with capable people.
   Action strategy:
   - Analyze and carefully study all jobs in the work unit to establish good job descriptions and job specifications.
   - Communicate clear, achievable performance expectations to persons working in every job, and do so on a regular basis.
   - Identify and maintain awareness of training and development needs for persons in every job.

- Take full advantage of any and all training and development opportunities available for staff, and create additional ones where necessary.
- Communicate staff limitations and capabilities, along with suggested transfers, promotions, and replacements, regularly to higher management.
- Treat each position vacancy as an opportunity to adjust job descriptions and recruit truly capable people.
- Build a positive image for the work unit so that capable people will want to work in it.
- Study and stay informed about relevant external labor markets and communicate labor market realities, including competitive wage and salary levels, to higher management.

Productivity agenda no. 2: Focusing on support

Action objective: Help people obtain the support they need to fully use job-related skills.
Action strategy:
- Clarify task directions for all subordinates.
- Provide regular and constructive performance feedback to persons in every job.
- Eliminate or change unnecessary or inconsistent rules, procedures, and policies.
- Treat everyone fairly and consistently in applying rules, procedures, and policies.
- Ask people regularly about their work and what resources and other support they need to do their jobs well.
- Work hard to make the best possible resource support (budgets, tools, technology, equipment, and facilities) available.
- Act in ways that help create a pleasant, socially satisfying, and dynamic work environment.

Productivity agenda no. 3: Focusing on effort

Action objective: Encourage people to work hard at their jobs.
Action strategy:
- Do everything possible to maximize ability and support as positive performance factors in the work unit.
- Involve subordinates in setting performance objectives and standards for evaluating results.
- Allocate rewards and reinforcements in a performance-contingent manner whenever possible.
- Use praise and public recognition regularly as rewards for a job well done.
- Train all supervisors in the work unit to be truly competent in dealing with their subordinates.
- Keep higher management informed about performance accomplishments within the work unit.
- Be a visible, confident, and enthusiastic leader in the work place.
- Act as a model of high performance.

Developing high-performance teams

Achieving health care productivity through managerial excellence requires more than developing individual managers. It is also important to create management teams with the same high-performance edge. A second application of the high-performance equation is a framework for performance-oriented team building. A recommended approach is to have groups of managers who work together meet and proceed through the following team-building steps.

1. Gather and discuss data on managers' perceptions of the causes of poor performance by subordinates and themselves.
2. Present and discuss the high-performance equation and its rationale.
3. Conduct and discuss a performance audit of the situational status of each factor in the equation.
4. Conduct and discuss a force-field analysis of situational forces affecting each factor in the equation.
5. Use results of the performance audit and force-field analysis to develop and implement plans for performance improvements, and to evaluate performance results.

The managerial team development process begins when managers, like those in the scenario presented earlier, confront their perceptions about the causes of poor performance. This unfreezes existing attitudes and raises questions that stimulate a climate of positive inquiry. Presentation and discussion of the high-performance equation uses this climate to focus attention on avenues for potential constructive action and change.

The managerial implications of the equation are best examined in the specific context of the organization and situations shared by the participating managers. The database for this self-assessment is obtained through group work on a performance audit and force-field analysis, both of which use the high-performance factors of ability, support, and effort as points of reference.

A performance audit examines current levels of accomplishment under each performance factor. Just like the individual action agendas, the basic outline for an audit (see Table 2) can and should be enhanced to accommodate the special needs of a given situation. Answers to the audit questions establish a needs assessment that initiates discussion on ways to improve each performance factor, such as dealing with special training needs for building and enhancing appropriate managerial skills.

TABLE 2

OUTLINE FOR PERFORMANCE AUDIT

Performance factors/action objectives	Performance audit questions
Ability: Staff jobs with capable people	How well are we ● analyzing jobs? ● recruiting and hiring capable people? ● ensuring good training and development?
Support: Help people use relevant skills	How well are we ● providing clear, challenging goals? ● giving feedback? ● providing useful procedures and policies? ● formulating good job designs? ● providing pleasant work environments? ● paying market-competitive wages or salaries?
Effort: Encourage people to work hard at their jobs	How well are we ● maximizing ability and support for all people in all jobs? ● using rewards and incentives well? ● acting as models of high performance? ● acting as enthusiastic leaders?

TABLE 3

SAMPLE FORCE-FIELD ANALYSIS OF HIGH-PERFORMANCE FACTORS

Element	Driving forces facilitate performance	Resisting forces inhibit performance
Focus on ability	Good in-house training Good orientation program Available cross-training Job flexibility Good job entry rules Good reputation	Job market shortages Poor work hours Understaffing Poor job descriptions Time pressures Poor recruiting
Focus on support	Good resources Clear, written goals Clear policy manual New technology Good performance reviews Good organizational structure	Budget constraints Unclear goals Poor policies and procedures Ineffective supervisors Poor equipment and facilities Poor benefits
Focus on effort	Good role models High expectations Immediate positive feedback Monetary performance rewards Participative goal setting Advancement opportunities	Union mentality Low performance norms Inflexible reward system Unenthusiastic staff Lack of recognition Lack of top management support

Data from health care managers attending a full-day workshop on the high-performance equation and its implications.

The next step is a situation-specific force-field analysis (see Table 3). This analysis examines what managers feel actually helps and hinders them in attempts to improve the performance factors. Participants learn a lot about what is going on in their organizations and make important decisions on how to change things for the better. Because they are involved in the process, everyone is likely to be highly committed to follow through and implement any agreed-on action plans.

Developing a high-performance management culture

Much is being said these days about the relevance of culture, a shared set of beliefs and values, to organizational success. When it comes to creating a strong and positive culture, the role of the manager as leader and sponsor is singled out for its importance.[5] Developing a high-performance culture requires a way to clarify relevant beliefs and values and to reinforce them throughout an organization. The high-performance equation offers a framework that can be used to clarify such beliefs and values; health care managers represent the best vehicle for their dissemination and reinforcement.

> *Developing a high-performance culture requires a way to clarify relevant beliefs and values and to reinforce them throughout an organization.*

An important part of a manager's job is to serve as coach or teacher for others. All health care managers must accept responsibility for being good educators who ensure that their subordinates are fully capable of performing their jobs with a high level of accomplishment. When the manager's subordinates are supervisors themselves, this means helping them develop and maintain appropriate managerial skills.

One appeal of the high-performance equation is its use as a planning and action framework that the informed manager can readily teach others. At the same time that a manager directly uses the insights of the equation, subordinate supervisors should be taught and encouraged to use them too. The key terms, ability, support, and effort, should ideally become a vocabulary shared by all persons working in a supervisory capacity within an organization. When these performance-oriented words are used frequently, when their meanings are clearly shared, and when their action implications are mutually agreed on, a common performance theme emerges to unify and reinforce people in their day-to-day work. This is a building block for the emergence of a strong and positive high-performance culture in a work unit or organization.

• • •

High-performance managerial development is a useful, timely, and appropriate theme for managers in all types of health care institutions. When the focus is on the human resource component in productivity improvement, the elements in the high-performance equation can be readily used as a practical framework for performance analysis and action planning that respects managers' realities. While not exhaustive, the high-performance equation offers useful insights to any manager regardless of his or her level of responsibility. From the standpoint of managerial development, the equation and its implications can be used to develop people-oriented productivity improvement strategies that are strong in their underlying fundamentals and appropriate to specific work settings. The equation can also be used as a framework for educating others about their supervisory responsibilities, and for creating a self-renewing and productivity-oriented team-development process. When approached comprehensively in an organization, the benefits of high-performance managerial development include a strong institutional thrust toward productivity gains through better use of human resources.

REFERENCES

1. Mitchell, T.R., Green, S.G., and Wood, R.E. "An Attribution Model of Leadership and the Poor Performing Subordinate." In *Research in Organizational Behavior*, edited by B. Staw and L.L. Cummings. New York: JAI Press, 1981, pp. 197–234.
2. Schermerhorn, J.R., Jr. *Management for Productivity*. 2d ed. New York: Wiley, 1986.
3. White, R.W. "Motivation Reconsidered: The Concept of Competence." *Psychological Review* 66 (1959): 297–333.
4. Grove, A.S. *High Output Management*. New York: Random House, 1983.
5. Deal, T.E., and Kennedy, A.A. *Corporate Cultures: The Rites and Rituals of Corporate Life*. Reading, Mass.: Addison-Wesley, 1982.

Career development of women in health care administration: A preliminary consideration

Cynthia Carter Haddock
and
Nancy Aries

As greater numbers of women enter health care administration, it is important to understand their career perceptions so that organizations can assist them in becoming effective leaders. This article reports findings from focused interviews of women administrators regarding their career development and expectations and suggests issues that require attention.

Since the mid-1970s the number of women entering the field of health care administration through professional masters' degree programs has increased dramatically. Although the number and percentage of women health care managers have increased, little is known about their career development. The purpose of this article is to examine why these women chose health care administration as a field of endeavor, what kinds of career strategies they have employed, what they have experienced, and what they expect in the future.

HEALTH CARE—A WOMAN'S FIELD

Health care is a major employer of women in the United States: One of every six women who work outside the home is employed in health care. Women represent approximately 75% of the total health care labor force, and in some job categories, such as nursing and dental hygiene, the proportion of women is greater than 95%.

After a long period of absence, women have once again begun to advance into nonnursing managerial positions. As recently as the 1920s a majority of U.S. hospitals were managed by women, many of whom were Catholic sisters.[1,2] However, the number of women administrators had decreased dramatically by the 1950s. The recent advance of women into managerial positions is demonstrated by the growing number of women entering graduate health administration programs during the past decade. Women now represent a majority of graduates from such programs.[3,4]

While women have recently entered the health care management ranks in large numbers, top administrative positions are still held largely by men. The membership of the principal professional organization for health care administrators is suggestive: In 1985, 14% of the affiliates of the American College of Healthcare Executives (ACHE) were women.[5] Of the male affiliates, 38% were chief executive officers (CEOs), but only 22% of the female affiliates were CEOs. Of the 341 female CEOs in this group, the majority (65%) were

---

Cynthia Carter Haddock, *Ph.D., is an Associate Professor in the Department of Health Services Administration, University of Alabama at Birmingham. This work was completed while Dr. Haddock was on the faculty of the Center for Health Services Education and Research, St. Louis University Medical Center.*

Nancy Aries, *Ph.D., is an Assistant Professor in the Department of Health Care Administration, Baruch College/Mount Sinai School of Medicine, City University of New York.*

members of religious orders. Only 121 (35%) of the female CEOs were laywomen.

Despite the growing presence of women in health care administration, few studies concerning women in the field have been published. And, while a fair amount of attention has been given to women as they leave professional education programs and begin their careers, much less attention has been given to women's career progress five to ten years after graduation.

WOMEN AND CAREERS

Much of the research on women in management looks at work as an independent variable. Therefore, explanations of women's differential status focus on two sets of concerns: The social/psychological barriers that women must overcome if they are to achieve success, and organizational barriers that prevent women from attaining top managerial positions. These paradigmatic presentations have been characterized as person-centered versus situation-centered explanations of women's place in the organizational hierarchy.[6]

The person-centered explanations focus on two sets of issues. The first is the ways in which women may have been detrimentally socialized in regard to professional achievement. Early work by Horner indicated that women fear success and therefore limit their own professional achievements.[7] More recent studies of women's achievement focus on aspiration. Some researchers argue that women have lower aspirations than men do, while others argue that women have different aspirations than do men in regard to professional attainment.[8-10]

The other set of issues raised by person-centered researchers concerns women's skills. The initial argument made by Henning and Jardim was that women lack the requisite skills to compete with men in the business environment.[11] This work has been more fully developed by researchers who examine the areas of differential expertise. In some studies women were found to exhibit different management styles, which were judged to be less effective than those employed by men.[12]

Situation-centered explanations of women's differential achievement stress the extent to which gender bias exists within organizations. One area of concern is the implicit ways in which women's opportunities are blocked within organizations.[13] Women tend to be typecast for certain types of positions that limit their chances for advancement over the long run.[14] As a result, women do not have the same access as men to internal resources. One area that has been stressed is that women have less access to mentors than do their male counterparts.[15]

A second set of issues in situation-centered studies is the differential treatment of women. Women are expected to have a lower professional commitment than are men. They may be considered to be a bad investment by industry and treated accordingly.[16] Women are evaluated according to different standards so that women receive less feedback from supervisors than men do.[17] Women continue to be paid less than men. Strober found that the difference in pay became most apparent when supplementary compensation was taken into account.[18]

The literature on women in health care management follows the paradigm described by Riger and Galligan.[19] Initial articles reviewed the personal experiences of women who held top administrative positions.[20,21] The authors described women's frustrations, concerns, and hopes for the future. More recent articles assess women's positions in health care management. O'Hara and Abramson[22] examined the determinants of success for managers in the industry. They developed the concept of individual performance versus organizational contributions and concluded that women's attention to individual performance was the primary obstacle to their advancement. In recent studies, Caplan et al.[23,24] surveyed the members of a voluntary organization in the Boston area called Women in Health Care Management. Their articles primarily provide descriptive data on the association's membership.

Most researchers do not take into account the fact that work is not strictly an independent variable that can be detached from other areas of life in which women participate.

The approach that has been taken to the study of women in management, health care management in particular, is limited in one significant respect. The framework employed by most researchers does not take into account the fact that work is not strictly an independent variable that can be detached from other areas of life in which women participate. Despite many changes in society, women still bear primary responsibility for home and family activities.[25] While research has acknowledged that differences between men's and women's careers may exist, it is widely assumed that women's career paths should resemble men's if they are to be considered successful.[26] Incorporating the concerns of women outside the workplace could further our understanding of women health care managers and their career development by extending the person-centered and situation-centered paradigms that have appeared in the literature.

METHODOLOGY

To explore the career development of women health care managers, three focused group interviews were conducted in the New York and St. Louis metropolitan areas. The 13 participants in these group interviews were questioned about their decision to enter the field of health care management, their own career strategies, the comparability of their expectations upon entering the field with their actual experiences since that time, and their opinions as to how health care organizations might better facilitate the development of women managers' careers.

The decision to use focused group interviews was a deliberate one. Numerous studies have looked at issues of women in management, but few have examined this material in a broader social context. Therefore, an open-ended research design was chosen for the initial study in order to develop an understanding of attitudes and perceptions about these issues. Responses to the focused interviews then become a powerful tool with which to generate hypotheses for further testing and to design a structured survey instrument.[27]

The interview participants ranged in age from the early thirties through the middle forties. All held professional masters' degrees, although the degrees varied (e.g., M.B.A., M.H.A., M.P.A), and all had graduated from three to nine years prior to the time of the interviews. They were currently or had been employed in a variety of work settings: health care consulting, hospitals, health maintenance organizations, and public agencies. The majority (10 of the 13 participants) had at some time worked in the not-for-profit, voluntary hospital sector; none had ever been employed in the for-profit hospital sector.

Although the women's titles varied, in part because of differences in their organizational levels and also because the organizations in which they worked differed, all of the participants could be considered to be senior-level managers in their organizations. The highest organizational level represented was that of chief operating officer. No one in the interview groups was a CEO.

Of the 13 participants, 11 were married at the time of the interview, and 2 were divorced or separated. All had children or were expecting a child. Ages of the participants' children ranged from preschool to college age.

The participants were almost evenly divided in terms of previous clinical work experience. Six of the women had entered health care administration without a clinical background, and seven had worked in clinical areas prior to obtaining their professional degrees. Most of those with clinical experience had been nurses, although physical therapy and occupational therapy were also represented.

RESULTS OF THE GROUP INTERVIEWS

The choice to enter health care administration

Most of the women in the interview groups had entered professional masters' programs in health care administration at a time when women students were clearly in the minority. For these women it was typical to be one of only three to four women in a matriculating class. Therefore, understanding their motivations for embarking on a career in an essentially male-dominated field seemed to be an appropriate starting point for assessing their career development.

Over one half of the women in our study had had clinical experience prior to entering health care administration, and this clinical experience was primarily responsible for their choice to pursue a career in health care administration. Administration presented itself as a way to build on their clinical experience and expertise while furthering both their careers and their ability to participate in organizational decision making.

> I decided to go back to school when I was 32 or 33. I had gone the middle management route [in physical therapy] and thought there were opportunities for me to do something to get past that.

> I started out as a business undergraduate and switched to nursing. While working in a hospital I saw that my business background could help with problems in the hospital where I worked. That's how I got into the field.

At the same time, many of these women experienced frustration in their clinical positions because of their rate of compensation and their inability to influence organizational decision making.

> I was a provider, and I decided I didn't want to provide any more. I wanted to supervise, and I wanted to make more money than I was making in private practice.

> *"For me, work has been a way to have some kind of impact on things that I think are serious."*

Many of the women interviewed cited the helping, humane nature of health care as a factor in their career choice. This had clearly influenced the choices of those who had already been active in a clinical area.

> I love working in a hospital. I think that there is a real sense of family. There's a lot of things you can do for people.

The nature of health care was also a factor for those without a clinical background who wanted to enter a management career and saw health care administration as an opportunity to be a manager in a socially beneficial industry. For them the field represented a combination of business and health care.

A third career choice motivation that administrators in our study groups gave was political in nature. Especially in the case of those women who were working in the public sector, their view of health care as a political endeavor had influenced their decision to work in health care administration.

> I wanted to work in the public sector—in a public hospital or community-based program—somewhere that had to do with people in need.

Health care was seen as part of the larger social and human service network, an area in which one could effect social change.

> For me, work has been a way to have some kind of impact on things that I think are serious.

Career strategies

Early career

The women administrators in our interview groups agreed that none of them had given much thought to a "career strategy" at the beginning of her career. While those who had previously worked in clinical areas were somewhat less naive about what it meant to be a health care administrator than those without clinical experience, all 13 women spoke of having no clear plans for their career as they graduated from their professional education programs.

> At the time I graduated, I had a general feeling about where I wanted to be—a hospital administrator at a certain size hospital—but only that much.

Most of the interviewees felt that this lack of strategy was not peculiar to women but was also true of their male classmates as they began their careers. However, they believed that they and their male classmates had shared an equal commitment and desire to succeed in health care administration.

> Everyone of us felt, the women as strongly as the men, that there was a place [in health care administration] for us out there, and we'd find it.

Several voiced the opinion that their male colleagues had been more likely to regard a CEO position as the ultimate career achievement, while women were more likely to be open to other career possibilities and objectives.

> In my class, it seemed like all the guys wanted to be CEOs. The women seemed to want different types of careers.

> There was a certain portion of the class that were CEO directed and probably should have been. Of the women, a significant portion were open to other opportunities whatever they might have been.

A lack of consensus on the question of aspiration is perhaps best summed up by the interviewee who saw it as a philosophical rather than a gender-based orientation.

> I would see it as philosophical differences rather than male/female differences. Lots of the women were similarly oriented towards the goal of becoming CEO.

Later career

Virtually all of the women interviewed agreed that a career strategy had evolved as their careers progressed and that this process was seldom at a standstill. Their aspirations, ambitions, and goals, as well as the paths that seemed to lead toward them, changed as their careers progressed.

> When we came back to New York I accepted the one job that I would classify as taken strictly for career reasons. I formulated that I was going to be in health care forever, and I was interested in programmatic areas. But I insisted that if given the chance I would detour into finance, because I thought it would be good to have financial skills. That opportunity presented itself as I was looking for a job. I went to work at what had to be the most boring job I ever had.

Most felt that they were moving toward their goals, although at times "sidesteps" and "detours" were taken. All expressed a general satisfaction with their careers at the time of the interviews.

Effect of family on career strategy

To broaden the person-centered and situation-centered views of women's career achievements, we questioned the 13 women administrators about the contexts of their careers. We were interested in the effects that home and family might have had on their career progress and career strategy.

All of the women acknowledged the significant effect of family on career strategy as well as on the day-to-day realities of their careers. For some this had meant basing career decisions on a spouse's career moves. For others it had meant a temporary withdrawal from career or part-time work. But whether women had been working or raising children, their efforts to combine family and career had not resulted from systematic decision making.

> I never gave a minute's thought to how I would negotiate a career and child. I was totally focused on a career and it just happened that I was lucky to find myself pregnant at a good point in my career.

> It didn't occur to me that I was going to combine a family and a career. I already had the kids and the only thing I had to deal with was my career. I don't think it was an active decision-making process.

The confidence that one can combine career and family is also reflective of the times. There were few role models to whom these women could refer when assessing what had influenced them to meet the challenge. Furthermore, the rhetoric of the women's movement asserted that women should work outside of the home.

> I think I was naive. It was the era when "I can do it all." I didn't have to make choices. I was going to get married, I was going to have children, and I was going to have a career.

Once pregnant, a number of the women had regarded themselves to be under extreme pressure to prove their professional commitment. They were being judged according to the actions of all of the women who had preceded them. They felt compelled to counter the stereotype of the professional woman who promised to return to work six weeks after her baby was born but changed her mind once she was home with her child.

> I was at the Medical Center four years working for the same people. I had gotten three promotions. I was the hottest thing since sliced bread, and all of a sudden when I got pregnant the assumption was that I was going to leave. I was astounded that they would assume that. All the other women who had ever worked there had left when they were pregnant.

> When people first heard I was pregnant they said, "M's pregnant—she's not coming back." I kept saying, "I'm coming back. I'm really coming back." I worked the nine months before I delivered. I was showing everybody I was serious, and that was part of my reason for going back so early.

A number of the women also felt that having children had helped them to become better managers by giving them a healthier perspective on what was required to manage effectively.

> I think that having children helped me mature as a manager. Part of managing well is making decisions and living with them. When I was younger, I had to make all the right decisions. Now I'm not as obsessed about work.

Children also enabled these managers to work more effectively with their own staffs.

> I'm a better boss because I'm a mother. Before I had a child, if somebody came into work an hour late because they had to take their baby to the pediatrician, the first thing that crossed my mind was "Oh God, can't you do that on Saturday?" Now I understand, and I'm a little bit more liberal.

All of the women agreed that the responsibility for home and children had to be balanced with work obligations. This extra set of activities was perceived to

---

Once pregnant, a number of the women had regarded themselves to be under extreme pressure to prove their professional commitment.

---

weigh disproportionately more heavily on the women than on their own spouses and most of their male colleagues with families.

> My husband is most cooperative. If I say, "Our son called from the Tetons. His camera dropped down the mountain. Buy a camera and send it to him." "No problem, honey. Glad to do it!" But it is still my responsibility to figure it out. That's the point. I have a staff assistant called my husband. He will take care of certain specific things but he doesn't have to think about it.

Women also expressed regret about activities that they had to forgo. Some considered working in health care to be equivalent to community work, but others saw a need for broader social involvement and found time to be a constraining factor.

You didn't ask about other obligations besides work and family. I have been able to maintain these obligations, but minimally. I'm active in a housing group, but I'm beginning to say I just can't do it all. It's with great reluctance, too, because I could have a much bigger impact on that housing group than I can on a lot of other things.

All of the women agreed that the most significant sacrifice that had resulted from the pressure to balance work and family obligations was their loss of personal time.

You can't be a wonderful mom and a wonderful wife and a wonderful executive and a wonderful friend and all the rest. There aren't enough hours in the day.

The issue wasn't so much whether I could do it, but that I was always tired. I felt that the person who suffered was me. My family was always fed; the house was always clean; the job was always done. There just wasn't enough time to take care of myself.

Even with the stresses and strains of work and home, several interviewees mentioned the satisfaction of making it all balance. (One woman added that this balancing of work and family was perhaps not so important to male colleagues' satisfaction levels.)

[Work] balances and my family balances, so I'm satisfied. And I'm happy right now. I'm satisfied with where I've been and where I'm going.

Career expectations

The highest organizational level represented among the 13 women administrators interviewed was one step from the top management position—chief operating officer. All of the women were committed to their work, including the individuals who had chosen to take a temporary leave from career or to work at home. Long hours were common to all of the women.

My day runs from 8:30 to 6:00. Kick in an hour or two at home sometimes.

I work 70 to 80 hours a week. Without the paperwork I do at home, I put in about 55 hours a week.

Despite their efforts, virtually all of the women believed that there were barriers that would prevent them from reaching the very top level of management.

Despite their efforts, virtually all of the women believed that there were barriers that would prevent them from reaching the very top level of management. Talent and hard work could not overcome existing administrative biases. While none used the term, they all agreed that there was a "glass ceiling".[28]

I've got an MBA. I've been working for 12 years. I have been a vice president of two major institutions. What else do you want? I'm not saying I should get the job, but I should get in the door.

The barriers that most of the women identified were related to gender.

I had expectations that the world would be a little fairer—that the bias against women was out there, but that I was not going to find it. It was a harsh, cruel slap in the face.

In addition to general perceptions, accounts were given of more subtle obstacles. One concerned the stereotypes of female managers.

I sat down with the president of the hospital and asked why I didn't get a certain job. He said, "You were one of the two final candidates. But you have a reputation for being aggressive, and we weren't sure that we were ready for someone who was as aggressive as you."

Another obstacle involved simultaneously satisfying the expectations for an executive and those for a woman. This was seen as a particularly formidable obstacle in organizations where many spouses of male executives had chosen the more traditional role of working only in the home and always being there to support husband and family.

Governing boards composed primarily of older men with more traditional spouses present one more source of misunderstanding for women administrators. Having to look to such individuals for support and organizational rewards, reinforces women's sense of being at a disadvantage and apparently raises doubts about the likelihood of fair treatment in the employment situation.

Even though the composition of boards is changing, I have seen no impact of their inclusion of women on the hiring of the CEOs.

Interestingly, the women interviewed regarded this situation as a fact of life, and several doubted that the situation would improve from a general perspective or from a woman's perspective.

I don't think the chances of getting a CEO position in an urban medium-size hospital are nearly as great as I did five years ago. That's the reality of an oversupply of administrators and cutbacks in the number of hospitals and administrators.

In our generation I think it's going to get worse, not better, because there will be lots of women at one level and then that will be it.

While most of the interviewees saw little likelihood that many women would achieve CEO positions in the near future, there was some hope that women would attain these positions eventually. It was not clear whether or not these women believed that the increased number of women health administration graduates would be enough per se to change career achievement expectations for women. However, they concurred that to ignore the influence of women as leaders, service providers, and consumers in the health care field would be a mistake.[29]

CONCLUSIONS

Our interviews with 13 senior-level women health care administrators made it is apparent that these women did not perceive themselves to be different from male health care administrators in terms of motivation, commitment, or talent. Those who had chosen to leave their careers temporarily or to work part-time understood clearly the trade-offs and sacrifices with respect to ongoing involvement in organizational decision making and career advancement.

Despite similar initial intentions, these women perceived two key areas of difference between their career experiences and those of their male colleagues. An understanding of these perceptions is crucial to any study of career development, because individual perceptions play a key role in the definition of "career."[30]

The first area of difference is the effect of family and home on career. Women in our interview groups clearly felt that family responsibilities fell disproportionately to them. Thus they were presented with a set of obligations and activities not faced by most males. While most of the women perceived that they had derived benefits from combining career and family, the stress that it added to their lives should not be minimized.

A second area in which women saw themselves as differing from male colleagues was the ultimate organizational possibilities available to them. All 13 women believed that definite limits to their potential career advancement exist. The possibility of attaining CEO positions, for example, was perceived to be severely restricted by a variety of gender-related obstacles. Others have reached the same conclusion concerning women's corporate possibilities in general.[31]

If health care organizations are to meet the challenges posed by an increasingly turbulent and at times hostile environment, they must seek out and develop top management talent. In doing this, they will have to address the issues that women face, because the pool of new management talent now includes many women.[32]

Although women have entered the managerial ranks of U.S. businesses in considerable numbers, there is evidence that their careers do not advance at the same rate as those of men.[33] It is unclear whether or not this is the case in health care administration specifically; however, there are indications that barriers may exist, especially as women approach the top level of organizations. In view of our society's commitment to equality of opportunity, we cannot allow this issue to be obscured by the increasing number of women entering the field. It is not at all clear that women's advancement to upper-level management positions will accelerate simply because more women enter the field.

As a starting point, we must become more responsive to the issues of work and family. The women we interviewed had made very clear choices to balance managerial careers with family responsibilities, asking little help from the organizations in which they worked. Because not all women will make a similar choice, there is a risk that capable women managers will be lost at the point in their careers when they are ready to move into top managerial positions. Although the women interviewed were skeptical about the feasibility of alternative work forms for health care administrators (e.g., job sharing, part-time work), especially at higher managerial levels, there must be a commitment to explore their use. Assistance with obtaining safe, affordable child care is an option that should also be explored.

The challenges faced by women in health care are not an issue for women alone but for men as well and for society as a whole.[34] The Public Health Service Task Force on Women's Health Issues recommended that a national goal be established to "increase, as rapidly as possible, the number of women in key positions in health practice, administration, research, and education."[35] Attention to women's career progression in health care administration is critical as large numbers of women enter the field and as other women weigh their career options. In fairness to the profession and to those it serves, we must ensure that health care organizations have the best leaders available.

REFERENCES

1. Rosner D. "Heterogeneity and Uniformity: Historical Perspectives on the Voluntary Hospital." In *In Sickness and in Health: The Mission of Voluntary Health Care Institutions*, edited by S.D. Seay and P. Vladeck. New York: McGraw Hill, 1988.
2. Neuhauser, D. *Coming of Age*. Chicago: American College of Hospital Administrators, 1983.
3. Korn-Ferry & Association of University Programs in Health Administration. *Health Administration Employment: A Survey of Early Career Opportunities*. New York: Korn-Ferry International, 1987.
4. Plant, J. "MHAs and the New Hospital Job Market." *Hospitals* 59, no. 5 (1985): 80–86.
5. American College of Healthcare Executives. *A Biographical Dictionary of the Membership*. Chicago: ACHE, 1986.
6. Riger, S., and Galligan, P. "Women in Management: An Exploration of Competing Paradigms." *American Psychologist* 35 (1980): 902–10.
7. Horner M., "Toward an Understanding of Achievement-Related Conflicts in Women." *Journal of Social Issues* 28 (1972): 157–76.
8. Harlan, A., and Weiss, C. "Sex Differences in Factors Affecting Managerial Career Advancement." In *Women in the Workplace*, edited by P. Wallace. Boston: Auburn House Publishing Co., 1982.
9. Strober, M. "The MBA: Same Passport to Success for Women and Men?" In *Women in the Workplace*, edited by P. Wallace. Boston: Auburn House Publishing Co., 1982.
10. Fox, M.F., and Faver, C. "Achievement and Aspiration: Patterns Among Male and Female Academic-Career Aspirants." *Sociology of Work and Occupations* 8 (1981): 439–63.
11. Henning M., and Jardim, A. *The Managerial Woman*. Garden City, N.Y.: Anchor Press, 1976.
12. Rosen, B., and Jerdee, T. "The Influence of Sex-Role Stereotypes on Evaluations of Male and Female Supervisory Behavior." *Journal of Applied Psychology* 57 (1973): 44–48.
13. Kanter R.M. *Men and Women of the Corporation*. New York: Basic Books, 1977.
14. Epstein, C.F. "The New Women and the Old Establishment: Wall Street Lawyers in the 1970s." *Sociology of Work and Occupations* 7 (1980): 291–316.
15. Harlan and Weiss, "Sex Differences."
16. Rosenfeld, R. "Women's Occupational Careers: Individual and Structural Explanations." *Sociology of Work and Occupations* 6 (1979): 283–311.
17. Harlan and Weiss, "Sex Differences."
18. Strober, "The MBA."
19. Riger and Galligan, "Women in Management."
20. Appelbaum, A.L. "Women in Health Care Administration." *Hospitals* 49, no. 18 (1975): 52–59.
21. Friedman, E. "Women CEOs: They're Good for the Field, but Is It Good for Them?" *Hospitals* 54 (1 February 1980): 45–48.
22. O'Hara, C., and Abramson, F. "Women in Health Care Management." *Health Care Supervisor* 1, no. 3 (1983): 35–44.
23. Caplan, D.L. et al. "Women Health Care Managers: An Economic and Employment Profile." *Health Care Management Review* 9, no. 2 (1984): 29–38.
24. Caplan, D.L. et al. "Women Health Care Managers: An Economic Update." *Health Care Management Review* 13, no. 1 (1988): 71–79.
25. Geerken, M. and Gove, W. *At Home and At Work: The Family's Allocation of Labor*. Beverly Hills, Calif.: Sage Publications, 1983.
26. Strober, "The MBA."
27. Hisrich, R., and Peters, M. "Focused Groups: An Innovative Marketing Research Technique." *Hospital and Health Services Administration* 27, no. 2 (1982): 8–21.
28. Morrison, A.M., White, R.P., and Van Velsor, E. *Breaking the Glass Ceiling: Can Women Make it to the Top in America's Largest Corporation?* Greensboro, N.C.: Center for Creative Leadership, 1987.
29. Dempsey-Polan, L. "Once and Future Leaders in Health Administration." *Hospital and Health Services Administration* 33, no. 1 (1988): 89–98.
30. Hall, D.T. *Careers in Organizations*. Pacific Palisades, Calif.: Goodyear Publishers Company, 1976.
31. Morrison, White, and Van Velsor, *Breaking the Glass Ceiling*.
32. Spruell, G. "Making It, Big Time—Is It Really Tougher for Women?" *Training and Development Journal* 39, no. 8 (1985): 30–33.
33. Ekstrom, R.B. "Women in Management: Factors Affecting Career Entrance and Advancement." *Selections* 2, no. 1 (1985): 29–32.
34. Navarro, V. "Women in Health Care." *New England Journal of Medicine* 292 (1975): 398–402.
35. Public Health Service Task Force on Women's Health Issues. "Women's Health." *Public Health Reports* 100, no. 1 (1985): 73–106.

High-performing managers: Leadership attributes for the 1990s

Montague Brown
and
Barbara P. McCool

Entrepreneurial, innovative, energetic, community-oriented leaders are needed for health care enterprises. Successful leaders work on a variety of challenging projects and are able to play multiple roles.

Wanted: entrepreneurial, innovative, energetic, community-oriented leaders for health care enterprises. If we are to prosper as a profession, the leaders for tomorrow need to be identified and developed today.

What kind of superman or superwoman is needed to tackle successfully the growing complexities of our health care enterprise? Can we find talented people who have the energy, conviction, vision, and attention to detail necessary to get the job done while remaining flexible about the future?

Over the past four years in both informal and indepth interviews we have talked with today's health care leaders about the attributes and characteristics needed by future leaders. We interviewed people whose accomplishments we had read about, and we received recommendations from members of the *HCMR* Editorial Board, journalists, and people interviewed at earlier stages of the project. In this process many more people were identified than we could include in our project.

Our method of selection and time and energy allotted to each interviewee represents more the methodology of the journalist than of the scientific investigator. Our selection was also biased toward people in their early thirties and forties. We deliberately set out to get as much information as possible from people we thought were the high performers of today and most likely to be the leaders of tomorrow. While we consulted many older leaders in health care our major focus was on those likely to move into top leadership positions in the industry, not those currently there. We also focused heavily on people working in hospitals and hospital systems.

People working in psychiatric care, long-term care,

Montague Brown, *M.B.A., Dr.P.H., J.D., is Chairman & CEO of Strategic Management Services, Shawnee Mission, Kansas, and Editor of* Health Care Management Review.

Barbara P. McCool, *R.N., M.H.A., Ph.D., is President of Strategic Management Services, Shawnee Mission, and Associate Editor of* Health Care Management Review.

This article was distributed to the participants in the roundtable discussion on which the next article is based. They were asked to prepare additional remarks that would expand the discussion beyond the issues presented here.

The authors thank David K. Hamilton, C.P.A., President of Strategic Ventures Group, and Shawnee Mission, Kansas, and Sherry R. Larson, M.T.(ASCP), M.P.A./H.S.A., Consultant for Strategic Management Services, Shawnee Mission, for their assistance in preparing this article.

managed care, and other aspects of health care are underrepresented in our interviews. However, we are confident that excellent leaders are developing there. In fact, they have much in common with the people in more traditional hospital roles with one exception: Because they are often in the new modes of care, these leaders are less likely to defend the status quo of community hospitals in the overall scheme of health care delivery.

Our interviews benefited from years of reading health care and business articles on management as well as recent and classic books on general topics, and, more recently, from working with high-performing people in business, law, engineering, and science.

We set out to address the issue of leadership in health care in a simple yet comprehensive fashion. In this article, we have logically grouped together the key attributes while providing more labels for related attributes. We describe the personal characteristics of high performance and the most likely attributes of tomorrow's health care leaders. While few individuals are likely to possess all of them, most high performers do strive to move in this direction. We have also identified some of the reasons these attributes appear desirable in the context of health care.

TRAITS OF A SUCCESSFUL LEADER

Healthy, energy giving, hard working

The health care environment is constantly bombarded with ideas, issues, and pronouncements of massive change. This can produce inertia and psychological shutdown if administrators do not keep pace with the cognitive overload. It is essential that administrators present themselves to employees, physicians, corporate officials, and others as being in charge and on top of things. Corporate changes, including the development of new major marketing functions, managed care programs, and other innovations add complexities to the job of knowing where to get information, support, and coordination within and outside the system.

All this complexity makes it essential that administrators have sufficient energy to present a strong countenance to all the parties who depend on them for leadership. This is extremely time consuming because of the number of people who need assurance, information, and decisions. It means hard work, often concentrated on separating the important from the trivial. All this must be done amidst the high noise level from numerous information sources including ever higher levels of dissonance within the medical community.

The health care culture should stress good physical conditioning. Like big league athletes, administrators must be fit for action in the ninth inning or the last 30 seconds in the fourth quarter. They cannot expect a fresh substitute in midstream. After the game, they must preside at a department head meeting to go over new changes in strategy. Then they will freshen up before going to dinner to help sell a prospective board member on joining a local group or heading up a new program for the hospital.

At the same time, administrators must be patient like the angler who knows what works and is not tempted to try every new lure offered by an overzealous salesperson. Keeping calm while others panic is helpful and essential.

One might draw on the analogy of the poker player who moves through the game knowing when to bet, when to hold, and when to fold. Health care is not a game of chance, but there are times when it is similarly difficult to sort the good moves from the bad and to know when to hold instead of shifting strategies precipitously in the middle of the game. The energy level has to be high to properly analyze options and sometimes to have the strength to withstand the temptation to jump with every new noise in the dark.

People who have such traits can be recognized and can be developed by health care leaders who possess these traits themselves. These potential leaders will need seasoning in a variety of situations before taking on the big role. This is the true meaning of mentoring.

The successful administrator will be calm in the middle of a stormy environment. This person will be making decisions, working long hours, and seeking the right combination of moves to keep the organization on target.

The successful administrator will be calm in the middle of a stormy environment. This person will be making decisions, working long hours, and seeking the right combination of moves to keep the organization on target.

These traits must be blended with a sense of time and history. The leader may be identified and rewarded in the short term but his or her ultimate success will be a result of working meaningfully in a 3-, 5-, 10-, and 20-year time frames. People with vision, rooted in the future, win many accolades for short-term specific performance because they are prepared to recognize opportunity everyday and they realize the need for changes as the environment shifts.

Creative, intuitive, innovative

The best managers are open to new possibilities, especially in highly complex situations. They engage in scenario-building mental exercises, asking themselves and others many "what if" questions. They have the ability to see many possibilities and the patience to keep their options open until they literally envision the appropriate choice of action. Clearly, these leaders are open to the less obvious, the idea from left field, the "eureka" factor.

They are also calm in the presence of the ambiguous situation before the light clicks on and every piece of the puzzle is clear. Tolerating dissonance is the hallmark of creative managers. They conceptualize well and draw from their previous learning experiences.

Mission oriented, entrepreneurial, visionary

Innovation, creativity, and similar attributes alone can lead to disruptive and even destructive behavior. At times new ideas do come out of the woodwork to energize an organization. But, for the long haul, it is creative thinking within the control of a truly mission-oriented person that moves organizations forward.

This thinking apparently starts from a sense of where the executive wants to go or what he or she wants the organization to be when it is fully mature. This is mission-oriented behavior. It starts with the fundamentals and analyzes alternative futures and actions in terms of how they can contribute to the more general and fundamental concept of the total enterprise. This touchstone in vision guides the successful executive. But it is a sense of the possible linked to a willingness to act, to bet the store, that separates idle dreaming from wholehearted commitment.

In many health care organizations, this often means going with one's gut feeling, telling others about it if it works and burying it when it does not. However, a few buried failures do not diminish the organization's level of energy. The overall batting average is the true test of success, not performance during a single time at the plate.

TYPES OF LEADERS

The successful leader integrates various forms of leadership into an effective style.

Networker, boss leader, resource builder

Hospital administrators operate with less formal authority than almost any managers in the world. Hospital managers are guided by professional norms other than their own. Their biggest customer-worker is the physician who reports to no one and maintains competitive hospital privileges as an ongoing strategy. In voluntary hospitals, the administrator will lead and manage the board. In investor-owned hospitals, where the administrator does not have a local fiduciary board to direct his or her actions, he or she has a corporate body with many centers of power and influence guarding the resources needed to do the job well. Even health professionals who are employees affiliate strongly with their unique professional norms and goals.

Health care, like no other field, rewards the diligent network builder: the man or woman who cultivates relationships and who seeks and provides assistance to multiple consultants. The network builder is one who weighs and balances perspectives to ensure that all relevant stakeholders know he or she cares and will be there for them just as he or she expects they will be there as needed. As the field becomes more complex and as corporate relationships and power shift, the skill of the diligent networker will be essential. He or she understands what is going on and is unlikely to be surprised.

Today's successful hospital leaders excel at networking upward with the boss. They know their bosses' needs and approaches and work to keep their bosses informed about their own needs. Put another way, good subordinates manage their bosses. Bosses must be told, diplomatically perhaps, when something does not fit, but they must be told simultaneously what will achieve the desired end. Since each boss must do this for his or her boss or board or the board's constituents, one can characterize health care as management from the bottom up, while breadth of vision is shared from the top down. The vision leads the manager while the manager leads the visionary. It is hard to specify this situation as a trait or characteristic, but, as Supreme Court Justice Potter Stewart once said about obscenity, "I cannot define it for you, but I know it when I see it." So too, as chief executive officers read this, they will remember those who have worked for them and have managed them well. Those who manage their bosses well will ultimately become the leaders of organizations and will work to manage their various constituencies in a similar fashion.

The ideas, resources, and power to get something done in complex situations come from an increasingly wide variety of sources. The successful leader-administrator is acutely aware of that fact and spends seemingly endless hours cultivating those whose energy and acceptance will be necessary to get the work done. This is not a one-shot task. In fact, many of those interviewed have spent years working to build, maintain, and utilize personal relationships throughout their hospitals and communities, and in the field. This fact underscores the need to develop leadership from within rather than continuously bringing in talent from the outside. On the other hand, new developments in tertiary care centers, marketing, insured products, and other such factors make it essential to bring in outside experts who have new ideas and can bring insiders up to speed more quickly.

The successful manager knows whom he or she needs and has the contacts and networks to bring them in as needed. He or she can orchestrate a mind-boggling array of varied interests in much the same way that a skilled golfer senses the opportunities to make a difficult shot and executes it, while all the spectators wonder how he or she planned such a complex maneuver. The professional executes the shot but only after hard work, study, careful cultivation of interest, and practice. Managers who perform well in networking do so in large part because they are trusted and are expected to be fair in their dealings with other members of the networks.

In talking to leaders, one often gets a sense that what they are talking about is present reality, when in fact it is often a rich imagery of the future. Their constant attention to day-to-day reality reminds us of the work of the sculptor or painter who tinkers, dabs, selects paints, works over, and brings life to a dream. But even as he or she works, the dream shifts and takes on newer realities. Thus we see not a completed master plan but a dream that in its unfolding makes leaders often look like folks just muddling through. In fact, management literature over the years has reflected advice to construct master plans while behavioral studies of reality look like messy muddling through. Master planning may indeed feed dreams but it is dreams that feed these muddling processes and ultimately lead to successful management outcomes.

These skills come from hard work, unceasingly considering alternatives, and having the courage to act when the timing is right. With the market changing dramatically, the dreams of leaders in the field need to be fed from a broader array of ideas and resources than have traditionally fed the hospitals and hospital chains of primary and secondary care hospitals.

Analyzer, synthesizer, evaluator

Not many leading administrators pore endlessly over statistics and number combinations. Still, they are not afraid of numbers: Indeed, those less comfortable with quantitative analysis tend to have quite capable financial and marketing analysts on their team. A number of good administrators have served as controllers in early jobs and retain a keen sense of the value to be derived from analysis of operations and potential returns on investment.

More than numbers, leaders seem to enjoy sampling and testing ideas, using conceptual and analytical constructs of various disciplines. This is a form of

---

More than numbers, leaders seem to enjoy sampling and testing ideas, using conceptual and analytical constructs of various disciplines.

---

play, exploration, and enrichment of one's own thinking. The integration and cross-fertilization of ideas, concepts, and data points lead to more creative approaches to problem solving and to path finding—the direction one takes to serve the organizational mission.

At times such leaders may seem to be out in left field compared to those wedded to a particular conceptual-analytical perspective. More likely, these leaders have discovered a synthesis that transcends any particular perspective and have incorporated that synthesis into their personal world view. These quantum leaps can likely be explained in a logical, sequential manner only after the fact, and then incompletely.

Implementer

Beautiful thoughts, well formed, seldom finish the job. Action must follow. More thought must follow initial action, planning, organizing, and directing. Management must continually be brought into play with and through other people. Openness to input from others throughout the implementation process contributes heavily to successful accomplishments.

People management, problem solving, and dogged follow-through to ensure correct action remain essential elements of management leadership today. Every successful manager we have met either does these well or has teamed up with someone who complements his or her own lack of skill or time to carry out these functions.

Successful implementers remain optimistic champions of ideas even while appearing to be hopelessly bogged down in transition problems.

People-oriented caregiver

Basically, people seek excellence in service when they use a health care facility. Ultimately, every new service and every change in operations occurs because people can identify with, accept, and believe in what we do because we care about other people, employees, patients, and visitors—the people of the communities we serve. This is implicitly and explicitly on every agenda. It is the touchstone of every action.

As multihospital systems have grown, the key issue raised by opponents has been whether this basic orientation to service will be lost. Leaders of high-performing health care organizations keep sight of this issue, even as they balance the interests of staff, top management, stockholders, regulators, and other important stakeholders. As one leader in the investor-owned sector aptly put it, his firm brought sound business practice to sound clinical practice provided by caring people serving patients. Any component without the others ultimately fails. Integrity and ethical behavior are integral to health care.

Every successful, high-performing manager in every industry knows that delivery of value to customers ultimately separates winners from losers. The high performers in our interviews know it too, and they keep that concept in the forefront of their vision of their firm for the future and for the day-to-day work of the organization.

HIGH-PERFORMING MANAGERS AT WORK

Besides the high-performing characteristics successful managers exhibit, these leaders have definite work tendencies.

Using all opportunities

Innovative, open to change, hard working, moving toward goals driven by an internalized vision of mission and commitment describe the successful manager. This individual probably has a degree in management, and has spent years building a knowledge and experience base working in the health care field and loves it dearly. This person respects the rights and integrity of others and knows how to stimulate people in conflicting situations and bring out the best ideas to fit the inner vision of excellence.

Advancement occurs fairly routinely, but major growth can and does come without any apparent change of place or formal position. Some yearn to travel and do a wide variety of things, while others never cease to grow in the soil of their origin. Growth takes many forms. Understanding this reality will not take the fact of being in one place as a mark against leadership ability and the potential for other things. Given the overall complexity of the field and the necessity for leaders to build the networks or root systems in the soil of their organizations, moving leaders around often can broaden, but it can also lead to an alienation from the deep rooting behavior that seems so much a mark of success of those interviewed in this process.

Not taking no for an answer

Leader-administrators are deeply concerned about the movement of market forces, while at the same time loyal to the traditional customers, the physicians. They have probably tried everything mentioned in the professional literature as well as business and health publications far afield from the more traditional sources. But they continue to scan the possibilities for other skills that will aid them in getting ahead in the marketplace. They have options and probably have exercised many, while holding onto the possibility of more successful moves within their current organizations. They wait while moving ahead. They understand that sometimes one takes one step forward and two steps back while trying to get something right. While leader-administrators may be disappointed that first attempts fail, they help others around them to stay the course and get ready for the next phase or stage of development.

By contrast, the less successful administrators remember the failed attempt but keep wondering when the magic answer will be sent down to resolve their local problems. In the meantime, they sympathize with the principal customers, the physicians, who then wonder if they should not jump ship to stay ahead.

Readying the hospital for all contingencies

Leader-administrators already have a sense of the business community. They build more ties and strive to be visible to this market segment, trying to serve this community's needs while increasing awareness of the potential for more direct dealings with the hospital. These leaders recognize that they are responsible for a major community enterprise and have aggressively inserted themselves into broader community organizations and networks. Again, by contrast, the more pedestrian administrator is not involved in the community, nor is he or she known much outside the hospital's direct areas of operation.

Increasingly, leader-administrators are knowledgeable about managed care systems. They know people who are managing such systems, keep up with their growth, and think about how such systems will eventually be integrated into a more holistic system of services. Moreover, they know that down the road they will have major responsibilities for this element of health care, just as today they have a major leadership role in the hospital sector. Leader-administrators will find ways to bring these components together when their market becomes ready for it.

Less adept administrators do not know much about these things and think either that managed care systems will not develop or that they will develop but will be outside their sphere of influence, perhaps resting in the insurance division of the company or some other arrangement.

Leading vertical integration efforts

Leader administrators are positioning their organizations to play a major role in the coming vertically integrated system of health services. They do not know just how their hospitals will approach all the issues involved, but they see elements of the system coming together and work continuously to bring their teams up to date on possibilities. Furthermore, they build new programs that make their organizations the natural focus of such efforts. They have vision; confidence that they can get things done; working knowledge of the latest movement toward their objectives; and a host of networks conditioned to help make it possible when the time comes.

These people have a far-reaching view. Current acronyms and maneuvers by competitors do not ruffle their feathers so much as they offer grist for scenario-building exercises. From time to time, they add programs, move programs around a bit, or call in some corporate or consulting talent to flesh out the options agenda. When opportunities exist, they act.

Others wonder about whether their hospitals can survive the downward pressures. They want new programs to provide relief from pressures. For example they want a new marketing focus, something directed at women, or a health assessment unit. They shrink from the reality that every new day brings at least ten new challenges.

Leaders dream great dreams. They lose some of their battles, but they can distinguish the battle from the war. Most of all, leaders are courageous. They are not like the weak administrators who become intimidated at the mere sound and fury of marketing rhetoric from competitors.

Welcoming competition

Competition makes leaders work harder, but external competition never seems to be a more powerful driving force than leaders' own dreams and ambitions, which come from an internal sense of mission.

Competition makes leaders work harder, but external competition never seems to be a more powerful driving force than leaders' own dreams and ambitions.

Administrators shaped by competition are not leaders. But administrators with a keen sense of the competition can and do deal effectively to stay ahead most of the time, and even when falling behind know how to regain lost ground. Leaders do not always win the round but they rarely lose the game. Nor are they likely to allow competitors to define the game. Performing well and aspiring toward excellence are uppermost in the thoughts of all high-performing managers.

Being responsible and acting accountably

Leader-executives do not ultimately point the finger at higher authorities. Instinctively and maturely they know and act as though responsibility stops at their desks. Similarly, leader-executives act in an accountable fashion with the many who have a stake in almost every major action of the health care enterprise—patients, employees, physicians, owners of the enterprise, and the community.

• • •

There is no single trait or characteristic that identifies an outstanding leader. However, the presence of several of the characteristics described probably reveal a leader in action.

The composite developed here is imposing. No one person has ever displayed all of these traits. But those who have most of these characteristics are individuals who consistently aspire to do better, and that striving for excellence in performance, perhaps more than all else, is critical in looking for outstanding performers in health administration today.

Women health care managers: An economic update

Debra L. Caplan,
Lisa LeRoy,
Jacqueline M. Rosenthal,
and
Linda J. Shyavitz

In 1980, a survey of women health care managers was performed that documented their career status at that time. This article reports the results of a follow-up survey, done five years later, demonstrating gains in salary and responsibility.

In 1980, a survey of women health care managers was conducted, the results of which were published in *Health Care Management Review*.[1] The article, an economic and employment profile of female health care managers in Massachusetts, showed that women health care managers were earning excellent salaries at a young age, thought that they had opportunities for advancement, and were clearly committed to careers in health care. A second survey was conducted five years later to gain further understanding of—and updated information about—the job characteristics, salaries, and professional status of women health care managers, and to review the progress they had made.

METHODOLOGY

The survey was mailed on January 25, 1985, to approximately 400 women health care managers. The respondents were identified from membership lists of three Massachusetts health care management organizations: Women in Health Care Management; the Health Care Management Association; and the Health Administration and Medical Care Sections of the Massachusetts Public Health Association. Respondents were given one month to complete the survey. The data collection instrument was similar to the one used in the 1980 survey with certain modifications that reflected the authors' insights after analysis of the original data. The questionnaire took five to ten minutes to complete and required only multiple choice, check-off responses, or short answers. The questions centered around the type of organization in which the respondent worked, the respondent's position, her professional qualifications, work experience, and compensation and benefits.

Debra L. Caplan, M.P.A., is the Vice President for Clinical Services at the Brigham and Women's Hospital, Boston, Massachusetts, and is also a Visiting Lecturer in Health Policy and Management at the Harvard School of Public Health.

Lisa LeRoy, M.B.A., is a Senior Planner in the Health Care Planning and Program Development Department at the Brigham and Women's Hospital in Boston.

Jacqueline M. Rosenthal, M.P.H., is the Director of Hospital Relations, Harvard Community Health Plan in Boston. She is also a Visiting Lecturer in Health Policy and Management at the Harvard School of Public Health.

Linda J. Shyavitz, M.S., is the President and Chief Executive Officer at Sturdy Memorial Hospital, Attleboro, Massachusetts.

One hundred fifty-five surveys were returned. This response rate (39%) is high, and can be attributed to a high interest level on the part of the survey respondents.

Almost all of the surveyed individuals worked in Massachusetts, with an overwhelming majority (87%) located in Boston or the immediate suburbs. Thus, while it might appear that the survey is of only local interest, the findings can be used by women in other parts of the country for comparative purposes.

Survey results were reviewed during summer 1985, again, using a standard statistical package for the statistical analysis. The 1985 survey was developed and managed by three authors of the original study and a 1985 woman graduate from the Health Care Management Program, Boston University School of Management.

BACKGROUND

Although there continues to be an increasing number of articles on female managers in the corporate setting, the 1980 study by these authors and the follow-up study are unique in studying women health care managers as a class.

The survey findings are relevant in three contexts ranging from broad to narrow. First, it is interesting to compare this group of women to working women nationwide, bearing in mind that the women health care managers surveyed are clearly a white collar professional group. Ideally, this group would be compared to other professional women nationally, rather than to working women on average. Unfortunately, there is a dearth of information on comparable professional populations.

Secondly, the survey results serve as an employment profile of Boston area women health care managers. Third, the findings can be compared to the 1980 survey results in order to measure the progress that women in health care have made over the five-year period.

The number of women managers and administrators has grown over the last decade, and in no field has this occurred more quickly than in health care. In 1980, women accounted for 46.2% of health care managers. In 1985, it is estimated that women account for 57.0%. This represents tremendous growth, and it is predicted that health care services administration will be one of the fastest growing occupations (for men and women) over the next decade.[2]

Women have been joining the ranks of management in ever increasing numbers. Among managers and administrators in all industries, women hold about 30% of jobs.[3]

SALARY SURVEY FINDINGS

Education and salaries

The Women in Health Care Management group is an elite group by national employment and education standards. Both salary levels and education levels were extremely high compared with national averages. Of the women surveyed, 74.9% held one or more master's degrees and an additional 9.0% held doctorates. Only about 6% of women nationally have advanced degrees. The high education rate may be due, at least in part, to the concentration of educational institutions in the Boston area, and the resulting concentration of individuals with advanced degrees. This contrasts sharply with national averages for women in the labor force that show the median education being completion of high school.[4]

Salaries were high. The mean annual salary range was $30,000 to $39,999, about twice that of the national average salary ($17,600) for women managers and administrators.[5] Of the survey respondents, 65.8% made $30,000 to $59,999. Nationally, approximately 5% of full-time working women make $30,000 to $59,999 (30% of men fall into this range).[6] About 13.0% of the women surveyed made $50,000 and above.

Employment profile

Type of organization

A great variety of health care organizations were represented in the survey. However, approximately one half of the women surveyed worked for hospitals (47.7%). The next three most common employers were: consulting firms (8.4%), public health agencies (6.5%), and universities (5.2%). Each of the other types of organizations accounted for less than 5% of the survey population (see Table 1). It was somewhat surprising that a greater number of health maintenance organization (HMO) staff were not represented, given the recent tremendous growth of this type of organization.

Individuals worked mostly for private organizations (83.2%), as opposed to public organizations (16.8%). Of the 129 private organizations represented, 106 were not-for-profit, while 23 were for-profit (82.2% and 17.8% respectively).

TABLE 1

TYPE OF ORGANIZATION

Organization type	% of women
Hospital	47.7
Consulting firm	8.4
Public health agency	6.5
University	5.2
Health insurance agency	4.5
Home care, hospice	4.5
Other	23.2

Ownership of organization	% of women
Public	16.8
Private nonprofit	68.4
Private for profit	14.8

Location of organization	% of women
Boston	57.4
Suburban Boston	29.0
Other	13.6

Boston was the predominant location of employment with 57.4% of respondents. Suburban Boston was the second most common location of employment with 29.0%. The remainder of respondents, 13.6%, came from elsewhere in Massachusetts.

Position in organization

A large proportion of respondents (22.6%) were in senior management positions. Mid-level managers comprised 52.9% of the entire group, while consultants, staff analysts, professors, and others made up 9.7%, 7.7%, 3.2% and 3.8%, respectively (see Table 2).

Almost one third (29.0%) of the women reported directly to the chief executive officer (CEO) of their organizations. Four and one half percent were the CEO of their organization. These were primarily women who ran their own consulting firms.

Women were fairly equally divided between staff and line positions. This was true regardless of seniority or organization type. No correlation was found between salary and staff versus line positions. This was true in the first Women in Health Care Management survey five years ago, as well.

Professional activities

Women were generally very active professionally. Many served as guest lecturers and on the boards of community organizations. Almost one fifth did freelance consulting and the same number wrote and published. More than 25% were involved in four or more professional activities in addition to maintaining their full-time jobs. This was true of women in all salary

TABLE 2

POSITION IN ORGANIZATION

Position in organization*	% of women
Senior manager—health care provider	21.3
Senior manager—nonprovider	1.3
Mid-manager—health care provider	42.6
Mid-manager—nonprovider	10.3
Consultant	9.7
Staff analyst	7.7
Professor	3.2
Other	3.8

Direct supervisees	% of women
0	18.2
1–10	67.5
11–20	10.4
21+	3.9

Years in Current Position	%	Years in health care	%
0–2	44.5	0–4	9.7
2–4	23.2	4–9	31.6
4–6	15.5	9+	58.7
6+	16.8		

Annual salaries	%
$20,000	3.2
$20,000–$29,999	24.6
$30,000–$39,999	36.1
$40,000–$49,999	21.3
$50,000–$59,999	6.4
$60,000–$69,999	1.9
$70,000+	4.5

*Non–health care delivery organizations were defined as those that did not provide clinical services to patients (e.g., regulatory agencies, planning agencies, and consulting firms). Health maintenance organizations were considered to provide clinical services.

ranges. Similarly, the number of children a woman had did not affect the number of professional activities she engaged in. The group of women surveyed was a highly motivated one.

Demographics

The majority of respondents were 30 to 39 years old (51.7%). Adding the 40 to 44 age category accounted for 68.5% of respondents (see Table 3). No correlation was found between age and salary.

One half of the women surveyed were married (51.6%); 34.8% were single; 11.0% were divorced; and 2.6% were widowed. It is interesting to compare these statistics with those for working women nationally. A greater number of the women health care managers were single than the national average, fewer were married, a comparable number were divorced or separated, and fewer were widowed. The age distribution for the survey population was similar to that for employed women nationally.[7]

A recent *Wall Street Journal* survey of women executives (in all industries) found similar differences in its survey population. A greater number of women executives were single, fewer were married, and a greater number were divorced (see Table 3).[8]

In the responding population, 61.3% of health care managers reported no children, while 38.7% had one or more children. Of those women who had children, 75.0% had one or two; 25.0% had three or more children. These statistics were very similar to those for executive women in the *Wall Street Journal* survey (see Table 4).[9]

Of working women nationally, 39.0% are working mothers. This is comparable to the 38.7% of women health care managers in the survey sample who had children.[10]

Benefits

Benefits were fairly standard and included: health and life insurance (with partial to total employer contribution); tax shelters; 16 to 20 vacation days; expense accounts in 41% of cases; and retirement plans. Only 5.8% of respondents received cars. Day care was offered by 14.2% of employers. In the majority of cases, the employer did not subsidize day care (see Table 4).

Some of the new findings were those concerning maternity benefits. Interestingly, many women did not know what type of maternity benefits were offered by their organizations. Nine percent did not know if their organizations offered maternity benefits at all, while 21.3% reported that their organization provided maternity benefits, but did not know what those benefits were.

Thirty-one women reported the leave status for their most recent pregnancy: 4 took no maternity leave but remained on the job; 4 were not working at the time; and 4 resigned. Sixteen took unpaid leaves of absence. Of these, 12 women took unpaid leaves of less than 4 months, while 4 took leaves of more than 4 months. Three received 1 to 3 months of paid maternity leave.

TABLE 3

DEMOGRAPHICS

Age	% of women
25–29	11.6
30–34	25.2
35–39	26.5
40–44	16.8
45–49	6.5
50–54	7.1
55+	6.5

Marital status	% of women	% of working women nationwide	*Wall Street Journal* women executives
Single	34.8	25.1	30.0
Married	51.6	59.4	46.0
Divorced/ separated	11.0	10.4	17.0
Widowed	2.6	5.2	7.0

Number of children	% of women	*Wall Street Journal* women executives
0	61.3	58.0
1 and 2	29.0	33.0
3+	9.6	9.0

Education	% of women
No college degree	2.6
BA, BS, equivalent	9.0
One or more masters	79.4
Doctorate	9.0

Equity and equality

Two opinion questions were asked of survey recipients. The first was the woman's opinion on whether or not she received a competitive salary. Of women responding, 61.3% thought that they did receive competitive salaries, while 36.1% thought that their salaries were not competitive. Women in the highest salary ranges were more likely to believe that their salaries were competitive than those in lower salary ranges. Women less than age 45 were more likely to think that

TABLE 4

BENEFITS (PERCENT)

Benefit	1980	1985	% change
Health Insurance			
Employer pays 100% premium	44.4	40.0	(4.4)
Employer pays part of premium	50.0	55.5	5.5
Employer provides no health insurance	5.6	3.9	(1.7)
N/A	0.0	0.6	0.6
Life insurance			
Employer pays 100% premium	65.2	63.9	(1.3)
Employer pays part of premium	21.3	22.6	1.3
Employer provides no health insurance	10.1	11.0	0.9
N/A	3.4	0.6	(2.8)
Tax shelter			
Employer contributes to tax shelter	21.2	22.6	1.4
Employer does not contribute	26.2	49.0	22.8
No tax shelter provided	44.7	27.7	(17.0)
Vacation (days)			
10 or fewer	16.7	11.0	(5.7)
11–15	26.7	20.6	(6.1)
16–20	42.2	45.2	3.0
21+	14.4	20.6	6.2
N/A	0.0	2.6	2.6
Expense account			
Eligible	47.8	43.2	(4.6)
Not eligible	52.2	56.1	3.9
N/A	0.0	0.6	0.6
Free car			
Yes	11.4	5.8	(5.6)
No	88.6	94.2	5.6
Retirement plan			
Employer contributes to plan	67.4	70.3	2.9
Employer does not contribute to plan	6.7	9.0	2.3
No retirement plan offered	19.1	20.6	1.5
N/A	6.7	0.0	(6.7)

continues

TABLE 4 *continued*

Maternity and child care benefits	1985
Day care	
Employer offers day care	14.2
Employer does not offer day care	83.9
N/A	1.9
Maternity benefits	
Employer provides 1–3 months unpaid leave of absence	44.1
Employer provides 1–3 months paid leave of absence	10.9
Employer provides maternity benefits, type unknown	21.3
Respondent did not know if maternity benefits are provided	9.0
Employer provides no maternity benefits	13.5
N/A	1.2
Situation for most recent pregnancy	
Resigned	2.6
Was not working	2.6
Stayed on job	2.6
1–8 months unpaid leave of absence	10.3
1–3 months paid leave of absence	1.9
N/A	80.0

their salaries were competitive than older women (see Table 5).

The second opinion question asked whether or not the respondent thought that women and men in her organization were afforded equal opportunities and benefits. Here, 59.4% reported that opportunities were equal while 32.9% thought that opportunities were not equal. The older a woman was, the less likely she was to think that she was afforded equal opportunities and benefits.

In the *Wall Street Journal* survey, 70% of women executives said that they thought they had been paid less than a man of equal abilities during their careers.[11]

It was hypothesized that women who thought that they received equal opportunity would also think that they received competitive salaries. This hypothesis proved not true. One half of the respondents thought they received both equal opportunity and a competitive salary. One fifth thought that they received neither equal opportunity nor a competitive salary. The remaining one third of women were equally divided between those who believed that they received equal opportunity but not a competitive salary, or vice versa. These findings suggest ambivalence about women's salaries and opportunities.

Opinion on equal opportunity did not vary according to the number of years a woman had worked in health care or in her current position.

COMPARISON WITH THE 1980 SURVEY

The mean annual salary for the 1985 survey was $35,000. However, only part of the 1985 survey population was the same as the 1980 survey population—the Women in Health Care Management members. Isolating the Women in Health Care Management members, the mean average salary in 1985 was $40,000, compared with an average salary of $25,000 five years earlier. Adjusting for inflation this represents a real gain of $7,600.

There may be several reasons for this. In the 1984 Coles survey of hospital managers, it was found that New England Hospitals gave the greatest percentage base salary increases of any American Hospital Association (AHA) region in the country in 1984.[12] In addition, these raises were above inflation. Although not all of the women health care managers were in hospitals, the half that were would have had a large effect on the results.

It is also interesting to note from the Coles survey that individuals in hospitals whose jobs were externally focused received the largest pay raises. Directors of marketing and business development were first, followed by directors of public relations and development.[13] A large number of women in the survey worked for hospitals in these capacities.

Comparisons between the 1980 and 1985 surveys indicate that some gains in job responsibility appear to have been made. More women have direct supervisory responsibility, and a greater number in 1985 report directly to the CEO of their organization than did in 1980.

DISCUSSION

The preceding findings strongly suggest that women managers are doing well in the health care industry in Massachusetts. Of these women, 22.6% were already in senior management with another 52.9% in middle management. Of the respondents, 65.8% earned annual salaries of $30,000 to $59,999 as compared to 5% of working women nationally; 6.4% earned $60,000 or more.

Women less than age 45 were more likely to think that their salaries were competitive than older women.

This concentration of women in middle or senior management jobs has interesting implications for the future. There are only a few lay women and not many more Catholic sisters in top executive positions (i.e., chief executive officer, chief operating officer, chief financial officer) in hospitals, HMOs, and preferred provider organizations (PPOs). Of the women surveyed, however, 48% work in the hospital industry. The senior managers there run the risk that their career momentum will be abruptly halted if top executive positions do not open up for them. The middle managers will in turn be stalled behind them. These women appear to be actively preparing for advancement. Not only are they working at meaningful jobs, but their educational preparation—74.9% hold one or more master's degrees—and their broader professional activities—over 25% are active in 4 or more professional activities—suggest that they are preparing for optimal career growth. The career growth suggested by this survey could become jeopardized if the top positions remain inaccessible.

Over the last year (1986) a group of Boston-area senior women managers, primarily employed in hospitals, has been working to constructively address the very limited opportunities for women in top executive positions in hospitals. They have talked with representatives of executive search firms, hospital chief executive officers, hospital trustees, and other hospital industry leaders about opening the doors for women to top executive positions.

The effort has so far yielded the following results and observations. The representatives interviewed agree that there are many outstanding women in Massachusetts capable and ready for promotion to chief executive officer, chief operating officer, and chief financial officer positions. The pool of qualified women for chief financial officer jobs is the smallest.

TABLE 5

OPINIONS

Opinion whether salary is competitive	%
Yes, salary is competitive	61.3
No, salary is not competitive	36.1
N/A	2.6

Opinion whether afforded equal opportunity and benefits	%
Yes, afforded equal opportunity	59.4
No, not afforded equal opportunity	32.9
All-women organization	1.9
N/A	5.8

Opinion on competitive salary and equal opportunity	%
Receive competitive salary and equal opportunity	47.1
Receive neither competitive salary nor equal opportunity	20.7
Receive competitive salary, but not equal opportunity	15.7
Receive equal opportunity, but not competitive salary	16.4

Despite this favorable assessment of the pool of female talent, those executive search firms with whom the group has previously worked have not proposed women as candidates for their top executive searches.

It appears that there is a reluctance to propose women for these positions. The cause or source of the reluctance is unclear. Those board members with whom the group spoke say that they want the best candidates for positions regardless of gender and say that they depend on the search firms to know the labor pool and propose the best candidates. Some of the search firms say that they are prohibited by boards from proposing women. Of course, both could be true. However, what emerged from those discussions was a sense that most of the search firms that are in positions to strongly influence governing board attitudes about women executives are uneasy about doing so.

One area search firm—new to the health care industry and accordingly without preconceived notions about it—has actively and successfully presented women candidates to governing boards. It can be done. It is unclear whether the uneasiness of other search firms is a result of their biases, the biases of their clients, or perhaps both. The recent *Harvard Business Review* article by Sutton and Moore on executive women suggested that in 1985 there was still considerable, albeit reduced from 20 years earlier, agreement with the notion that business will never wholly accept women as executives.[14] For greater acceptance of women executives to continue to grow over the next 20 years both executive search firms and hospital governing boards need to examine their attitudes about women in top executive positions. In fact, they owe it to their organizations to test the talents of outstanding women in top executive positions.

In addition to the constraints imposed by lack of access to the top positions, women in the survey also indicated serious ambivalence about whether they were paid competitive salaries and were afforded equal opportunity in their positions. That ambivalence may be a result of other serious frustrations. While this survey did not measure women health care managers' salaries as compared to men's, the aforementioned Sutton and Moore article found substantial salary differences between men and women executives. Furthermore, Sutton and Moore found that only 42% of their male respondents and 18% of the female respondents thought women had at least equal opportunity for advancement.[15] If one applies these findings to the health care industry, one can conclude that equal pay and equal opportunity are serious issues for these women health care managers. Again, it is up to health care delivery organization boards and senior executives to ensure that talented and motivated women managers are compensated equally to men and provided equal access to promotional opportunities.

The survey indicates that hospitals and other health care organizations have provided meaningful career ladders for women managers, despite the underrepresentation of women at the top of these ladders. It also profiles an extremely motivated and educated group of middle managers, capable and eager for career advancement.

Future changes in the status of the survey population will be important to measure. In another five years will this group achieve substantial representation in top executive positions? Or does the apparent reluctance of executive search firms and organization boards to consider women portend otherwise? Will it be up to those few women currently in executive management positions to provide career opportunities for other women, or will we observe greater acceptance of women executives on the part of men?

There are many other questions raised by the survey findings, as well. Of particular interest will be trends in day care and maternity benefits.

Women's opinions about their salaries and opportunities will continue to provide insight to their situation relative to men. It is the hope of the authors that future surveys of women health care managers will demonstrate continual gains in salary and responsibility with increased attainment of top executive positions.

REFERENCES

1. Caplan, D.L., et al. "Women Health Care Managers: An Economic and Employment Profile." *Health Care Management Review* 9, no. 2 (1984): 29-38.
2. Bodger, C. "Sixth Annual Salary Survey." *Working Women* (January 1985): 68.
3. U.S. Department of Labor, Office of the Secretary, Women's Bureau. *Time of Change: 1983 Handbook on Women Workers.* Bulletin 298. Washington, D.C.: U.S. Department of Labor, 1983, p. 52.
4. Ibid., p. 107.
5. Ibid., p. 93.
6. Ibid., p. 83.
7. Ibid., p. 14.
8. "Survey of Women Executives." *Wall Street Journal*, October 1984, p. 21.

9. Ibid., p. 9.
10. U.S. Department of Labor, Office of the Secretary, Women's Bureau, *Time of Change: 1983 Handbook on Women Workers*, 22.
11. "Survey of Women Executives," 29.
12. Coles, B. "Coles' Compensation Survey." *Modern HealthCare* 14 (November 1, 1984): 21.
13. Ibid., 38.
14. Sutton, C.D., and Moore, K.K. "Probing Opinions: Executive Women Twenty Years Later." *Harvard Business Review* 63 (September-October 1985): 42-66.
15. Ibid., 56.

Wage differences and the concentration of women in hospital occupations

Andreas Muller,
James J. Vitali,
and
Diane Brannon

Women earn substantially less money than men. Many argue that this situation is the result of wage discrimination. The effect of the concentration of female employees within selected hospital jobs on wage rates is examined.

It is a well-known and controversial fact that women earn substantially less money than men. In 1982 women earned 63 cents for each dollar earned by men, a difference that has only slightly decreased over the last decade.[1]

The earnings gap between the sexes has provoked competing interpretations. According to one view, women have been socialized to opt for work in occupations such as nursing, teaching, or secretarial work, which fit the nurturing and supportive image of the traditional woman's role. Presumably, the ample supply of labor relative to its demand keeps wages for those occupations lower than for comparable male occupations. This labor market interpretation has been challenged by the observation that men in female-dominated occupations receive higher pay than women.[2,3]

Another view suggests that the earnings gap is largely a reflection of wage discrimination, which can take two forms. First, women are paid less than men for the *same* work, a practice prohibited by the Equal Pay Act of 1963.[4] The law mandates equal pay for jobs requiring "equal skill, effort, and responsibility, and which are performed under similar working conditions. . . ."[5] This doctrine is also well known by the popular parallelism: "equal pay for equal work."

Another form of wage discrimination occurs when employees of one sex in sex-segregated jobs are paid less than those of the opposite sex for *comparable* work.[6] For instance, highway engineers, a predominantly male occupation, earn about one-fifth more than registered nurses, a predominantly female occupation. The two jobs are rated roughly equal on criteria such as education, experience, and skills.[7]

This article examines whether the concentration of female employees within selected hospital jobs has any effect on wage rates once differences in the "comparable worth" of various hospital jobs have been statistically controlled.

Andreas Muller, *Ph.D., is Associate Professor at the University of Oklahoma Health Sciences Center, College of Public Health, Oklahoma City, Oklahoma.*

James J. Vitali, *1Lt., is with the United States Air Force, Medical Service Corps, Sheppard AFB, Texas.*

Diane Brannon, *Ph.D., is Assistant Professor, Health Planning and Administration Program, The Pennsylvania State University, University Park, Pennsylvania.*

To determine wage discrimination when work is performed in different jobs is a complex and somewhat arbitrary task.

THE COMPARABLE WORTH STANDARD

Wage discrimination within job categories can be fairly easily proven, if men and women do the same work; it involves a comparison of sex-specific wages. However, to determine wage discrimination when work is performed in different jobs is a more complex and somewhat arbitrary task. It requires a basis of comparison known as the comparable worth standard.

Comparable worth means that employers should pay workers equally, if their jobs are of equal value to them.[8] However, this definition is unclear as to what constitutes "equal value" or "intrinsic worth" of jobs. Critics maintain that the worth of a job to an employer is simply the amount that must be paid to attract and keep its work force.[9] Thus wages are a reflection of the interplay of factors determining the demand for and supply of labor. According to this view, a wage differential between comparable jobs may be only a reflection of differences in the supply and demand conditions affecting these jobs and not the result of intentional wage discrimination.

Additional difficulties arise with the measurement of comparable worth. In practice, job evaluations are used to rate different occupations according to criteria such as experience, skills, responsibility, and comfort of work place among others. Different rating systems have been devised that permit calculation of a summary score representing worth of jobs.[10]

However, critics of such rating schemes point out with some justification that "the relative ranking of jobs depends upon a multiplicity of arbitrary and subjective judgments: which job evaluation system to employ, which weight to give to each dimension or factor, how many points to assign to each job on each dimension, and other judgments."[11] A study by the National Academy of Sciences concluded that it is unlikely that a single, completely reliable job evaluation system will be developed. Consequently, the Equal Employment Opportunity Commission has not established such a standard in enforcing the Equal Pay Act.

Despite these difficulties, employer-developed job evaluations have been used in equal pay and comparable worth litigations. So far, the *County of Washington v. Gunther*[12] is the only case to reach the U.S. Supreme Court involving wage discrimination for jobs that were comparable but not equal. In this case, prison matrons were paid only 70 percent of the wages of male prison guards, while a job evaluation conducted by the county determined that matrons' pay should have been 95 percent. The supreme court emphasized that its decision was not based on judicial determination of comparable worth, since the evidence cited was from the county's own job evaluation. However, the decision is significant in that the court accepted the validity of an employer-conducted job evaluation.[13]

Since the late 1970s there has been increased activism surrounding the comparable worth issue in the form of collective bargaining, the filing of charges by professional associations,[14] and state and local government initiatives. The National Committee on Pay Equity reports that as of September 1983, state legislatures had enacted or adopted 26 laws or resolutions on comparable worth and four other states had enacted new fair employment practices laws that broadly prohibit sex discrimination.[15] While these laws vary widely, most relate to civil service systems and require that job evaluation and salary surveys be conducted. While most do not affect private employers directly, these state and local initiatives have broad potential for setting precedents and pushing up salaries in other segments of local labor markets.

STUDIES BASED ON COMPARABLE WORTH

Eight nurses in Denver challenged their employer, the City of Denver, on the basis that predominantly male professional and administrative occupations requiring the same education, experience, and supervisory responsibility made between $39 and $138 more per month than comparable nursing positions.[16]

Even larger pay differentials were noted in studies of state government jobs in the states of Washington, Minnesota, and Illinois. Job evaluators found wage gaps ranging up to 90 percent in favor of predominantly male occupations that were ranked similarly in comparable worth.[17]

In Connecticut, National Union 1199 conducted a pilot study in which 123 jobs were ranked according to factors such as knowledge, skills, mental demand,

and responsibility.[18] The study found that health care employees in mainly female occupations earned between $1,000 and $4,000 less per year than males in comparably ranked occupations.

The University of California sponsored a comparable worth study of nonacademic staff salaries at the university.[19] Jobs were classified as either male- or female-dominated if over 70 percent of all employees of a particular job were male or female. Job worth was measured by a summary score of "minimum number of years of education and experience required for adequate performance of the particular job."[20] The multiple regression analysis showed that on the average, female-dominated occupations received $329 less per month than male-dominated occupations of comparable job worth and racial composition.

Data and regression model

In 1979 a private consulting firm conducted a wage survey of health care personnel in 52 hospitals located in 17 eastern Pennsylvania and 3 adjacent New Jersey counties. Wage data were tabulated for three market areas (regions) comprising a similar number of hospitals. From this wage study, 40 job classifications were chosen that corresponded with those evaluated by another consulting firm in 1980. The occupations selected for this study are presented in Table 1.

Job evaluation determines the relative worth of one job to another job, or to many others. The type of job evaluation most commonly used is the point method.[21] It breaks down a job into components, or compensable factors. The job evaluation used in this study contained 14 factors whose relative importance was expressed by a number of points assigned by a team of job evaluators working for the consulting firm. The details of this procedure are described in the Appendix.

In order to determine whether wage rates show a sex-specific bias favoring male employees, the following multiple regression analysis was performed.

Minimum Hourly Wage $(y) = a + b_1$ education $+ b_2$ experience $+ b_3$ physical skills $+ b_4$ mental skills $+ b_5$ social skills $+ b_6$ accuracy $+ b_7$ mental effort $+ b_8$ physical effort $+ b_9$ physical conditions $+ b_{10}$ social conditions $+ b_{11}$ psychological conditions $+ b_{12}$ public contacts $+ b_{13}$ responsibilities $+ b_{14}$ authority $+ b_{15}$ Region I $+ b_{16}$ Region II $+ b_{17}$ percent female $+ e$.

TABLE 1

HOSPITAL JOBS SELECTED FOR REGRESSION ANALYSIS

1. Nurse supervisor
2. General head nurse
3. Staff nurse
4. Licensed practical nurse (LPN)
5. Nurse aide
6. Orderly
7. Nurse anesthetist
8. Nurse instructor, BSN
9. Physical therapist
10. Physical therapy assistant
11. Respiratory therapist
12. Nuclear medicine technician
13. X-ray technician
14. EEG technician
15. ECG technician
16. Lead medical technician
17. Medical technologist
18. Chief pharmacist
19. Staff pharmacist
20. Pharmacy aide
21. Social worker, MSW
22. Social worker, BSW
23. Dietician
24. Food service supervisor
25. Cook
26. Food service helper
27. Medical records clerk
28. Medical transcriptionist
29. Laboratory aide
30. Maintenance mechanic
31. Maintenance worker
32. Security guard
33. Housekeeping attendant
34. Unit/ward clerk
35. Clerk/typist
36. Infection control nurse
37. Utilization review coordinator
38. Computer operator
39. Histology technician
40. Ultrasound technician

The dependent variable (y) of this study is the minimum hourly starting wage for each of the 40 job classifications in the three regions. Since the job evaluation was designed to measure the relative worth of different jobs excluding factors such as seniority, the minimum starting wage was thought to be the most appropriate dependent variable. However, maximum and mean hourly wages were also analyzed.

The b coefficients show the effect, in dollar terms, of an independent variable such as education on wages when the effects of the remaining independent variables have been statistically controlled. The coefficient b_{17} is of main interest, since it indicates whether differences in the sex distribution of hospital jobs affect wages. Wage discrimination may exist, if b_{17} is negative and statistically significant.

Since no data on the sex distribution were collected by the consulting firms, this information was obtained by survey from one hospital in each of the three regions. For 31 of the 40 job classes, the percentage of female employees varied little among the three hospitals. Therefore, the mean percentage was calculated and used as the measure of the occupational sex distribution. The sex distribution variable was also coded in dummy form; 1 representing jobs with 50 percent or more female employees; otherwise it was coded 0.

To control for regional wage differences reflecting differences in the cost-of-living standard and different labor markets, two dummy variables were included in the regression equation. Their effects are represented by coefficients b_{15} and b_{16}, respectively.

Findings

Tables 2 and 3 clearly indicate that the hospital industry is primarily an employer of women. Among the 40 occupations selected for analysis, 29 are predominantly held by female employees. It is of interest to note that among the predominantly male occupations (Table 2) the chief pharmacist commanded the highest minimum starting wage ($8.74) while among the predominantly female occupations the nurse supervisor received the highest minimum starting wage ($6.75). Despite the wage difference of nearly $2 favoring the chief pharmacists, the job evaluation considered the nurse supervisor's position substantially more demanding than that of chief pharmacist; the comparable worth scores are 1712 and 1258, respectively.

The comparison of average wages indicates that "male jobs" receive 71 cents more per hour than "female jobs." However, on the average, these jobs are also more demanding as is shown by the average job evaluation scores. Therefore, the purpose for the following analysis was to find out whether this wage differential favoring male jobs is merely a reflection of greater job demands, or whether the concentration of females in certain jobs, in fact, reduces pay, once job demands are taken into account.

Table 4 presents the results of various regression specifications. First, the variables assessing comparable worth and regional differences are regressed on minimum hourly wages, and then, the sex distribution variable is included in the analysis. Model 1 indicates that the summary job evaluation score accounts for over 73 percent of the variance in minimum hourly wages. However, using the 14 components (compensable factors) as independent variables proved to be superior (see Model 2); they explain jointly over 92 percent of the variance in minimum hourly wages. Two dummy variables were in Model 3 to account for regional wage differences that slightly improve the fit of the regression ($R^2 = .934$).

Models 4 and 5 are of particular interest because they include the sex distribution variables. The multiple regression coefficients pertaining to both the per-

TABLE 2

HOSPITAL JOBS HELD PREDOMINANTLY BY MALES

Job class	Minimum hourly starting wage	Job evaluation score	% female
Orderly	$3.61	328	0
Nurse anesthetist	7.52	1105	44
Respiratory therapist	5.85	683	40
Nuclear medicine technician	5.60	666	38
Chief pharmacist	8.74	1258	0
Staff pharmacist	7.92	1274	45
Cook	4.42	361	42
Maintenance mechanic	5.07	986	0
Maintenance worker	3.88	223	0
Security guard	3.85	378	0
Ultrasound technician	5.18	618	25
Mean	$5.60	716.4	21.3

TABLE 3

HOSPITAL JOBS HELD PREDOMINANTLY BY FEMALES

Job class	Minimum hourly starting wage	Job evaluation score	% female
Nurse supervisor	$6.75	1712	100
General head nurse	6.17	1220	100
Staff nurse	5.51	943	97
Licensed practical nurse	4.24	650	95
Nurse aide	3.91	308	99
Nurse instructor (BSN)	6.51	985	100
Physical therapist	6.36	739	54
Physical therapy assistant	4.11	274	69
X-ray technician	4.90	666	83
EEG technician	4.43	400	100
ECG technician	3.74	391	100
Lead medical technologist	6.64	1132	67
Medical technologist	5.66	914	90
Pharmacy aide	3.72	262	58
Social worker MSW	6.26	860	92
Social worker BSW	5.45	575	88
Dietitian	6.17	651	100
Food service supervisor	4.54	334	67
Food service helper	3.44	152	89
Medical records clerk	3.63	263	100
Medical transcriptionist	4.00	326	100
Laboratory aide	4.33	469	100
Housekeeping attendant	3.44	219	58
Unit/ward clerk	3.66	306	99
Clerk/typist	3.58	282	100
Infection control nurse	6.08	1236	100
Utilization review coordinator	5.87	675	100
Computer operator	4.05	334	67
Histology technician	4.52	401	100
Mean	$4.89	609.6	88.7%

cent female and the sex dummy variable are statistically significant. The regression coefficient for percent female can be interpreted as follows: hospital jobs predominantly held by females pay 27 cents less per hour than jobs predominantly held by males, when factors measuring comparable worth and regional differences have been statistically controlled. Model 5 shows similar results. Hospital jobs predominantly held by males pay approximately 24 cents more per hour than those predominantly held by females.

> *Hospital jobs predominantly held by males pay approximately 24 cents more per hour than those predominantly held by females.*

To examine further the reliability of the findings, the regression analyses were repeated for a subsample of 31 job classifications ($N = 93$) for which the percentage of females varied little among the three

TABLE 4

REGRESSION RESULTS FOR MINIMUM HOURLY WAGE ($N = 120$)

Independent variables and intercept	Model 1	Model 2	Model 3	Model 4	Model 5
Education	—	.007 (9.692)**	.007 (10.370)**	.006 (8.442)**	.006 (9.108)**
Experience	—	.009 (3.765)**	.009 (4.028)**	.007 (2.783)*	.007 (3.194)*
Physical skills	—	.009 (1.969)	.009 (2.107)*	.008 (1.904)	.008 (1.853)
Mental skills	—	−.006 (−1.247)	−.006 (−1.335)	−.006 (−1.457)	−.006 (−1.498)
Social skills	—	.008 (1.417)	.008 (1.516)	.011 (2.039)*	.010 (1.995)
Accuracy	—	.006 (3.208)**	.006 (3.433)**	.006 (3.818)**	.006 (3.781)**
Mental effort	—	.005 (1.424)	.005 (1.523)	.004 (1.089)	.003 (.911)
Physical effort	—	−.005 (−1.553)	−.005 (−1.662)	−.007 (−2.249)*	−.007 (−2.310)*
Physical conditions	—	−.008 (−1.393)	−.008 (−1.491)	−.009 (−1.615)	−.011 (−1.879)
Social conditions	—	.040 (2.464)*	.040 (2.636)*	.040 (2.628)*	.045 (2.987)*
Psychological conditions	—	.004 (.288)	.004 (.308)	.012 (.825)	.006 (.422)
Public contacts	—	.003 (1.269)	.003 (1.358)	.003 (1.343)	.004 (1.497)
Responsibility	—	−.001 (−1.074)	−.002 (−1.150)	−.002 (−1.296)	−.001 (−1.106)
Authority	—	−.003 (−1.714)	−.004 (−1.834)	−.003 (−1.258)	−.002 (−.925)
Region 1	—	—	.132 (1.573)	.132 (1.588)	.132 (1.601)
Region 2	—	—	−.213 (−2.539)*	−.213 (−2.564)*	−.213 (−2.584)*
Evaluation score	.003 (17.973)**	—	—	—	—
Sex dummy	—	—	—	—	−.235 (−2.164)*
Percent female	—	—	—	−.0027 (−1.739)*	—
Intercept	3.097 (24.170)**	3.061 (24.015)**	3.088 (24.009)**	3.322 (17.926)**	3.352 (19.075)**
R^2	.733	.924	.934	.937	.938
F	323.040**	91.143**	92.389**	88.840**	90.337**

t = statistics in parentheses; *p≤.05; **p≤.001.
Note: One-tailed tests were used for sex distribution coefficients since direction was predicted.

TABLE 5

REGRESSION RESULTS FOR MINIMUM HOURLY WAGE ($N = 93$)

Independent variables and intercept	Model 3	Model 4	Model 5	Model 6	Model 7
Education	—	.006	.006	.005	.005
	—	(6.550)**	(7.257)**	(5.991)**	(6.136)**
Experience	—	.007	.007	.005	.005
	—	(2.556)*	(2.832)*	(1.620)*	(1.592)*
Physical skills	—	.004	.004	.003	.003
	—	(.748)	(.828)	(.677)	(.675)
Mental skills	—	−.013	−.013	−.012	−.012
	—	(−2.109)*	(−2.336)*	(−2.177)*	(−2.193)*
Social skills	—	.011	.011	.012	.012
	—	(1.596)	(1.768)	(2.071)*	(2.044)
Accuracy	—	.008	.008	.009	.008
	—	(3.303)**	(3.659)**	(3.989)**	(3.793)**
Mental effort	—	.009	.009	.008	.009
	—	(1.479)	(1.638)	(1.468)	(1.688)
Physical effort	—	−.004	−.004	−.006	−.007
	—	(−1.162)	(−1.287)	(−1.891)*	(−1.953)
Physical conditions	—	−.011	−.011	−.013	−.014
	—	(−1.644)	(−1.821)	(−2.038)*	(−2.175)*
Social conditions	—	.045	.045	.044	.053
	—	(2.060)*	(2.282)*	(2.252)*	(2.650)**
Psychological conditions	—	−.002	−.002	.007	.006
	—	(−.124)	(−.137)	(.506)	(.394)
Public contacts	—	.003	.003	.002	.002
	—	(.740)	(.819)	(.781)	(.523)
Responsibility	—	−.001	−.001	−.001	−.001
	—	(−.537)	(−.595)	(−.849)	(−.727)
Authority	—	−.001	−.001	−.000	−.001
	—	(−.526)	(−.583)	(−.154)	(−.219)
Region 1	—	—	.095	.095	.095
	—	—	(1.043)	(1.054)	(1.053)
Region 2	—	—	−.294	−.294	−.294
	—	—	(−3.219)**	(−3.252)**	(−3.249)**
Evaluation score	.003	—	—	—	—
	(15.083)**	—	—	—	—
Sex dummy	—	—	—	—	−.272
	—	—	—	—	(−1.565)
Percent female	—	—	—	−.0028	—
	—	—	—	(−1.612)	—
Intercept	3.159	3.126	3.192	3.431	3.463
	(22.785)**	(21.479)**	(22.553)**	(16.824)**	(15.556)**
R^2	.714	.920	.936	.938	.938
F	227.508**	63.874**	69.836**	67.125**	67.125**

t = statistics in parentheses; *$p \leq .05$; **$p \leq .001$.

Note: One-tailed tests were used for sex distribution coefficients since direction was predicted.

hospitals (see Table 5; the deleted job classifications were: nurse anesthetist, 44% female; physical therapist, 54% female; respiratory therapist, 40% female; nuclear medicine technician, 38% female; staff pharmacist, 45% female; pharmacy aide, 58% female; cook, 42% female; housekeeping attendant, 58% female; and ultrasound technician, 25% female). The regression analyses show the following: (1) overall, the regression models explain a similarly high amount of variance in minimum hourly wages; (2) the regression coefficients pertaining to the sex distribution variables are of similar size and are negatively related to wages. However, because of reduced sample size, the sex distribution coefficients are not statistically significant, although they approach the .05 level of statistical significance. Similar results were obtained when the analyses were repeated using mean and maximum hourly wage as dependent variables.

Discussion and conclusions

The data and regression model study has two main findings. First, factors such as education, experience, and accuracy predict hospital wages well. Regression models excluding sex distribution accounted for over 90 percent of the variance in minimum hourly wages, which suggests that job evaluations do provide guidelines for determining hospital wages.

Second, the sex distribution variables still explained additional variance in wages. The multiple regression coefficients showed negative relationships with minimum hourly wages and either were statistically significant or approached statistical significance. This finding suggests that hospital jobs held mainly by women receive lower wages than those held mainly by men.

The wage difference between predominantly male and female jobs ranged between 24 cents and 28 cents per hour. These figures represent about 5 percent of the mean minimum starting wage ($5.08). Expressed on an annual basis, the wage difference ranged between $499 and $582 of 1979 minimum starting wages and cannot be attributed to either differences in job evaluation scores or differences in regional wage conditions.

The wage gap found in this study is smaller than that found in previous research using similar methods. Perhaps this is in part a reflection of the fact that the regression models explained more variance in wages *before* the effect of sex distribution was assessed.

Nevertheless, there are several reasons the findings of this study must be interpreted with caution. A larger sample of hospitals would be necessary to validate the representativeness of the sex distribution measure. However, the reanalysis performed on the most reliable job classifications (see Table 5) determined that the results did not change appreciably.

Also, the analysis assumed that the sex distribution within the 40 hospital occupations remained unchanged between 1979 when the wage data were collected and 1984 when the sex distribution data were obtained. The assumption was tested on the basis of eight hospital occupations reported in the 1980 Census.[22] The correlation between the percent female workers reported in the Census and that found in the authors' survey was $r = .803$, suggesting that, at least for the eight job classes, the sex distribution measure remained rather stable.

It is also possible that some variable(s) unaccounted for in this study may be the cause of the wage differential. For instance, unionization may account for some of the wage differential observed in the study, but no data were available to test this hypothesis directly. Further research might be needed to explore this possibility.

In summary, the findings of this study support two conclusions: (1) that hospitals do systematically apply job evaluation criteria in the assignment of salary levels to job categories; and (2) that small but significant pay inequities exist between hospital jobs held predominantly by men and those held predominantly by women. The findings agree with prior research on this topic. Although tentative for reasons stated, a wage discrimination interpretation is consistent with the findings and suggests the need for continued efforts to eliminate pay inequities. Those efforts may include the following.

1. *Clarifying the Hospital's Position on Pay Equity.* Pay equity between the sexes is a widely held social value, and the concerned hospital administrator understands well the costs of deviance from such values. While no legal precedent exists for comparing employees of one system with those of another or with the market in general, hospitals are liable for their own inequities, with the trend moving toward comparison of comparable worth. Implicit in the use of job evaluation systems is the intent to avoid discrimination among both individual workers and job categories. It is reasonable to expect that hospitals will follow through with this commitment. The pro-

cess of contract negotiation appears to be a suitable forum for expression of pay inequities.

2. *Exploiting the Power of Job Evaluation Systems.* While legally somewhat of a "Catch-22" in that having a job evaluation study may provide evidence of inequities and not having one may imply the intent to discriminate, a comprehensive job evaluation system is integral to effective human resources management. Developments in automated job analysis systems[23] have dramatically increased employers' capacity to rationalize personnel systems with payoffs not only in terms of pay equity but also in improved structuring of span of control and accountability, better defined career paths, and specification of training needs. The design of a job evaluation system is critical and should involve both outside experts and employee representatives. The evidence from this study suggests that job factors such as the ones used in this study do determine hospital wages to a large extent. Coincidentally, the small amount of variance left unexplained by job evaluation or sex distribution factors calls into question the preeminence of uncontrollable labor market factors in the determination of hospital wages.

3. *Assessing the Costs and Benefits of Achieving Pay Equity.* Perceived inequity has been shown to result in job dissatisfaction accompanied by increased turnover and absenteeism and lowered performance.[24] Estimates of these effects should enter into a calculation of costs and benefits of correcting pay inequities within systems, as should costs resulting from successful litigation calling for retroactive adjustments.

If systematic job evaluations can account for all but a small percentage of hospital wage differences as indicated in this study, real inequities may be less than perceived inequities. Likewise, the apparent costs of nonretroactive adjustment may be in part offset by the tangible benefits of increased job satisfaction. Contrary to popular practice, criteria for job evaluations, survey results, and salary scales are best made explicit, as secrecy breeds exaggerated perceptions of inequity. Furthermore, employee participation in the hospital's efforts to address the issue of comparable worth will contribute to fairness at the work place.

REFERENCES

1. Gest, T. "Battle of the Sexes over 'Comparable Worth.'" *U.S. News & World Report* 96 (February 20, 1984): 73–74.
2. Belon, C.J., and Gould, K.H. "Not Even Equals: Sex-Related Salary Inequities." *Social Work* (November 1977): 466–471.
3. Gulack, R. "Just Beginning to Catch Up." *RN* 44 (August 1981): 40–43.
4. Equal Pay Act of 1963, 29 U.S.C. §206 (June 10, 1963).
5. Id. at 449.
6. Thomas, C. "Pay Equity and Comparable Worth." *Labor Law Journal* 34, no. 1 (1983): 3–12.
7. Gest, "Battle of the Sexes."
8. Thomas, "Pay Equity and Comparable Worth."
9. Nelson, B.A., Opton, E.M., and Wilson, T.E. "Wage Discrimination and Title VII in the 1980s: The Case against 'Comparable Worth.'" *Employee Relations Law Journal* 6 (1980–81): 381–403.
10. Brady, R.L., Persson, L.N., and Thompson, S.E. *Comparable Worth Compliance Handbook.* Stanford, Conn.: Bureau of Law & Business, Inc., 1982, Chap. V.
11. Nelson, Opton, and Wilson, "Wage Discrimination and Title VII," 384.
12. County of Washington V. Gunther, 452 U.S. 161 (1981).
13. Thomas, "Pay Equity and Comparable Worth."
14. "ANA Assembles Litigation Team to Win Pay Equity for Nurses." *American Nurse* 15, no. 10 (November–December 1983): 3, 20.
15. Comparable Worth Project, National Committee on Pay Equity and National Women's Political Caucus. *Who's Working for Working Women: A Survey of State and Local Government Pay Equity Initiatives.* Washington, D.C.: National Committee on Pay Equity, 1984, p. 35.
16. Schrader, E.S. "Nurses Challenge Women's Pay for Women's Work." *AORN* 23, no. 2 (1976): 169–170.
17. Gest, "Battle of the Sexes."
18. Stiefel, N. "Equal Pay for Women Workers." *1199 News* 15 (August 1980): 8–9.
19. Norris, B.A. "Comparable Worth, Disparate Impact, and the Market Rate Salary Problem: A Legal Analysis and Statistical Application." *California Law Review* 71 (1983): 730–75.
20. Ibid., 761.
21. Brady, Persson, and Thompson, *Comparable Worth Compliance Handbook.*
22. Commonwealth of Pennsylvania, DOL and Industry Office of Employment Security. *Pennsylvania Equal Employment Indicators.* Harrisburg, Penn.: Labor Market Analysis Section, 1983, pp. 7, 8, 12, 14.
23. Gael, S. *Job Analysis: A Guide to Assessing Work Activities.* San Francisco: Jossey-Bass, 1983.
24. Lawler, E.E. *Motivation in Work Organizations.* Belmont, Calif.: Wadsworth, 1973, pp. 83–86.

APPENDIX

CALCULATION OF JOB EVALUATION SCORE

Job evaluators selected 14 compensable factors and assigned weights (points) representing their relative importance in the hospital industry (see Table A-1). Then each job description was inspected and ranked on each factor on a scale from 1 to 6 according to the level of qualification or job demands. For instance, for the educational factor, up to 8th grade education was assigned level 1 requiring the ability to read and write simple messages. In contrast, level 6 requires the completion of a graduate degree, i.e., MSN, or MS in hospital administration. Each job was ranked from 1 to 6 on the other factors in a similar fashion. The summary job evaluation score was calculated by the following formula:

$$JES = \Sigma F_i \times 2_i^{(L_j - 1)}$$

where F_i = points of factor i
L_j = level of factor j
JES = Job Evaluation Score

TABLE A-1

POINTS ASSIGNED TO COMPENSABLE FACTORS AT LEVEL 1

Factor	Points	Total points
1. Education		12
2. Experience		12
Knowledge/skills		12
3. Physical	4	
4. Mental	4	
5. Social	4	
6. Accuracy		7
7. Mental effort		7
8. Physical effort		6
Working conditions		9
9. Physical	3	
10. Social	3	
11. Psychological	3	
12. Public contacts		8
13. Nonsupervisory job responsibility		12
14. Management responsibility and authority		15

Empowering middle managers in hospitals with team-based problem solving

Robert Dailey,
Frederick Young,
and
Cameron Barr

A team-oriented problem-solving procedure using management project teams was developed to improve quality of care and productivity in a private, nonprofit hospital. A focus group of managers developed the procedure, endorsed by top management. The work of four manager teams that followed the procedure saved the hospital more than $500,000.

Middle managers are responsible for sustaining the quality of clinical and support services, but their authority for recommending solutions to resource allocation and interunit coordination problems has not kept pace with their technical decision-making authority. When middle-level hospital managers have low decision-making authority for attacking and solving problems of interunit effectiveness, their job scope and depth are reduced. This leaves their managerial skills undeveloped or underutilized.

In Methodist Hospital, top managers found themselves often intervening in departmental operations because middle managers were unable to solve intraunit cost and interunit coordination problems. The effect of this was the shrinking of the authority of middle managers to routine, technical decision making in clinical and support services. They had little authority to address staffing problems, budgeting issues, and interunit coordination problems. The erosion of their authority had undermined their beliefs about the extent to which they could aggressively address service quality problems occurring in and between their units. This article presents a change process that required middle managers to follow a problem-solving procedure to solve operational productivity and quality-of-care problems in Methodist Hospital.

EMPOWERING MIDDLE MANAGEMENT

Methodist managers participated in a decentralized process to improve productivity and service quality in its units. The program rested on the classical assumptions of organizational development and principles that underlie the construction of unit-level productivity indicators in hospital settings.[1,2] The empowerment process began with a middle-management focus group that defined problems in the hospital and developed an acceptable intervention procedure to solve them.

Robert Dailey, *Ph.D., is Professor of Management and Organizational Behavior at the College of Business, Drake University, Des Moines, Iowa.*

Frederick C. Young, Jr., *M.H.A., is the President and Chief Executive Officer of Methodist Health System Foundation, Inc., and Pendleton Memorial Methodist Hospital.*

Cameron C. Barr, *M.H.A., is Executive Vice President for Hospital Operations of Pendleton Memorial Methodist Hospital.*

Launching the Empowerment Process

Hospital executives identified 10 middle managers with effective problem-solving skills. The managers came from lab, pharmacy, nursing administration, security, radiology, respiratory therapy, dietary, business operations, surgery, and ambulatory surgery. In a focus group, they concentrated on identifying a straightforward problem-solving process that fit the hospital's organizational culture. The middle managers stressed that the procedure should

- improve expertise in problem-solving skills among department heads, directors, and head nurses;
- enhance coordination between hospital units;
- build teamwork throughout the hospital;
- establish unit productivity, quality of health care, and quality-of-service indicators for hospital units;[2,3] and
- concentrate on internal hospitalwide quality and productivity problems.

Executives did not participate in the focus group. They agreed that their presence would have made middle managers reluctant to honestly convey their feelings about existing problems and administrative systems. The problem-solving procedure was refined and presented to the executive team for its approval.

Benefits of the Focus-Group Procedure

The focus-group procedure successfully communicated the need for expanded middle-manager involvement in solving unit performance and productivity problems to the hospital's key managers. Participants agreed that they could use business problem-solving practices in their units without compromising the units' quality of health care. They also learned that they could freely express their concerns about the hospital to each other in a safe setting. Finally, they realized that they could successfully discuss coordination problems without upper management's being present to function as referees. The nine steps in the problem-solving procedure developed by the focus team are presented below. The management training subjects that support the various elements of the intervention are shown in parentheses.

1. Meet with unit staff to identify symptoms and causes of productivity and quality-of-care problems (brainstorming).
2. Determine the most important facets of the problem from the unit(s) perspective (nominal group technique).
3. Determine underlying causes for the problem (cause-and-effect analysis).
4. Do a baseline audit to assess the magnitude of the problem in terms of dollars, work hours, error rates, etc. (Pareto analysis, basic statistical analysis).
5. Report results of analysis and recommendations to the executive team (action-planning and management-presentation skills).
6. Evaluate recommendations by top management.
7. Implement solution and track results.
8. Develop productivity indicators (PIs) and cost-of-quality ratios to track solution savings.
9. Publicize results to entire hospital.

The subjects presented during the training matched the problem-solving intervention procedure developed by the middle-manager focus group. The procedure was endorsed by top managers who did not design it, so the training had to correspond closely to the steps in the procedure. All hospital managers and medical administration attended the training.

The primary effect of designing an intervention procedure and presenting it to the executive team was to begin delegating more authority to teams of middle managers who were prepared to attack and solve service-quality problems at their work level. In the larger scope of hospital operations, it was crucial that productivity and quality-of-care problems be solved more often by middle managers. If this could be achieved, then executives could concentrate more on the strategic positioning of the hospital in its competitive market.

Constructing and Presenting a Program to Strengthen Managerial Problem-Solving Skills

A training program was developed to support the focus team's intervention procedure. It emphasized business problem solving, teamwork, quality-circle concepts, employee participation in decision making and formal feedback to top management from problem-solving teams. The training was conducted in groups of 15 managers in four training cycles, each lasting one and one-half days. It emphasized quality-circle methods to identify the true causes of problems and the magnitude of their contributions to problems under study. Excellent references are available describing the application of quality-circle techniques in health care settings.[4-6]

Advanced concepts introduced in the training were productivity tracking and the determination of the

> *The emphasis on measurement is important; it is the only way unit personnel can keep score for the health care work performed by their units.*

cost of quality (see Appendix A) in the delivery of health care services.[3,7] Productivity tracking systems are made up of indicators of work unit activity that specify a clear relationship between valid quantitative and qualitative measures of unit work output and unit-level goals for cost control, productivity, and the quality of health services delivered. The emphasis on measurement is important; it is the only way unit personnel can keep score for the health care work performed by their units. Productivity and quality indicators can be used to involve unit staff in decision making to increase the quality and cost effectiveness of the unit's health care services. When staff can clearly see the link between core work activities, measurements of work outcomes, and unit goals, then they are

- aware of how their efforts affect unit performance,
- prepared to be involved in work unit decision making,
- committed to the success of the hospital,
- likely to address problems that surface in work unit performance data,
- interested in quick feedback on work unit performance, and
- convinced that rewards are connected to excellence in productivity and service quality.

Table 1 shows the properties of PIs and quality indicators (QIs).[2] A unit's indicators should be related to

TABLE 1

PROPERTIES OF QUALITY AND PRODUCTIVITY INDICATORS FOR HOSPITAL UNITS

Features of the indicators	Hospital work unit examples
1. Developed through staff involvement	1. Number of errors in patients' bills
2. Represent the central aspects of team-based performance and health care service quality	2. Formulary carrying costs for each physician admitting patients to the hospital
3. Measurable and presented in terms which all unit staff can understand	3. The rate of unit compliance for posting lab reports to patient charts prior to physicians' rounds
4. Not the same things as individual employee accountability standards	4. Value and number of lost medications which cannot be charged to patients
5. Sensitive to changes in employee effort and team-based problem solving	5. Number of same-day surgery patients who are correctly processed per day
6. Related to work methods in the unit	6. Total variance in temperature for food served by meal
7. Can be altered or changed as the unit's strategic goals change	7. Number of cold meals served to patients over 70 years of age
8. Meaningful to upper management	8. Percentage of supplies which arrive on the units no more than six hours before they are needed
9. Reflect interdependence among hospital units which must coordinate health care services	9. Reductions in supply carrying costs created by J.I.T. methods
10. Reflect the needs of the hospital, patients, physicians and staff	10. Complaint rate about quality of care from patient family members
11. Not necessarily the same as quality assurance monitors	11. Response time to call lights
12. May be linked to merit-based bonuses received by unit managers	12. Rate of transfer to the nursing pool by unit RNs

the division's strategic plan. They are the primary link between the everyday aspects of a unit's health care work to the strategic goals for the unit. They chart unit-level outcomes rather than individuals' job performance. They are not substitutes for effective performance-appraisal practices at the staff level. Instead, they are meaningful indicators of unit success and failure.

PIs and QIs are sensitive to changes in a unit's teamwork. Staff should develop their unit's indicators, because they understand the goals of the unit and the way the unit's progress toward those goals is measured. A unit's indicators require top managers to share unit performance information with unit staff so they can influence their unit's indicators. PIs and QIs pull unit staff into the decision-making process to solve productivity and service-quality problems at their level. The construction and maintenance of effective indicators further empowers middle managers as they convert their goals in the strategic plan to indicators supported by the unit staff.

The Cost of Service-Quality Errors in Hospitals

Determining the cost of quality in hospital units requires the accurate measurement of the cost of errors in health care work. A quality:productivity ratio (QPR) is[3]:

$$QPR = \frac{\text{Number of service items not rejected}}{(\text{Total number of items} \times \text{processing cost per item}) + \text{Number of error items} \times \text{Reject processing cost per item}}$$

This ratio allows for changes in the quantity of work produced while considering quality changes in the quantity of work produced. The ratio implies that each rejected unit of health care service (e.g., poorly developed X-ray, cold meal served, blood sample lost) must eventually become a "good" item for health care services to successfully proceed. The ratio quantifies the relationship between productivity and the quality of service, which is measured as the cost of correcting deficient service units. Its elements capture the relationship between units of health care service and the cost of producing them correctly.

An Organizational Development Observation

Some organization development specialists may object to the use of a focus-group procedure to develop a method to attack and solve operational problems in organizations. It could be argued that all employees should have a chance to participate in a diagnostic phase that reveals all symptoms of problems confronting a hospital. Such an approach is consistent with classical survey feedback methodology. The hospital needed to reduce costs without the morale, motivation, and performance losses associated with a reduction in force. Top managers wanted to avoid a layoff because they believed that the staff was at the center of the institution's mission. Management wanted a fast answer to this question: "Where can we have the most impact on cost and productivity in our operations in the shortest period of time without losing good employees or ruining morale?" The answer was to quickly and deeply involve middle managers in the operational problems of the hospital and train the managers to solve them. Time-consuming survey feedback work would only have added to the cost of the project and delayed the involvement of middle managers in substantive team-based problem solving. The focus-group procedure was an efficient and meaningful way to get middle managers involved in changing the way the hospital made its operational problem-solving decisions.

Implementing the Intervention Procedure

Top managers selected four problems from those identified during the training. Each problem was assigned to a team of five to seven middle managers. Top management had previously agreed to an intensive pilot test of team-based problem solving. Five working days after the training was completed, the problem-solving teams were set up. The rapid transition from training to problem solving was important because it built on the good will, openness, and urgency developed during the training phase. The four problems selected by administration were

1. How can we standardize the departmental special ordering system which could then be used throughout the hospital?
2. How can we shorten patient waiting time in the emergency department up to the decision point to admit or send the patient home?
3. How can we effectively use pharmacy preparations in nursing units and in recovery?
4. How can we ensure that lab reports are posted to patient charts in nursing units in a timely manner?

Each team had two coordinators from the original focus group. Groups were created to reflect the interunit problem shown above. Thus top management created teams with managers who needed to cooperatively solve problems they had not solved in the past. The goal of creating heterogeneous problem-solving teams of middle managers was to build problem-solving effectiveness in the middle management level. The hospital's traditional decision structure had not created a fertile environment for team-based problem solving. Breaking these traditions in the existing organizational culture was essential. Vice presidents did not serve on problem-solving teams during the pilot test for this reason.

To sustain program momentum, teams had to successfully complete their projects in five weeks. In the past, problem-solving teams took up to a year to make recommendations, often based on flawed information and analysis. The expectation of unlimited time to complete a team problem-solving assignment was changed, and the teams readily accepted the very tight time limits.

RESULTS OF THE PILOT STUDY

Standardizing the Process for Special Ordering

A team worked on standardizing the departmental special ordering system using surgery and purchasing as test units for a new, hospitalwide system. The team learned that employees were baffled by the absence of a standardized form and ordering system for special-order items. They found that the departments negotiated their own procedures with purchasing. No one in purchasing was responsible for informing surgery staff about the status of their special-order items if they had to be back-ordered by the vendor.

Special orders in surgery averaged 333 occurrences per month during the first quarter of 1989. The labor costs for handling each special order in surgery were $9.20. The annual cost estimate for this activity in surgery was $36,700.

The team developed a new requisition form to eliminate the three forms previously required for special ordering. A card system was developed for surgery so a designated employee could track the status of special-ordered items. The team also set policies for form use, card system use, and notification of back-order status. Personnel in surgery and purchasing were trained to follow the new procedures. The pilot test was conducted after administration approved the recommendations. The team expanded the pilot test to include ambulatory surgery and the emergency department. The team reasoned that they should have a validity check for the action plan shown above.

During the pilot test, surgery's rate of special orders fell by 520 percent. With a cost per order of $9.20, the team determined that the new system would save $2,475 per month or $30,000 per year in surgery. Adopting the new form and method could save $50,000 on a hospitalwide basis.

DECREASING PATIENT WAITING TIME IN THE EMERGENCY DEPARTMENT

A second team addressed how patient waiting time in the emergency department (ED) could be minimized up to the decision point to admit or send the patient home. ED personnel indicated that the problem was exacerbated by overutilization of ED services, ineffective scheduling of physicians during high-census periods, slow turnaround for lab test results, and scheduling problems with radiology.

The team conducted work-flow analysis in the ED and decided to alter procedures for registering and discharging ED patients. Analysis indicated that patient registration required an average of 19.31 minutes (national norm is 10.00 minutes), while discharge time averaged 24.25 minutes in the ED (national norm is 20.00 minutes). The team recommended that ED patient registration be transferred to the admitting department for evening admissions. At other times, ED would handle its own registrations. The ED cashiers would be responsible for flagging the records for those patients who should return to the ED cashier for discharge. ED also developed a set of productivity indicators to track registration, waiting, and discharge times for patients.

Estimating the full cost of clerical time at $14.50 per hour, the team arrived at an annual savings of $76,800.

During a second measurement period, 94 percent of all registrations and discharges were done within the national norms. With an average of 236 patients per week being admitted and discharged in the ED, the team estimated that 4,960 work-hours in registration and discharge activities would be saved during the first year. Estimating the full cost of clerical time at $14.50 per hour, the team arrived at an annual savings of $76,800.

UTILIZATION OF PHARMACY PREPARATIONS IN NURSING UNITS AND RECOVERY

The third team worked to improve the utilization and collection system for pharmacy preparations used in nursing and recovery units. The team's analysis isolated numerous features to the problem. They found that
- nurses were "borrowing" drugs from one patient and giving them to another;
- documentation was inadequate for drug use in the pharmacy and the recovery unit;
- drug transportation was inadequate owing to patient room transfers and slow deliveries; and
- piggybacks and IV solutions were not used, not returned to pharmacy, thrown away, or allowed to become warm.

The team determined that 61 instances of lost drug and IV charges per week occurred in pharmacy and the hospital recovery unit. The team found an average of 20 minutes was spent in pharmacy and recovery to identify the origin of each error. The analysis indicated that $21,150 per year was spent to track such errors. Analysis of underbillings for pharmacy preparations distributed to all nursing units indicated that annualized underbilling errors were costing the hospital $292,000 for drugs and $77,700 for IVs and piggybacks. When the cost for tracking down errors was added, the total annualized estimated losses were $390,850.

The team's recommendations were numerous. Pharmacy couriers were to make rounds to all nursing units to pick up and deliver medications, IVs and IV/piggybacks. Pharmacy and nursing would develop the courier's schedule. Pharmacy would verify the adjustment printout daily against the previous day's cart fill list. Nurses' notes and the IV flow sheet were combined and required to show when new IVs were hung. Nursing designed a procedure to cover the return of all unused medications. Pharmacy analyzed IV solutions and admixtures for their true length of stability rather than using 48 hours for all solutions. A newly created pharmacy chart audit committee composed of six nurses and pharmacists would meet monthly to audit randomly selected patients' charts.

Top management authorized the installation of the team's recommendations. The problem-solving team is developing PIs to track
- the rate of returns of IV piggybacks;
- the number of IVs and piggybacks that must be destroyed;
- the ratio of dispensed medications to medications that appear on patients' bills (chart auditing productivity indicator); and
- the turnaround time for IVs and piggybacks ordered but not dispensed.

The team will develop a QPR to track the costs of preparing an IV or piggyback that is prepared but not used or thrown out.

POSTING LAB REPORTS TO PATIENTS' CHART

The fourth team addressed the problem of ensuring that lab reports were posted to patient charts in nursing units in a timely manner.[8] The team found that too many reports were being produced; the results were not being posted to patients' charts in a timely manner; one and sometimes two reports per patient per day were thrown away because the reports were delivered to the nursing units at the wrong time; excessive report production was wasting resources and making extra work for nurses; and physicians had to make their rounds without complete patient charts.

The team developed a new process for compiling and distributing laboratory reports. The lab supervisor was made responsible for ensuring that lab results were delivered to the units within 30 minutes of printing time for all routine reports and within 10 minutes of printing time for all STAT requests. Head nurses were responsible for ensuring that all lab results were charted within a half-hour of report delivery. The lab was required to run a cumulative report by 2:00 P.M. and stopped running the interim report. Ward clerks charted the cumulative report by 5:00 P.M. before the physicians' rounds. The lab quit printing reports at 8:00 P.M., while it continued to print the interim report at 2:00 A.M. and deliver it to units by 2:30 A.M.

The team implemented its steps in a pilot test. By eliminating one interim report, $16,826 in materials and labor could be saved annually. The team developed a QPR to track the service-oriented effectiveness of its action plan (see Appendix A). The QPR shown in the appendix is a first approximation because the team will build in more accurate estimates for the cost of the doctors' waiting time during rounds when reports are not posted. Given the short time covered by the pilot test, the team's QPR adequately illustrates how methodical a problem-solving team can become. The team calculated that the hospital could save $6,424 per year by reducing the average number of daily report reprocessing from its current level of 15 to 5 (see Appendix A). The anticipated total annual savings for the hospital are $23,250. The team found that compliance with the recommended action-plan steps for the production, distribution, and charting of lab test results rose to 95 percent across all nursing units by the end of the pilot test.

Opinion survey results confirmed the success of the pilot test. Nurses, ward clerks, lab personnel, and physicians stated that the new system was more effective and that far fewer patient charts were incomplete. These effects were consistent with increases in the team's QPR during the pilot test period.

• • •

The intervention was designed to increase middle managers' involvement in hospital problem solving. It helped Methodist Hospital reduce costs and improve the quality of services by delegating more authority for problem solving to teams of middle managers. It changed managerial reporting relationships in (1) planning problem-solving efforts; (2) managing the work of problem-solving teams; (3) reporting results to upper management; and (4) implementing the results of problem solving.

The intervention differs from quality-circle programs in several important ways. Highly competent middle managers participated in a focus group process to identify and develop a team-based, problem-solving procedure, which they followed once they were trained in its procedures. Most quality-circle programs do not target a specific managerial level in the organization. Quality-circle programs do not give middle managers more authority to solve interunit problems. The intervention also altered the nature of managerial reporting relationships. In traditional quality-circle methodology, the organization's decision-making hierarchy is unchanged. Under this arrangement, the work of quality circles does not require the alteration of managerial reporting relationships nor does management have to provide budget information or productivity data to the quality circles. Top management does not alter its work relationships with middle managers. The present intervention enriched the scope and depth of middle managers' jobs by drawing them into a hospitalwide problem-solving process that required them to work on highly motivated teams.

The pressure created by the short problem-solving cycle ensured that teams of middle managers would be highly motivated to act quickly on information with a bearing on the problems assigned to them. The information-sharing and time-urgency features of the intervention are not regular features of quality-circle programs.

The intervention produced lasting solutions to problems the hospital had not adequately solved by conventional, hierarchical means.

The intervention produced lasting solutions to problems the hospital had not adequately solved by conventional, hierarchical means. There were several reasons for this success. First, managers had substantially improved abilities to apply business problem-solving skills to problems they formerly ignored or wrote off as being insoluble. Second, they had to function effectively as members of a multidisciplinary team, which had to quickly understand and solve an assigned problem. By strengthening the problem solving skills of middle managers and assigning responsibility for solving problems with hospitalwide implications to them, managers saw their role in hospital problem solving in a different light. They concluded that management was moving away from centralized practices to more decentralized delegation of authority. The intervention again differs from a quality-circle program because it altered and expanded the responsibility and authority of middle managers. At best, volunteer-based quality-circle programs help employees to become more involved with their work groups as they find ways to make production systems more efficient and focused on quality. Traditional quality-circle interventions have less impact in this area.

REFERENCES

1. Friedlander, F., and Brown, L. "Organization Development." *Annual Review of Psychology* 25 (1974): 313–16, 320–31, and 336–41.
2. Dailey, R. "Productivity Monitoring Systems In Hospitals: A Work Group Focus." *Health and Health Services Administration* 33 (1988): 75–83.
3. Adam, E., Jr., Hershauer, J., and Ruch W. *Productivity and Quality: Measurement as a Basis for Improvement.* Englewood Cliffs, N.J.: Prentice Hall, 1981.
4. Orlikoff, J., and Snow, A. *Assessing Quality Circles in Health-Care Settings.* Chicago, Ill.: American Hospital, 1984.

5. Lawler, E.E., III, and Mohrman, S. "Quality Circles: After the Honeymoon," *Organizational Dynamics* 15 (1987): 42–54.
6. Adair, M. *Quality Circles in Nursing Service: A Step-by-Step Implementation Process.* New York, N.Y.: National League for Nursing Publications, 1982.
7. Dailey, R. "Redefining Hospital Performance: A New Commitment to Competitiveness Through Employee Involvement." *DRG Monitor* 6, no. 3, (1988): 1–8.
8. Dailey, R. "Strengthening Hospital Nursing: How to Use Problem-Solving Teams Effectively." *Journal of Nursing Administration* 20, no. 7/8 (1990): 24–29.

APPENDIX A

Team four worked on reducing costs and increasing the timeliness of reporting laboratory test results for posting to patients' charts. Using the quality: productivity ratio (QPR) methodology for determining the cost of quality, the team produced a working ratio. This appendix illustrates the team's reasoning for its QPR. The numbers are estimates developed by the team after observing report-processing activities for a period of five days.

Cost to Produce an Original Lab Report

1. Materials
   a. 2800 report forms per case @ $49.00.
   b. Average report is 10 pages in length. $ .18
   c. Average number of daily reports.   .65
   d. Daily cost of materials.   11.70
2. Work force
   a. Lab technician on computer for 2 minutes.   .37
   b. Lab secretary prints, tears, sorts, and delivers (7 minutes per report).   .09

   TOTAL COST PER ORIGINAL REPORT $ .64

Cost to Reproduce a Report (late or lost)

3. Doctor cannot locate report during rounds
   a. Nurse searches for report (approx. 2 minutes). $ .48
   b. Nurse telephones lab secretary to reprocess report.   .24
   c. Lab secretary converses with nurse about the report (reason for being misplaced or lost).   .13
   d. Lab secretary locates and prints hard copy in 3 minutes.   .39
   e. Lab secretary delivers to unit and returns to lab in 4 minutes.   .52

   TOTAL COST TO REPRODUCE AN ORIGINAL REPORT $ 1.76

$$QPR = \frac{\text{No. of reports not reprocessed}}{(\text{Tot. no. of reports} \times \text{processing cost per report}) + \text{No. of reprocessed reports} \times \text{Reprocessing cost of incorrect report}}$$

$$= \frac{45}{65 \times \$.64 + 15 \times \$1.76} = .662 \text{ reports/\$}$$

The team reported preliminary data showing that the QPR is rising because fewer reports are being reprocessed. The team attributes this result to the fact that one interim report has been eliminated. Also, employees cited more realistic production and distribution policies as reasons for fewer occurrences of report reprocessing. The team estimates that the current figure of 15 reprocessed reports per day can be cut to 5. This would represent an additional savings of $6,424 per year.

Part II

CULTURE, BEHAVIOR AND MOTIVATION: TOOLS FOR CHANGE

Cultural change versus behavioral change

Richard D. Nordstrom
and
Bruce H. Allen

Health care culture is a powerful force in hospitals that must be taken into consideration in attempting to effect changes in employees' behaviors. A survey can be taken to assess the cultural climate of a health care institution.

Few management concepts are more difficult to come to grips with than the term *culture*. Yet the word is readily used and is considered to be widely understood. The variety of definitions from scholarly sources is immense. The work of Kroeber and Kluckhohn uncovered some 164 definitions of culture.[1] One thing that is clear from these definitions is that culture consists of both behavior and thought. It is interwoven throughout our day-to-day activity. Behaviors are rewarded or sanctioned according to cultural values. The question posed by this article is: Can one change the element without changing the whole? Or is it possible to make a behavioral change in an organization without first changing the cultural values within which this change will take place?

There cannot be culture without a group. Culture is owned by the group and is embedded in the group. Groups that have been together long enough to share problems and solutions, feel the effects of solutions, and see people come and go from the group will have a culture.

Culture proves to be both beneficial and detrimental to groups.[2] Shared thoughts and behaviors make groups more efficient[3] in that they work more economically in terms of both time and money. Stability is generated because newcomers are soon taught to conform and thus perpetuate the cultural norms.[4] The liability of culture is found in the fact that efficiency does not mean effectiveness. When the strategic achievement needs of the organization are not being met by the group's shared values, effectiveness suffers. Cooperation may be good for individual groups but the organization may require a different level of cooperation or commitment to the task between groups rather than within groups.

IS THERE A HEALTH CARE CULTURE?

Organizations are groups and have groups within them. A culture and component subcultures exist in

---

Richard D. Nordstrom, *Ph.D., is a Professor of Marketing at California State University–Fresno. He is active in many health care and marketing organizations and has participated in health care marketing consulting for the past 12 years.*

Bruce H. Allen, *Ph.D., is Professor of Marketing in the School of Business at San Francisco State University. He recently served as Vice-President for Marketing Strategy at the Community Hospitals of Central California and is a marketing consultant for hospitals and multihospital systems.*

all organizations. Hospitals are no exception. However, one should not assume that all health care institutions have the same culture. A given hospital will be drastically different from others. The reasons for differences lie in variations in management, local economy, technology and specialization, and socialization needs.[5]

As with all cultures, health care cultures are dynamic, evolving constructs. As management changes over time, new personalities arrive; new means of solving problems arise as new technology is acquired; new demands on communication develop as the organization grows; rewards for behavior change as the economy changes; and the needs of the group shift as the group grows. Therefore, as with all corporate cultures, health care culture is heavily weighted by the influence of organization leaders.[6,7]

An ideal health care culture is not a relevant normative standard. Culture based on what people should be doing rather than on what they are doing would be virtually impossible to attain. Similarly, simply stating what a culture should be does not mean people will shift in that direction.[8] Therefore, wisdom suggests that managers gain an understanding of cultural values that influence group dynamics and group behavior. This information can be obtained through research or by intuition. Normally, research will prove more reliable.

CAN A CHANGE IN BEHAVIOR CHANGE CULTURE?

Cultures evolve as shared value beliefs and behaviors become widely accepted. Temporary behavior changes (e.g., a minor change in routine, such as a change in the order a janitor cleans floors) need not reflect the culture. A long-term behavioral change (e.g., a change in procedure, such as a change in the equipment a janitor uses) must reflect the culture.[9] Behavior changes without consideration of cultural impact are doomed to constant monitoring and require constant use of punishment and reward to sustain the change. If culture changes, too, behavior change can be enduring because it is self-sustaining.

In short, corporate culture transformations in health care organizations require a combination of behavior modification and a shift in values and beliefs. Without the shift in values and beliefs, the culture remains unchanged and behavior modification is temporary. Hospital managers must assess the type of behavior change desired, the length of time over which the change is to be obtained, and the fit with present culture.

HOW DOES CULTURE INFLUENCE BEHAVIOR IN A HOSPITAL?

Behavior modification is influenced by the culture in that efficiency is defined by culture, i.e., peers may define what is considered efficient. This does not relate to effectiveness, however, because a person who is viewed by peers as efficient may not be effective to the organization. Thus it follows that the better understood and the more widespread the shared values become, the more effective the organization as a whole will become.[10]

First, the corporate or health care culture must be clearly presented to the employees. This will demonstrate management's awareness and concern for employees and their widely held values. It will also help

---

The corporate or health care culture must be clearly presented to the employees. This will demonstrate management's awareness and concern for employees and their widely held values.

---

employees see how their own attitudes fit or do not fit the set of values held by the group as a whole.

Second, and virtually simultaneously, the value of the desired cultural philosophy must be reflected by management actions, by rewards, by communications, and by selection and indoctrination of new employees.[11]

Perhaps the critical aspect of seeking change in any health care culture is a thorough understanding of the existing culture. Hospitals are currently embracing guest relations programs designed to encourage their employees to be more customer, people, or guest oriented. This is another way of saying that hospitals are trying to change the health care culture. Care must be taken if these organizations are to be successful. The starting point is an understanding of the existing culture.

HOW CAN HEALTH CARE CULTURE BE UNDERSTOOD?

Employee climate or culture studies provide one means of assessing the health care culture in a hospital. However, exploring the culture is of no value if top management is not interested in an objective analysis of the data or is unwilling to make plans based on the data. Given that the chief executive officer wants to know about the status of cultural values, a carefully designed survey instrument will help her or him learn what shared values guide the organization.

An appropriate instrument for the cultural survey in health care organizations should be designed and tailored to fit the needs of the particular institution. This can be a topic for extensive discussion. Outside consultants can be retained for the research. If time and money do not permit, several adequate instruments can be purchased and adapted for use.

A properly designed instrument will show clearly what areas are of importance to employees. Rather than look at these as problems, management should focus on the results as employee concerns. Once management knows what the employees feel is important, the next step is to analyze how management can or should proceed either to capitalize on these concerns or to modify them for greater corporate effectiveness.

Communication of selected results to the employees is a must. A cultural climate survey helps employees understand that their concerns are widely held. Recognition of this fact can be beneficial if management wants the shared values to work toward greater effectiveness. Justification of behavioral change is easier when it is in the context of contributing to reduction of group concerns. A change in priorities is easier to accept when the evidence upon which the priorities are founded is widely known.

Action plans are also necessary. It does little good to isolate a concern of the group if nothing is to be done. Inaction sends a message as clearly as action.

In conducting a cultural climate survey, privacy of respondents should be protected to encourage response. One should keep in mind that the purpose of a culture study is to learn about widely shared values. Why a person or thing is liked or disliked is more important than who or what is disliked or liked. It is suggested that data be handled in as broad a context as possible to permit focus on the total organization rather than on just a few individuals or departments. However, should an obvious trouble spot be disclosed, it may call for direct managerial attention.

In one multihospital system study that was conducted, since culture is multifaceted, each question in the survey was asked using three dimensions. First, respondents were asked to indicate "how pleased" they were with the item. This question and all other similar questions (type A) served as the basis for the appreciation dimension. Next they were asked to say "how helpful" the item is for them in doing their job. This and all other similar questions (type B) served as the basis for the job performance dimension. Finally, employees evaluated the item's value to them in providing job satisfaction. This and all other similar questions (type C) served as the basis for the satisfaction dimension.

Broadly speaking, employee attitudes reflect:
- some aspect of the individual's means of evaluating group and organizational effectiveness (appreciation);
- some aspect of the individual's means of evaluating group and organizational efficiency (performance); and
- some aspect of the individual's means of evaluating group and organizational agreement with work, working conditions, and peers (satisfaction).

These dimensions apply to various aspects of the job, such as training.

RESULTS FROM A MULTIHOSPITAL SYSTEM'S CLIMATE STUDY

Using data obtained from a survey conducted by the authors in a system of three hospitals and three convalescent facilities in Fresno, California, factor analysis was run to reduce, simplify, and gain a better understanding of the data. By examining the statistical relationship between and among variables it was possible to distill large numbers of survey data into clusters with similar underlying dimensions. If a study has been properly designed and responded to by those surveyed, the clusters will be logical.

From the analysis of data obtained from responses to the type A questions it was possible to uncover four underlying factors (clusters of questions) that contributed to the understanding of the cultural value system. The number of underlying factors was determined by including only factors with an eigenvalue of one or more. These were evaluative norms reflecting appreciation. It does not matter how good or bad something is in reality. What is shown from asking people how pleased they are with an item is how the corporate culture evaluates or considers this item's importance.

Appreciation factors

Responses indicated that employees felt these areas greatly influence their appreciation of work, peers, and work environment.

1. supervision and decision making;
2. group cohesion;
3. tangible rewards and recognition; and
4. nurses, physicians and patients.

These are for the system studied, not for all hospitals.

Factor 1: supervision and decision making. The most important element of appreciation consisted of a group of topics related to supervision and decision making. These topics covered:

- supervisors' ability to help with problems;
- employees' freedom to give ideas and their possession of the knowledge that supervisors would listen to them;
- supervisors' encouragement of employees to be superior;
- supervisors' ability to interact with employees on a personal basis;
- supervisors' communication to employees of how well they do their job;
- work groups' decision-making processes; and
- management's concern for what employees think.

Factor 2: group cohesion. Topics included in this facet of appreciation were:

- how people in the group get along;
- how people in other groups get along with a specific work group;
- what type of people are in the group;
- how everyone tries her or his best; and
- how a specific work group thinks quality is important.

Factor 3: tangible rewards and recognition. In this set of hospitals, personal elements of the job were third in priority as shown by the cultural climate study. Within this factor topics show that employees in this health care culture share beliefs such that appreciation is based on the following topics:

- information received from management;
- procedures and policies of management;
- equipment;
- benefits;
- pay;
- opportunity for promotion; and
- surroundings.

Factor 4: nurses, physicians, and patients. Finally, factor analysis shows that the last group of appreciation values were nurses, physicians, and patients. Employees showed appreciation for these groups and thought that these were groups to appreciate.

Performance factors

The second type of response assessed widely held values relative to performance. These factors emerged as cultural justifications for job-related behavior:

- teamwork;
- problem solving;
- information;
- training; and
- nurses, physicians, and patients.

These were the elements employees used to evaluate job and organizational efficiency.

Satisfaction factors

Satisfaction on the job is affected by four factors, which are the groups of responses that indicate how the corporate culture assesses job satisfaction. The satisfaction factors were:

- ability to discuss ideas and problems;
- rewards (pay, training, etc.);
- people; and
- nurses and physicians.

These were the elements used to evaluate what makes the job or organization good or bad. In this health care culture, satisfaction values were also shared. Satisfaction was a function of being able to discuss views, rewards, fellow employees, and nurses and physicians. Patients were not included in the structure of values that contributed toward satisfaction; because in this system, patients were not included in the factors that contributed to the job's worth or the organization's effectiveness.

The Appendix shows the complete set of responses in the clusters developed from the use of factor analysis. The health care culture has formed over time. The cultural values in a particular hospital may not be the same as the values indicated from the data obtained in the six institutions used for this study. What is certain is that understanding of the climate in health care is vital.

HOW CAN KNOWLEDGE OF CULTURE BE USED TO INFLUENCE CHANGE?

Culture serves as a justification of behavior. People will use the shared values of culture to justify what they are doing. "No one else . . . . Why should I?" These and similar responses indicate when an employee feels cultural values do not support a requested behavior.

When management wants to introduce a desired change that will require employees to change behaviors, two avenues must be followed simultaneously. The new behavior should not be reinforced by external justifications alone. In short, financial or tangible rewards should not be used as the justification for short-term behavior modification. It is through use of intrinsic forms of motivation that people see the inherent value of the requested behavior. Second, managers must communicate new beliefs that support the need for the desired change. When people accept these values (new beliefs) then the desired behavioral change will take place.

Cultural values that are deeply held, that have been in place for a long time, or that create strong emotional responses are more difficult to modify and have a greater influence on the behaviors of employees. The cultural system analyzed above does not rank patients high in influencing appreciation, job performance, or satisfaction values. In this organization it would be impossible to gain full adoption of a guest relations program in a short time span. Rewards and recognition should be designed to support a desired change, new employees should be recruited with desired new talents, and management should communicate by word and deed their commitment to the new values.

Communication of cultural norms can be accomplished by both direct and indirect means. Direct pronouncements, letters, and memoranda are necessary to avoid confusion. However, indirect means such as rituals, ceremonies, stories, heroes, logos, decor, dress, and actions, are also necessary. The credibility of the communicator must be good before communication is accepted. It is not until management demonstrates with action that the desired behavior is truly desired that cultural change will be effected.

Information obtained from this survey alerted management to the challenge it faced in instituting a program aimed at focusing more attention on patients, guests, and others who are in contact with the hospital. It was clear management would have to set the tone. Some of the many ways to help set the tone are:

- Praise desired behavior by using an internal publication or newsletter. This is simply a means of telling a story showing that the behavior is desired and will be noticed (rewarded). Stories help people see that "others are doing something," too. They help take away justifications for not acting as directed by the new policy.
- Redo patient rooms to show commitment to patient attitudes and comfort.
- Get top managers on the floors to show real concern for what is going on with patient care.
- Publicize good patient interactions other than those with nurses.
- Provide sensitivity training for admissions people, housekeeping staff, and other personnel to show the importance of their role.

Hospitals must use the "try it, you'll like it" approach to gaining behavior change or cultural change. After inducing people to try the new behav-

---

Hospitals must use the "try it, you'll like it" approach to gaining behavior change or cultural change.

---

ior, they must be shown how they will like it. There are many variations on this theme. "Try it and you'll find it is just about the same as what you are now doing." "Try it and you'll find it makes your job easier."

Sometimes removal of cultural deviates is also necessary. Old behavior builds on old behavior. Sometimes there are people who refuse to adapt to desired changes. In many instances, long-time management may itself be the culprit in perpetuation of the old culture. Management cannot expect others to change if the managers are not going to lead the way. If management is new or is leading the way, then it is sometimes necessary to remove people or to remove their authority should they be retarding the movement toward the desired cultural change. This is but another means of demonstrating the value of the change.

Management must also recognize the fact that as new people are employed, they will be indoctrinated to the old cultural values by the existing employees unless they are given a clear understanding of the new or desired values top management means to reward. Although it is not possible to have a perfect fit between culture and people, one should not hire those who cannot fit the desired behavioral posture. If management wants employees to be more people oriented, all new hires must be capable of being people oriented. (This assumes management has taken the first steps in communicating the desired change, has taken the lead in showing others how to be more people oriented, and has found ways to demonstrate the value and reward of the new behavior.)

IS THIS ETHICAL?

Understanding what or how people think is not likely to impose on personal or religious values. It is the nature of a manager's job to guide and influence. Ethics is concerned with widely accepted mores. People bring their culture to work and a work culture is formed. Understanding and shaping organizational beliefs and values is appropriate as a means of leadership. Organizational goals can be reached only if employees understand them and do what is necessary to reach them.

HOW CAN CULTURAL CHANGE BE MAINTAINED?

Thoughtful planning is the key element in maintaining positive shared values or in changing a set of shared values that has become detrimental to the advancement of the organization. Obviously, a culture could be compatible with the needs of the institution and therefore not require change. Managers who think and act as though their organizations were systems see the interaction of people as a part of the system.

Careful hiring of employees, willingness to remove those who present problems, and thorough indoctrination of new hires all promote the plan. Praising success and being on constant alert for stories to tell about those who are supporting the new action are all needed.

Recent attention to corporate culture from books like *In Search of Excellence*[12] have focused attention on the fact that organizations that help employees see that the quality of their life and the quality of the service they provide are one and the same seem to be the more successful firms. Leadership in product innovation does not guarantee leadership with people.

• • •

If health care institutions desire behavioral change, they must not ignore the managerial potential of culture. Culture can be assessed through the use of survey techniques. This article reported results of cultural assessment from an instrument that has been used in a number of hospitals. The data from this study showed management which shared cultural values were directing employee behavior prior to introduction of an extensive new program.

Factor analysis was performed on data from six health care facilities. These data showed that properly designed questions can help evaluate the corporate health care climate. In the data used, the employees evaluated four factors as distinct areas of appreciation that helped shape their activity: supervision, work group, rewards, and nurses, physicians, and patients. Job performance was influenced by concerns about teamwork, supervision, information, training, and nurses, physicians, and patients. Job satisfaction was influenced by concerns about problem solution, rewards, people at work, and nurses and physicians.

It would seem that if the managers of these institutions wanted to introduce a change, they would have to address the credibility of communications. All efforts would need to be carefully developed to show employees that management had changed, that behaviors were being rewarded, and that the desired behavior would be consistent with their old values. The probable starting place would be the demonstration of management's new posture by new actions, activities, and communications from management.

Culture exists. Culture must be considered in management. Change cannot be undertaken unless culture is considered and simultaneously altered. Culture may or may not be what one would like but it is a relevant factor in operations, planning, and strategy.

REFERENCES

1. Kroeber, A.L., and Kluckhohn, C. *Culture: A Critical Review of Concepts and Definitions.* New York: Vintage Press, 1952.
2. Service, E. *Cultural Evolutionism: Theory in Practice.* New York: Holt, Rinehart & Winston, 1971.
3. Ulrich, W.L. "HRM and Culture: History, Ritual and Myth." *Human Resource Management* 23, no. 2 (1984): 117–128.

4. Koch, D.L., and Steinhauser, D.W. "The Changing Corporate Culture." *Datamation.* (October 1983): 247–256.
5. Service, *Cultural Evolutionism.*
6. Donnelly, R.M. "The Interrelationship of Planning with Corporate Culture." *Managerial Planning* 29 (June 1984): 8–12.
7. Fierman, J. "The Corporate Culture Vultures." *Fortune* 108 (October 17, 1983): 66–72.
8. Short, L.E., and Ferratt, T.W. "Work Unit Culture: Strategic Starting Point in Building Organizational Change." *Management Review* 73 (August 1984): 15–19.
9. Albert, M. and Silverman, M. "Making Management Philosophy a Cultural Reality, Part 1: Get Started." *Personnel* 61 (January–February 1984): 12–21.
10. Schein, E.H. "Coming to a New Awareness of Organizational Culture." *Sloan Management Review* 25 (Winter 1984): 3–16.
11. Sathe, V. "Implications of Corporate Culture: A Manager's Guide to Action." *Organization Dynamics* 12 (Autumn 1983): 5–23.
12. Peters, T.J., and Waterman, R.H. *In Search of Excellence.* Cambridge, Mass.: Harper & Row, 1982.

APPENDIX

EXPLANATION OF FACTORS INFLUENCING CULTURE IN SIX HEALTH CARE FACILITIES

Appreciation factors (obtained from all A responses)

Factor 1: supervision	Factor 2: cohesion	Factor 3: rewards	Factor 4: nurses and physicians
Supervisor helps with problems	Cooperation between groups	Policies of management	Nurses
Supervisor listens to your problems	People you work with	Equipment you work with	Physicians
Your supervisor	Everyone is trying best	Benefits	Patients
You receive enough information	Quality is important	Pay	
Management wants to know what you think	Promotion		

Job performance factors (obtained from all B responses)

Factor 1: teamwork	Factor 2: supervision	Factor 3: information	Factor 4: training	Factor 5: nurses and physicians
Teamwork	Supervisor	You get enough information	Equipment	Nurses
Cooperation with other groups	Supervisor helps with your problems	Pay	Surroundings	Physicians
Others try their best	Supervisor encourages	Opinions are heard	Training	Patients
The people I work with	Decision process			
Group thinks quality is important				

Job satisfaction factors (obtained from all C responses)

Factor 1: problems	Factor 2: rewards	Factor 3: people	Factor 4: nurses and physicians
Supervisor can help with your problems	Equipment	People with whom I work	Nurses
Your ideas are encouraged	Benefits	Cooperation	Physicians
Your supervisor values your opinions	Surroundings		

Using managerial role motivation theory to predict career success

Max G. Holland,
Cameron H. Black,
and
John B. Miner

Managerial role motivation theory has proved to be useful for understanding executive performance in a wide range of highly structured organizational environments. Consistent results of studies indicate that the theory may be useful for understanding managerial behavior and predicting performance in health care organizations.

While some people exhibit outstanding performance as staff members or first-line supervisors, they fail to achieve comparable success when elevated to executive positions. It is important to understand the differences in managerial performance and the specific characteristics or attitudes associated with the career success of a manager. Research addressing such issues has been conducted across a wide range of organizational settings, but little has been conducted within the context of health care organizations. The research on which this article is based looks at attitudes toward certain managerial roles and their relationship to career achievement in hospital administration.

MOTIVATION TO MANAGE

Three important ingredients for success in any endeavor include appropriate skills, understanding of goals and expectations, and motivation or the desire to succeed. None of these is more important than the others. In fact, if one factor is lacking, the others may not compensate, no matter how strong they are. For example, a person may be highly motivated to perform a job, but without the right skills or a clear understanding of goals and objectives, that person is unlikely to succeed. Likewise, a skilled person who lacks motivation may do poorly. While all three factors are important to the managerial job, managerial motivation is of particular significance. It may be expressed as the desire or willingness to carry out specific role requirements associated with managerial success.

Miner's[1,2] managerial role motivation theory is based on the principle that certain role requirements can be associated with managerial success. Those who have more motivation to manage are more successful in managerial roles. Miner's original research

---

Max G. Holland, *Ph.D., is a Professor of Management and Health Administration at Georgia State University, Atlanta, Georgia, and serves on the faculties of the Institute of Industrial Relations and the Gerontology Center.*

Cameron H. Black, *Ph.D., is a consultant to numerous organizations. She was previously an Assistant Professor of Business Administration at Dalhousie University in Halifax, Nova Scotia.*

John B. Miner, *Ph.D., is a Professor in the School of Management, Jacobs Management Center, State University of New York at Buffalo.*

on this topic began while he was associated with a major oil company.[3] In studying research and development managers, he found that managers who focused on scientific and professional pursuits were less effective than those with broader organizational interests.

This research led to the development of a theoretical framework that associates specific role requirements with managerial performance within the structured environments of large organizations. Although managerial jobs vary from one organization to another, certain roles appear to be generic across various environments. The foundation of managerial role motivation theory is based on the role requirements expressed in the following assumptions[4]:

1. Managers are expected to deal effectively with their superiors and to obtain support for actions at higher levels. Thus managers should have positive feelings toward those holding positions of authority over them.
2. Managers are expected to accept the challenge of the job and achieve results for themselves and their subordinates. This expectation requires managers to have a favorable disposition toward competition, as a negative attitude will likely result in behaviors that fall short of role expectations.
3. Managers are expected to take charge and behave in an active and assertive manner. Those who prefer passivity or who dislike assertiveness probably lack the motivation to perform some of the roles of a manager.
4. Managers are expected to exercise power over subordinates. Those who dislike imposing their wishes on others probably will not be as effective in managerial roles as those who get satisfaction from directing others.
5. Managers are expected to be unique and to assume a position of high visibility. To carry out this role, managers must be willing to deviate from subordinate groups and behave in a manner that may invite attention, discussion, and criticism from subordinates. People who enjoy being the center of attention will more likely be effective in this way.
6. Managers are expected to perform routine administrative tasks that accompany the job. People who are uncomfortable with administrative realities are less likely to fulfill this role than those who face these tasks directly and positively.

Role motivation theory has stood the test of validity when applied to a wide range of managerial positions in numerous large, hierarchically structured, bureaucratic organizations. These organizations represent private industry, the military, and government. A recent view of 26 validation studies in such organizations provides ample evidence to support the theory,[5] but studies conducted outside the theory's appropriate domain (i.e., in less structured environments) have consistently failed to yield significant findings. Such research outside the hierarchical domain included university faculty members, management consultants, sales personnel, small entrepreneurs, research scientists, and the like.

At an earlier time hospital management may have fallen outside the bureaucratic domain and within the ideological domain of religion or the professional domain of medicine. But hospital executives have increasingly assumed roles comparable to those of top-level managers in other high technology industries.[6] The modern administrator is no longer simply a business manager or a coordinator of resources, but is increasingly recognized as a top-level corporate executive. This is reflected in changing organizational structures, changing executive titles, increasing compensation, and increasing status.

It may be assumed that hospital executives with higher levels of managerial motivation will have experienced greater career success than their counterparts with lower managerial motivation.

As a starting point, it may be assumed that hospital executives with higher levels of managerial motivation will have experienced greater career success within their work domain than their counterparts with lower managerial motivation. Assuming that career success may be reflected by advancement to top positions in larger hospitals, by higher compensation levels, and by more rapid advancement to the position of chief executive officer (CEO), the following hypotheses were examined in this study:

1. There is a positive relationship between managerial motivation and the size of the hospital; therefore, higher managerial motivation should be associated with larger facilities. This hypothesis is based on the fact that interorganizational

movement from smaller to larger facilities is often perceived to reflect career advancement and is often based on job performance.
2. There is a positive relationship between managerial motivation and executive compensation; therefore, higher managerial motivation should be associated with a higher salary. This hypothesis is based on the principle that outstanding performance is recognized through financial rewards.
3. There is a positive relationship between managerial motivation and rapidity of movement to the chief administrative position; therefore, more managerial motivation should be associated with less time spent in hospital employment prior to becoming chief executive officer. This hypothesis assumes that high performance in lower-level management positions will result in a faster promotion to a chief executive position.

METHODOLOGY

Testing these hypotheses requires a valid instrument for measuring managerial motivation based on the concepts underlying managerial role motivation theory. It also requires an assumption that managerial motivation can, at least in part, be described by the variables considered in the hypotheses. In addition, it necessitates a representative sample of executives to provide the data for the study.

Measurement instrument

The Miner Sentence Completion Scale (MSCS) has been used for over 25 years to measure managerial motivation. It is intended for use within the context of hierarchical organizations as specified by Oliver[7] and is designed to measure attitudes that reflect motivation variables underlying managerial role motivation theory.

The MSCS is designed to measure emotional responses to specific role-related variables by converting a set of defined stems (sentence beginnings) into complete sentences. Some examples follow:
- Federal judges (*are all incompetent*). This reflects a negative opinion toward an authority figure.
- When one of my staff asks me for advice (*I tell him or her what to do*). This reflects a positive response or a willingness to impose wishes on others.
- When running a race (*I always wear running shoes*). This response toward competition does not reflect a positive or a negative response; it simply states a fact.

The MSCS is based on seven subscales with five items each. The items are scored to reflect positive (+1), neutral (0), or negative (−1) responses. Each subscale has a potential range of +5 to −5. The aggregate MSCS score is computed by summing the subscales and has a potential range of +35 to −35.[8,9] There are two versions of this instrument. The original MSCS requires participants to write responses to each stem. The version of the MSCS used for this research allowed participants to select one of six possible alternatives for completing the sentences.

The subscales of the MSCS are designed to measure the attitudinal predisposition of the subjects toward the specific motivational variables of the theory. In the case of competition, there are two subscales to differentiate between competition in games and competition within work-related situations. There is one subscale for each of the other five characteristics. The seven subscales of the MSCS are therefore defined as
1. authority figures,
2. competitive games,
3. competitive situations,
4. assertive roles,
5. willingness to impose wishes on others,
6. desire to stand out from the group, and
7. routine administration functions.

It would be ideal to relate MSCS scores of the subjects to precise measures of performance as managers. However, obtaining comparable performance measures across multiple organizations presents a difficult, if not insurmountable, task. Therefore, proxy measures were used. These measures included hospital size, financial compensation, and rate of advancement to a chief executive position.

Hospital size was considered to be a reasonable measure of performance since it can be shown that career advancement in hospital administration often involves interorganizational mobility as one assumes the top position in a larger, more prestigious institution. Hospital size was measured by the number of beds, according to figures published by the American Hospital Association.

Financial compensation was also considered to be a measure of performance. It was recognized that income may be related to variables such as seniority and age, but it is also a reasonable measure of performance. For example, high-performing managers within a specific age or experience classification are expected to receive higher financial rewards than those with lower levels of performance. Salary information was provided by the respondents through a biographical data sheet that accompanied the questionnaire. Each administrator was requested to check a specific income category as opposed to providing actual salary.

The biographical data sheet also requested information related to work experience. This included the total number of years as a chief executive officer and the number of years in other positions. The criterion for success was considered to be number of years in health care organizations prior to promotion to a chief executive position. This assumes that those who exhibit superior performance in lower-level management positions will be rewarded by more rapid advancement to a chief executive position.

As might be anticipated, age was found to be significantly related to certain personal characteristics, such as income and total work experience, but measures that would correct for these age differences were not considered since a negative relationship was found to exist between age and the total MSCS score. That is, older subjects tended to score lower than their younger counterparts. However, correlations between age and MSCS scores were low and were not statistically significant. Furthermore, other research with practicing managers has failed to establish consistent age relationships with MSCS variables.[10]

Sample

The data for this study were obtained by mail from 668 chief executive officers of acute care hospitals in 16 states. The response rate was approximately 36 percent of the administrators who were originally asked to participate in the study. A Kolgomorov-Smirnov goodness-of-fit test was used to compare the hospital characteristics of respondents to the total population. The sample was found to be representative of the population.

The study focused on the South, Southwest, and West Coast because these are the regions of the country in which investor-owned hospitals are concentrated. An initial objective of the research was to compare administrators in investor-owned hospitals with those in nonprofit organizations. In this connection, Black[11] concluded that there were no significant differences in MSCS scores between the administrators

TABLE 1

CHARACTERISTICS OF STUDY SAMPLE
(N = 668)

Category	Percent
Region	
South Atlantic (Florida, Georgia, North Carolina, South Carolina)	24.3
Mid-South (Alabama, Arkansas, Louisiana, Tennessee)	26.6
Southwest (Arizona, New Mexico, Oklahoma, Texas)	23.1
West Coast (California, Oregon, Washington)	26.0
Hospital ownership	
State and federal	8.8
Local government	36.7
Voluntary	42.7
Corporate for-profit	11.8
Sex of hospital administrators	
Male	92.1
Female	7.9
Age of hospital administrators	
Under 40	31.9
40–49	31.0
50–59	26.0
60 and over	11.1

in investor-owned hospitals and those in voluntary nonprofit facilities or government facilities. Thus, differentiations along ownership lines were not included in the current study. A profile of the sample is shown in Table 1.

The data were analyzed in two ways. Correlations were used to compare MSCS scores with the criteria. In addition, one-way analyses of variance were carried out to compare MSCS scores between meaningful criteria categories. Because of missing data, these analyses typically used less than the entire sample. None of the analyses, however, represented samples of fewer than 646 respondents.

RESULTS

All three hypotheses were confirmed with analyses of variance and correlations. The data in Table 2 indicate relatively weak but consistent support for the hypotheses.

The data clearly show a positive relationship between hospital size (number of beds) and total MSCS score. The mean scores by hospital size categories are presented to demonstrate in practical terms that ad-

TABLE 2

RELATIONSHIPS OF MSCS TOTAL SCORE TO CRITERION VARIABLES

	N	Mean MSCS score	Analysis of variance	Correlation
Number of beds			3.19*	.13†
Fewer than 100 beds	287	6.18		
100–299 beds	270	6.90		
300 beds or more	110	7.73		
Administrator's compensation			7.82†	.11†
Less than $60,000	514	6.41		
$60,000 or more	141	7.92		
Time in health care employment prior to promotion to CEO			4.89†	−.13†
4 years or less	226	7.84		
5–9 years	199	6.80		
10 years or more	221	5.92		

* $p < .05$
† $p < .01$

ministrators of larger hospitals reflected higher scores than those in smaller facilities.

Although the correlations are weak, a significant relationship between mean MSCS score and financial compensation supports hypothesis two. These findings indicate that executives with the highest levels of managerial motivation as measured by the MSCS are compensated accordingly. It may be presumed that highly motivated executives are rewarded for performance within their current positions, or that they move to other organizations that offer higher financial compensation.

Hypothesis three is also confirmed through these statistical tests. Those with the most rapid rates of promotion to CEO positions (fewer years of lower-level experience) reflect higher MSCS scores (resulting in a negative correlation coefficient). The mean MSCS score of those achieving a CEO position in four years or less is 32 percent higher than those working ten years or more prior to promotion to CEO. These findings support the concept that managerial motivation is related to the rate of advancement. It could be assumed that rapidly rising, long-term administrators develop managerial motivation on the job, but this interpretation is not consistent with the fact that a negative relationship was found to exist between age and MSCS scores. In addition, there is no significant relationship between MSCS scores and total work experience. These findings are consistent with studies in other industries, which yield similar conclusions in comparing age and total experience to managerial motivation.[12]

All three factors used to measure performance were shown to be independently related to MSCS scores, and thus it may be concluded that motivation to manage is associated with managerial performance as measured by each factor. It may be of more interest, however, to consider the effects of combining these factors and to contrast those subjects with the highest MSCS scores to others.

The data presented in Table 3 present the mean MSCS scores by various two-way analyses. Although other statistical techniques were used to achieve and verify these findings, the data presented in Table 3 are easy to understand and to interpret. The classifi-

TABLE 3

MEAN TOTAL MSCS SCORES BY TWO-WAY CLASSIFICATION

Executive compensation	Hospital size Larger	Hospital size Smaller	Analysis of variance
Higher	7.38	6.77	1.50
Lower	7.83	6.35	

Hospital size	Promotion rate Faster	Promotion rate Slower	
Larger	9.69	7.44	4.66*
Smaller	7.01	5.95	

Executive compensation	Promotion rate Faster	Promotion rate Slower	
Higher	10.51	6.46	6.32*
Lower	6.21	6.84	

* $p < .01$

cations are based on hospital size (200 beds or larger versus smaller), income ($60,000 or more versus less), and rate of promotion to a CEO position four years or less versus more time).

The data in Table 3 clearly indicate that although hospital size and executive compensation are each independently related to MSCS scores, in combination they are not adequate predictors of these scores. On the other hand, combining promotion rate with each of the other factors reveals some significant findings. Combining promotion rate with hospital size shows that administrators of larger facilities who achieved a CEO position most rapidly demonstrated significantly higher scores on the MSCS than did others. Of greater significance was the combination of promotion rate and executive compensation. Those with higher salaries and more rapid rates of promotion exhibit extremely high MSCS scores.

An analysis that considered combinations of all three factors was also conducted. It was found that administrators who have greater incomes, who are affiliated with larger hospitals, and who advanced to a CEO position most rapidly demonstrated significantly higher MSCS scores than those with any other combination of these factors. The mean values for the total MSCS score and subscale scores are presented in Table 4.

The subscale scores are included in Table 4 to compare specific characteristics of the highest scoring groups to others. The most significant finding is that those in the highest scoring category have a stronger willingness to impose wishes on others (power motivation) and are more competitive (in-work related situations as well as in games). They also show a stronger willingness to stand out from the work group (to be unique) and have a more favorable attitude toward authority figures (acceptance of organizational realities). The higher score on assertiveness was not statistically significant. The mean score on routine administrative functions was lower for those in the highest scoring group, reflecting a possible desire to deal with broader organizational issues as opposed to detailed tasks.

These data are presented in the hope that they will be enlightening and will reinforce the concept that managerial motivation is an important component to managerial success. It is clear that there are differences in the performance measures used in comparing those with high MSCS scores and those with lower scores. The data show that those who have achieved greater success in managerial roles demonstrate higher MSCS scores that are primarily influenced by attitudes toward engaging in competition, using power, being unique, and respecting authority.

TABLE 4

COMPARISON OF TOTAL AND SUBSCALE SCORES*

	Mean scores of administrators	All other administrators	Analyses of variance
Total MSCS score	11.88	6.57	13.14†
Subscales			
Authority figures	1.38	0.64	3.98‡
Competitive games	2.69	1.53	4.99‡
Competitive situations	1.94	0.62	7.97‡
Assertive roles	1.00	0.33	2.37(n.s.)
Willingness to impose wishes on others	1.37	0.43	7.62§
Desire to stand out from the group	2.50	1.76	4.23‡
Routine administrative functions	1.00	1.26	0.51(n.s.)

* Comparison between administrators in the largest hospitals, with the highest salaries and the most rapid promotions, and all other administrators.
† $p < .001$.
‡ $p < .05$.
§ $p < .01$.
n.s. = not significant.

DISCUSSION

The results of this study consistently support the hypotheses, indicating that managerial role motivation theory is applicable to hospital administration and that the MSCS may be useful in providing understanding and predicting outcomes within that context. It should be noted, however, that the general

> *The results of this study indicate that managerial role motivation theory is applicable to hospital administration and that the MSCS may be useful in providing understanding and predicting outcomes within that context.*

results, although statistically significant, are not as strong as in some research based on the MSCS. This appears to be a consequence of several factors. First, studies conducted across a large number of organizations typically result in weaker findings than those conducted within a single organization or a small number of organizations, because decisions, such as those on promotion and compensation, are influenced by a wide range of individuals, organizational cultures, and value systems. In other words, each organization has a unique set of criteria for measuring performance and this article represents 668 individuals, all from different hospitals, thus compounding the measurement problem.

Second, the criteria used in this research were intended as substitute measures for actual job performance. They are rough approximations at best, influenced by considerations such as financial constraints, seniority, market conditions, and the availability of managerial talent, as well as actual job performance. Although the relationships shown here are interesting and theoretically important in their own right, more circumscribed motivation–behavior relationships would prove stronger than those involving these global indexes, and thus would yield even stronger findings. For example, comparable performance evaluation measures between all subjects would be highly desirable for such a study, but gathering such data would be impossible across the wide range and large number of organizations represented in this study.

Finally, the analysis of variance results are stronger than the correlations, simply because of the skewing that occurs on some of the criteria. The mean differences noted for these analyses are substantial and meaningful, indicating there are significant differences that offer strong support to the managerial role motivation theory.

A final point relates to the level of the hospital administrators' MSCS scores. The overall mean total in the sample was 6.7. Is this high or low relative to other relevant groups? Although the multiple-choice version of the MSCS has been used much less frequently in research than has the free-response version, there are some comparative data. In a sample of U.S. Army officers attending the Command and General Staff College at Ft. Leavenworth, Kansas, the mean multiple-choice score was 7.0.[13] Public school principals in a large school district had an average score of 5.7.[14] Officers in a metropolitan police department had a mean score of 3.1,[15] and supervisors in a state agency had a mean score of 2.7.[16] Among supervisors in a major corporation the mean score was 5.4.[17]

Relative to these groups the sample of hospital executives scored high. The mean score of administrators of larger hospitals who are paid particularly well and who moved into top positions relatively rapidly was 11.9 (more than 80 percent higher than the rest of the sample). Scores of these participants are comparable to those of CEOs in very large private corporations,[18] and are substantially higher than the typical managerial population that has been measured by the MSCS. The data consistently support the conclusion that managerial role motivation theory is applicable to the field of hospital administration.

● ● ●

This research is the first known effort to use managerial role motivation theory within the context of hospital administration. While other studies have been carried out in the health care industry,[19] none have dealt specifically with CEOs. The results of this study are sufficiently promising to encourage further research and to stimulate an interest in managerial motivation as an important consideration in selecting and developing managerial talent. Of course, performance in the managerial job depends on a number of factors. Motivation should not be viewed as the only, or even the most important, consideration for assessing managerial potential. But it is a factor that should not be overlooked in recruitment, selection, promotion, and performance evaluation.

This research is not intended to suggest that a specific psychometric instrument, such as the MSCS, be used as the sole basis for personnel decisions. It is intended to show that managerial role motivation theory and its underlying assumption related to managerial behavior are applicable to hospital administration and that motivation can be associated with certain performance measures. The findings of this study should be of special interest to those responsible for recruiting, selecting, and developing managerial talent, such as hospital boards, corporate human resource personnel, executive search firms, and hospital executives.

The research clearly implies that managerial motivation, as measured by the MSCS, can be associated with specific measures of career achievement within hospital organizations. The predisposition toward certain role prescriptions was shown to be significantly related to financial compensation, organizational size, and rate of promotion to a CEO position. If these are accepted as reasonable criteria, this research leaves little doubt that managerial role motivation theory is applicable to managers in hierarchically structured health care organizations.

REFERENCES

1. Miner, J.B. *Studies in Management Education*. Atlanta: Organizational Measurement Systems Press, 1965.
2. Miner, J.B. "Limited Domain Theories of Organizational Energy." In *Middle Range Theory and the Study of Organizations*, edited by C.C. Pinder and L.F. Moore. Boston: Martinus Nijhoff, 1980, pp. 273–86.
3. Miner, J.B. "Sentence Completion Measures in Personnel Research: The Validation of the Miner Sentence Completion Scales." In *Personality Assessment in Organizations*, edited by H.J. Bernardin and D.A. Bownas. New York: Praeger, 1985, pp. 145–76.
4. Miner, J.B. "Twenty Years of Research on Role-Motivation Theory of Managerial Effectiveness." *Personnel Psychology* 31 (1978): 739–60.
5. Miner, "Sentence Completion Measures."
6. Schultz, R., and Johnson, A.C. *Management of Hospitals*. New York: McGraw-Hill, 1976.
7. Oliver, J.E. "An Instrument for Classifying Organizations." *Academy of Management Journal* 25 (1982): 855–66.
8. Miner, J.B. *Scoring Guide for the Miner Sentence Completion Scale*. Atlanta: Organizational Measurement Systems Press, 1964.
9. Miner, J.B. *1977 Supplement-Scoring Guide for the Miner Sentence Completion Scale*. Atlanta: Organizational Measurement Systems Press, 1977.
10. Miner, J.B. *Motivation to Manage: A Ten Year Update on the "Studies in Management Education" Research*. Atlanta: Organizational Measurement Systems Press, 1977.
11. Black, C.H. *Managerial Motivation of Hospital Chief Administrators in Investor-Owned and Not-for-Profit Hospitals*. Ph.D. diss., Georgia State University, 1981.
12. Berman, F.E., and Miner, J.B. "Motivation to Manage at the Top Executive Level: A Test of the Hierarchic Role Motivation Theory." *Personnel Psychology* 38 (1985): 377–91.
13. Lardent, C.L. *An Assessment of the Motivation to Command among U.S. Army Officer Candidates*. Ph.D. diss., Georgia State University, 1979.
14. Lovett, M.C. *The Efficiency of the Miner Sentence Completion Scale for School Administrator Selection*. Ed.D. diss., University of Florida, 1984.
15. Holland, A.M. *Comparative Analysis of Selected Predictors of Police Officer Job Performance*. Ph.D. diss., Georgia State University, 1980.
16. Quigley, J.V. *Predicting Managerial Success in the Public Sector: Concurrent Validation of Biodata and the Miner Sentence Completion Scale in the Georgia Department of Human Resources*. Ph.D. diss., Georgia State University, 1979.
17. Miner, *1977 Supplement-Scoring Guide for the Miner Sentence Completion Scale*.
18. Berman and Miner, "Motivation to Manage at the Top Executive Level."
19. Holland, M.G. "Can Managerial Performance be Predicted?" *Journal of Nursing Administration* 11, no. 3 (1981): 17–21.

Problem solving by hospital managers

Teresa M. Steffen
and
Paul C. Nystrom

When managers confront complex problems, their attitudes toward problem solving affect their behavior. The problem-solving attitudes of over 100 women and men who manage six hospitals are analyzed.

People's attitudes about problem solving differ dramatically. For instance, some problem solvers value a rational approach devoid of any emotionality or subjectivity. Other problem solvers prefer using intuition to solve problems. Some individuals seem to be more concerned about reaching a solution than are others. Some people gain more enjoyment from problem solving than do others. People with a high need for certainty spend considerable time attempting to define problems clearly and they seek to avoid ambiguous situations. Ethical considerations seem more important to some individuals than to others engaged in problem solving.

WHEN ATTITUDES INFLUENCE BEHAVIOR

Individual differences in attitudes may not trigger important differences in behavior in certain situations.[1] When people work in routine jobs, their behavior is largely controlled by bureaucratic rules, standard operating procedures, and close supervision directed at compliance.[2] However, when people work in jobs requiring the exercise of substantial discretion, their attitudes can directly and strongly influence their behavior.[3] Organizational designs typically route complex problem-solving tasks to those employees whose jobs require discretion.[4] These employees usually are managers and other professionals.

Managers and other professionals are assigned a subset of problems known as ill-structured problems. Problems can be arrayed along a continuum that ranges from well-structured to ill-structured. This concept of an ill-structured problem originated in research on artificial intelligence.[5,6] Problems can be ill-structured because they lack clear definitions of desired goals, or lack clear assessments of current conditions, or lack clear indications of the processes that would enable the achievement of goals.[7] Ill-structured problems provide more opportunities than do well-structured ones for problem-solvers' attitudes to affect their behaviors.

Teresa M. Steffen, *P.T., M.S., in Health Care Management, is a doctoral student in Management Science at the University of Wisconsin-Milwaukee, and Director of Program Development at Share Therapeutic Services in Wisconsin.*

Paul C. Nystrom, *Ph.D., is a Professor of Organizations and Strategic Management in the School of Business Administration at the University of Wisconsin-Milwaukee.*

PRIOR EVIDENCE

Past research produced a questionnaire that measures attitudes about problem solving.[8] Statistical analyses yielded six separate factors, as shown in Table 1. Factor names reflect the basic content of the items that had been found to cluster together empirically. Reading this list of 21 items reveals that some statements describe beliefs and other statements describe values. Beliefs address descriptive issues that explain the whys and hows, known technically as cause–effect relationships. Values address prescriptive issues that enunciate the shoulds, which indicate preferences and normative standards.[9]

TABLE 1

FACTOR LOADINGS FOR ITEMS ON PROBLEM SOLVING[10]

Rationality
- Problems should be solved without any emotional involvement. 1* .74
- Problems ought to be solved more scientifically than by intuition. 2 .60
- Problems need to be viewed unemotionally. 3 .72
- Problem solving is a rational process. 7 .54
- If mathematical techniques were used more in problem solving, the overall quality of the solutions would improve. 11 .55
- I try not to let my feelings interfere when I am solving a problem. 17 .71

Need for intuition
- Many problems are first solved intuitively and data are gathered to support the solution. 5 .42
- There are some problems that cannot be solved. 16 .62
- Not all problems have solutions. 18 .58
- I tend to jump to conclusions when solving a problem. 21 .56

Solution concern
- Sometimes we ought to use the solution to a problem even though it is harmful to the environment. 10 –.44
- The most important aspect of problem solving is reaching the best solution. 12 .61
- Getting people to use the solution is the most important aspect of problem solving. 13 .70
- Solutions should always lead to a better world. 14 .60

Enjoyment
- Problem solving can be aesthetically pleasing. 4 .54
- I enjoy solving problems. 8 .50
- It is important that the problem solving be approached with a step-by-step procedure. 9 –.49

Need for certainty
- Problems that are abstract are a waste of time. 19 .55
- Computers can solve problems better than people. 20 .53

Ethics
- Solutions to problems ought to be evaluated on moral grounds as well as technical. 6 .52
- The best solution to a problem should be implemented even if it is legally questionable. 15 –.55

*Numerals immediately following items indicate sequential position in questionnaire.

TABLE 2

MEANS FOR OTHER PROFESSIONAL GROUPS' ATTITUDES TOWARD PROBLEM SOLVING[11]

Attitudes	Research engineers N = 25	Engineering students N = 28	Business managers N = 30	Pre-law students N = 26
Rationality	13.38 (2.42)*	13.28 (1.97)	14.38 (2.60)	11.99 (2.63)
Need for intuition	7.54 (1.41)	7.20 (1.86)	6.58 (1.75)	7.69 (1.81)
Solution concern	4.24 (1.74)	5.43 (1.47)	4.89 (1.46)	5.70 (1.44)
Enjoyment	2.50 (.91)	1.66 (1.02)	2.31 (.83)	2.16 (.88)
Need for certainty	2.30 (.95)	2.72 (.72)	2.11 (.80)	2.18 (.74)
Ethics	.58 (.65)	.16 (.77)	1.60 (.62)	1.24 (.61)

*Standard deviations reported in parentheses below means.

When four samples of professionals completed the questionnaire, they tended to display predictably different attitudes toward problem solving.[12] These expected differences in attitudes supposedly reflect training and socialization practices that differ between professions. The underlying assumption is that a profession's selection, education, on-the-job training, and reward systems tend to create relatively homogeneous groups of people.[13] Results in Table 2 illustrate the directions of these differences in problem-solving attitudes between four professions. For instance, student engineers expressed the highest need for certainty whereas business managers expressed the lowest need for certainty among the four groups studied.

Within a profession, individual differences in attitudes may stem partially from organizational determinants. Managers working in higher level jobs tended to express stronger attitudes favoring rationality than did managers in lower level jobs.[14]

Attitudes toward problem solving also can affect the ways that people recognize and define problems. A study of university students in a management course revealed that people who enjoyed problem solving also demonstrated a willingness to consider new feedback that contradicted their initial conceptualization of a problem.[15] On the other hand, people with a strong need for certainty seemed unwilling to alter their problem conceptualizations in the face of disconfirming feedback.

UNANSWERED QUESTIONS

The literature does not report any studies of health care managers' attitudes toward problem solving. In what ways do the problem-solving attitudes of health care managers differ from those of business managers? Previous studies have not investigated any potential differences in problem-solving attitudes based on gender. In what ways do women's attitudes toward problem solving differ from those of their male colleagues? Finally, nobody has analyzed the possible effects of organizational cultures on problem-solving attitudes. Do managers in different organizations express different attitudes toward problem solving? This study provides answers to all three of these questions.

A STUDY IN SIX HOSPITALS

The first author discussed this research idea with a top executive in each of six hospitals. After obtaining their willingness to participate, three-page questionnaires were provided for distribution to the managerial personnel in each hospital. All managers responded anonymously; they mailed their completed questionnaires directly to the university, and the questionnaires did not ask for a name or use any identification number.

The managers

The study gathered completed questionnaires from 103 managers. All six hospitals operate in one metropolitan area. Two hospitals are large, one medium, and three smaller in size. Two of the smaller hospitals emphasize specialities. These hospitals differ considerably in their financial well being.

Participants averaged 40 years of age, 11 years of managerial experience, and 5 years in their current job. The study included 49 women and 51 men (3 managers did not indicate their gender).

The questionnaire

A major part of the questionnaire contains 21 items that measure problem-solving attitudes. Each item uses a 5-point Likert scale ranging from (1) disagree strongly to (5) agree strongly. Each item response has been weighted by the factor loading reported in Table 1, and then summed across all items in a factor to yield a person's scores for these six factors.

The questionnaire originally had been developed by using a statistical procedure that identified maximally independent factors. Pearson correlation coefficients reported in Table 3 indicate that these factors also exhibit substantial independence in this study of hospital managers. As desired, most correlations do not differ significantly from zero, meaning no associations whatsoever. Only 2 of the 15 correlations reached a statistically significant level (p < .05). Rationality correlated directly with solution concern (r = .22) and inversely with enjoyment (r = .18).

MAJOR FINDINGS

Hospital managers compared with other professionals

In what ways do hospital managers possess problem-solving attitudes that differ from those of business managers? The 30 business managers referred to in Table 2 worked at vice-presidential or general management levels within major corporations. The t-tests re-

TABLE 3

CORRELATIONS BETWEEN PROBLEM-SOLVING ATTITUDES

Attitudes	Rationality	Need for intuition	Solution concern	Enjoyment	Need for certainty
Rationality					
Need for intuition	.05				
Solution concern	.22	.04			
Enjoyment	–.18	.12	–.01		
Need for certainty	.15	.03	.07	–.07	
Ethics	.07	.11	.04	.05	–.11

TABLE 4

DIFFERENCES IN PROBLEM-SOLVING ATTITUDES BETWEEN HOSPITAL MANAGERS AND BUSINESS MANAGERS

Attitudes	Hospital managers N = 103	Business managers N = 30	Difference t	Significant p
Rationality	13.12 (2.36)*	14.38 (2.60)	−2.49	.02
Need for intuition	7.07 (1.62)	6.58 (1.75)	1.42	
Solution concern	5.91 (1.48)	4.89 (1.46)	3.13	.002
Enjoyment	2.40 (.64)	2.31 (.83)	.63	
Need for certainty	2.04 (.76)	2.11 (.80)	−.44	
Ethics	1.10 (.67)	1.60 (.62)	−3.63	.002

*Standard deviations reported in parentheses below means.

ported in Table 4 indicate that those business managers place more emphasis on rationality; that is, they more strongly prefer to use mathematical techniques and to avoid emotionality than do the hospital managers. Hospital managers exhibit a higher concern for finding solutions than do the business managers. Business managers express a higher interest in the ethics factor than do the hospital managers. Hospital managers do not differ significantly from business managers in terms of need for intuition, enjoyment of problem solving, or need for certainty.

Comparing hospital managers with earlier findings for other professionals (refer back to Table 2), the hospital managers respond about the same on the rationality factor as did the research engineers and the engineering students. Hospital managers responded about the same on solution concern as did the pre-law students, but substantially higher than the research engineers. Finally, hospital managers do not exhibit a particularly low interest in ethical issues, as Table 4 might initially lead one to conclude. It appears that business managers responded unusually high regarding ethics, rather than hospital managers having responded unusually low regarding ethics. Hospital managers expressed a much stronger attitude regarding the ethics factor than did research engineers or engineering students.

Gender and other background characteristics

In what ways do female hospital managers possess problem-solving attitudes that differ from those of their male colleagues? Two problem-solving attitudes exhibit statistically significant differences based on gender (Table 5). Women express a higher concern for finding solutions; men express a higher need for certainty.

A comparison of the 27 upper-level managers with the 75 middle-level hospital managers yields only one significant difference. Upper-level hospital managers

TABLE 5

GENDER DIFFERENCES IN PROBLEM-SOLVING ATTITUDES OF HOSPITAL MANAGERS

Attitudes	Women N = 49	Men N = 51	Difference t	Significant p
Rationality	13.28 (2.27)*	13.03 (2.47)	−.51	
Need for intuition	7.31 (1.67)	6.85 (1.57)	−1.42	
Solution concern	6.26 (1.72)	5.59 (1.16)	−2.27	.03
Enjoyment	2.34 (.65)	2.44 (.61)	.79	
Need for certainty	1.87 (.72)	2.13 (.72)	1.78	.08
Ethics	1.13 (.70)	1.06 (.66)	−.50	

*Standard deviations reported in parentheses below means.

express greater enjoyment of problem solving (t = 1.94, p < .06).

On the whole, hospital managers' attitudes about problem solving are not related to their other background characteristics. Most correlation coefficients do not reach statistically significant levels. The few exceptions indicate moderate relationships of age with rationality (r = 14, p < .08), age with solution concern (r = .15, p < .07), years of education with enjoyment (r = .22, p < .02), and years of experience with rationality (r = .20, p < .03).

Similarities between hospitals

Do managers' attitudes about problem solving differ significantly between hospitals? Analysis of variance (ANOVA) tests comparing averages for each of the six hospitals indicate that four of the six problem-solving attitudes do not differ between hospitals (Table 6).

In order to isolate the precise nature of the two observed differences between hospitals that involved enjoyment and need for certainty, the authors utilized the Newman-Keuls procedure for testing pairwise contrasts between means. Managers' problem-solving attitudes in five hospitals do not differ from each other in terms of enjoyment or need for certainty. But one particular hospital stands out from the rest. This hospital was in the midst of a major reorganization accompanied by layoffs, including the elimination of some managerial jobs. Managers in this one hospital registered significantly lower enjoyment of problem solving as well as significantly higher need for certainty. Thus, the statistical analyses lend support to the questionnaire's validity.

Table 6

ANALYSIS OF VARIANCE COMPARING SIX HOSPITALS

Attitudes	Mean square between hospitals	Mean square within hospitals	F Ratio	Significant p
Rationality	8.65	5.42	1.59	
Need for intuition	3.40	2.60	1.31	
Solution concern	2.32	2.18	1.06	
Enjoyment	.86	.38	2.26	.05
Need for certainty	1.35	.53	2.55	.03
Ethics	.50	.44	1.14	

The t-tests indicate that business managers place more emphasis on rationality; that is, they more strongly prefer to use mathematical techniques and to avoid emotionality than do the hospital managers.

• • •

Hospital managers' attitudes about problem solving differ in some ways from those of business managers. Within the group of hospital managers, women's attitudes differ in some respects from those of men. Hospital managers' attitudes about problem solving are relatively independent of various background characteristics such as age, education, and years of experience. Finally, and perhaps most importantly, hospital managers' attitudes about problem solving typically do not differ significantly even though they manage very different hospitals.

These findings suggest that managers of professional organizations such as hospitals share similar beliefs and values. These beliefs and values, including attitudes about problem solving, probably take shape during the college years and then become reinforced by socialization processes directed at new job holders. It appears that an occupation's culture exerts more influence than does an employing organization's culture.[16] A similarity in problem-solving attitudes among managers in different types of hospitals implies an ease of managerial mobility between hospitals. Managers who approach ill-structured problems in a similar way undoubtedly can communicate and cooperate more easily than can managers who approach problems in ways that differ and conflict.

REFERENCES

1. Nystrom, P.C., and Starbuck, W.H. "Managing Beliefs in Organizations." *Journal of Applied Behavioral Science* 20 (August 1984): 277–87.
2. Nystrom, P.C. "Designing Jobs and Assigning Employees." In *Handbook of Organizational Design*, vol. 2, edited by P.C. Nystrom and W.H. Starbuck. New York: Oxford University Press, 1981.
3. Jaques, E. *Equitable Payment*. New York: Wiley, 1961.
4. Mintzberg, H. *The Nature of Managerial Work*. New York: Harper & Row, 1973.
5. Newell, A. "Heuristic Programming: Ill-structured Problems." In *Progress in Operations Research*, edited by J. Aronofsky. New York: Wiley, 1970.
6. Simon, H.A. "The Structure of Ill-structured Problems." *Artificial Intelligence* 4 (1973): 181–201.
7. Reitman, W.R. "Heuristic Decision Procedures, Open Constraints, and the Structure of Ill-defined Problems." In *Human Judgment and Optimality*, edited by M. Shelly and G. Bryan. New York: Wiley, 1964.
8. Lyles, M.A. "An Empirical Investigation of Interdisciplinary Differences in Problem Solving Attitudes." *Proceedings*. Kent, Ohio: Midwest Division, Academy of Management, 1978.
9. Beyer, J.M. "Ideologies, Values, and Decision Making in Organizations." In *Handbook of Organizational Design*, vol. 2, edited by P.C. Nystrom and W.H. Starbuck. New York: Oxford University Press, 1981.
10. Lyles, M.A., and Mitroff, I.I. "Organizational Problem Formulation: An Empirical Study." *Administrative Science Quarterly* 25 (March 1980): 102–19.
11. Lyles, "An Empirical Investigation of Interdisaplinary Differences."
12. Ibid.
13. Van Maanen, J. "Breaking In: Socialization to Work," In *Handbook of Work, Organization, and Society*, edited by R. Dubin. Chicago: Rand McNally, 1976.
14. Lyles and Mitroff, "Organizational Problem Formulation."
15. Herden, R.P., and Lyles, M.A. "Individual Attributes and the Problem Conceptualization Process." *Human Systems Management* 2 (1981): 275–84.
16. Van Maanen, J., and Barley, S.R. "Occupational Communities: Culture and Control in Organizations." In *Research in Organizational Behavior*, vol. 6, edited by B. Staw and L. Cummings. Greenwich, Conn.: JAI Press, 1984.

Quality circles: the myth and reality of hospital management

Naoki Ikegami
and
Seth B. Goldsmith

Although quality circles have been successfully introduced into U.S. hospitals, they have not fared well in Japan. In light of the fact that Japanese hospitals have within their organizational structures many elements considered a precondition for the success of quality circles, their hesitance in adopting this innovation deserves attention. It is useful to examine the reasons quality circles may be more feasible in the United States than in Japan.

Although there have recently been reports of the successful introduction of quality circles (QCs) into U.S. hospitals,[1,2] QCs have not fared well in Japanese hospitals. In light of the fact that Japanese hospitals inherently have within their organizational structures many of the elements that must be considered a precondition for the success of QCs, their hesitance in adopting this innovation deserves attention.

First, the organizational features of Japanese style management that seem to be instrumental in the development of QCs must be examined. Then, problems must be identified in the application of QCs to the hospital setting. The authors suggest organizational changes that would seem to make the introduction of QCs more feasible in the United States than in Japan.

QCs: DEFINITION AND BACKGROUND

QCs, known in Japan as Quality Control Circles, are one way of achieving quality control. Quality control has been defined as "a methodological system for economically producing products to match the customers' needed quality."[3] The term *product* has been gradually enlarged to include not only manufactured goods but also services. The term *customer* does not mean just the buyer but also the workshop or team involved in the next process on the assembly line.

Original concept

The initial idea of quality control came to Japan from the United States after World War II as part of a methodology of scientific management that focused on the objective, statistical approach. However, this original concept has undergone a considerable metamorphosis by incorporating the prevailing cultural and ethical traditions of Japan. The objective, quanti-

---

Naoki Ikegami, *MD, D.Med.Sc., is Assistant Professor in the Department of Hospital and Medical Administration at Keio University School of Medicine in Tokyo, Japan. He received his medical and doctoral degrees from Keio University and his master's degree from the University of Leeds in England. In 1984 he was awarded a McGaw Fellowship by the Association of University Programs in Health Administration.*

Seth B. Goldsmith, *Sc.D., is Professor of Health Administration at the School of Public Health, University of Massachusetts at Amherst. He is the author of many articles and books on health care, Editor of the* Journal of Ambulatory Care Management, *and a columnist for a Japanese health care journal.*

tative approach has been combined with a moral, holistic philosophy.

This can best be illustrated by quoting from the official definition of QCs in Japan. QCs are defined as "small groups to perform voluntarily quality control activities within the same workshops. This small group carries on continuously as a company-wide quality control activities, self-development, control and improvements within the workshop utilizing quality control techniques with all members participating."[4]

The primary goal of the QC is "to contribute to the improvement and development of the enterprise."[5] The words *voluntarily*, *self-development*, and *all members participating* have clearly been developed within the Japanese context, as has been the emphasis on the company as a whole.

Contemporary developments

When analyzed from a theoretical viewpoint, the contemporary QC can be understood as a combination of the objective, quantitative methodology of scientific management and the participative, cooperative philosophy of industrial humanism. It may be considered as an optimal mixture of MacGregor's Theory X and Theory Y in which the organization's economic efficiency becomes the focal binding theme for meeting the need of the worker's social and psychological satisfaction in performing tasks.[6]

Without the quantitative perspective, the movement would lack objective criteria for assuring an increase in productivity and would remain as a moral exhortation for better workshop relationships. The statistical evidence is needed to demonstrate to both QC participants and management that actual improvement has occurred as a result of their activities.

Without the participative, cooperative perspective, there would be little hope that QCs could ever be voluntarily and even enthusiastically welcomed in the workshop. It is this philosophy that deserves attention, as it appears to have been the Japanese organizational environment which has led to its emergence.

The contribution that the Japanese have made in introducing the participative, cooperative perspective can be best illustrated by the divergence between Demming, the original advocate of quality control, and his Japanese disciples.

This divergence was pointed out by Williams, who noted that Demming's thesis seems to be that "*managers* are more responsible than *production workers* for quality of products. According to Demming, quality assurance is synonymous with a reliable process, and quality control requires sophisticated technical and statistical procedures."[7] While Demming proposed a managerial approach, the Japanese have turned it to a bottom-up, workshop-oriented, and voluntary activity.

The key turning point in this transformation would seem to be the founding of the journal *Workshop and QC* in 1962. It focused on the first assembly line foremen and was to serve as a practical and inexpensive textbook for QC activities. The inaugural edition of the journal called for the registration of QCs with the Japanese Association of Science and Technology; this could be regarded as the formal founding of QCs.

Since that time, over 80,000 QCs have been registered. Local and regional meetings of QCs have been organized for the purpose of presenting their experiences and exchanging information. The meetings are attended with almost religious fervor, and there appears to be little doubt that the successful QC activities are the result of voluntary contribution from QC members.

Thus there is a need to first study the organizational structure of Japanese management that has fostered this transformation. Japanese management style, which includes a cooperative spirit, a nonspecialized homogeneous personnel, lifetime employment, and the emphasis on long-tem objectives, will be analyzed from the viewpoint of explaining the underlying dynamics of QCs.

Although the authors are aware of the fact that sociohistorical tradition unique to Japan has made its contribution in determining these features, an attempt has been made to minimize their explanation in esoteric terms and to describe these aspects so that they can be understood and assimilated by westerners. As the development in QCs shows, it is possible to assimilate desired features to fit required ends.

JAPANESE MANAGEMENT—SALIENT FEATURES

Not just a corporation but a cooperation

One feature of the Japanese company that would awe the foreign visitor is the annual ceremony held in

April to welcome the new employees. Several hundred young people just graduated from college or high school, all in new suits, assemble to undergo an elaborate ritual. It begins with the opening words of the chief executive officer (CEO) stressing the ethical duties of members of the organization and concludes with a proclamation by a representative of the new employees vowing to abide by this code.

It is, in fact, an initiation to join a cooperative body and not just employment in a corporation. The new employees all start at the same line at the bottom to begin their long struggle to the top of the pyramid. Very few have postgraduate degrees, and their baccalaureate degrees usually have little relationship to their new tasks. The primary merit of this system is that although it takes a large effort in training new employees, they are open-minded and malleable. They can be indoctrinated to the values and customs of the corporate *Gemeinschaft* (community culture). Any expertise henceforth acquired can only be used in conjunction with the operation of the organization.

Thus there is less attachment to the acquired skill, and the organization meets with little resistance when transferring the individual to a new field or requesting that he or she handle auxiliary tasks as the needs of the organization change. It would be far more traumatic for the employee to quit and try to develop a new set of working relationships in a different organization. In most cases, it is better to remain in the cooperative workshop and to retain seniority rights and social standings.

What tends to be overlooked by outsiders who see the apparent harmony of the workshop is that the homogeneous nature of the personnel leads to a bitter struggle for promotion, as there are no distinguishing external features that mark new employees. The one way for the energetic young person to realize his or her ambitions is to participate actively in the immediate group in which he or she is associated.

Collective responsibility is taken by the work section for any improvement or mishaps, so it is of benefit for each member to advance the position of the section as much as possible. If the individual has made an exceptional contribution to the section, the section chief may strongly urge a symbolic increase in salary at the time of the annual advancement. This effectively places the individual ahead of those who had entered the company at the same time.

Because the number who can advance is limited, there is a fierce struggle between section chiefs as to which of their protégés should qualify for a position. Thus an interdependent concentric force maintains the dynamism of the organization.

The Japanese are just as ambitious as westerners, or even more so, as they are not conscious of any fundamental differences in ability or background. Furthermore, each member of the starting class has an equal chance to rise to the top of the organization with its cherished perquisites. The difference lies in the methods employed to realize their ambition; the Japanese tend to rely on a two-step process.

The first step is to promote the goal of his or her section, and the second step is to rely on an interdependent feeling of obligation with the section chief. The need to develop this mutual feeling of dependence leads to an emphasis on after-hours socializing between superiors and subordinates. Both have a vested interest in developing this relationship, for the superiors must be assured of commitment and good communication if they too wish to advance.

The Japanese concept of collectivism does not imply a total submergence of individual interests for the benefit of the group, but rather pursuit of self-actualization using the dynamics of interpersonal relationships.[8] It does not necessitate a complete shift from the western perspective, but rather a development of tendencies that have been relatively latent.[9] The driving force behind the energy displayed in those participating in QCs could be regarded as resulting from this basic cooperative and competitive environment, which, in turn, is derived from the homogeneous character of the workshop personnel.

Lifetime employment

Lifetime employment has been cited as one of the distinctive features of Japanese management. However, it is better to regard it as a necessary component arising from, and strengthening the cooperative nature of, that management. To function as a cooperative body, all members, especially core members, need to have a binding commitment to the organization. (Temporary workers and those employed in the small subcontracted companies typically do not have lifetime employment.)

Working for a rival company is regarded as betrayal by all concerned. The only time that commitment to the organization can be terminated with mutual good will is when the employee reaches the age of retirement.

For lifetime employment to be efficient, there is first a need to undergo an initial indoctrination on institutional goals and values. Without the readiness of the worker to undertake new assignments as dictated by the changing organizational needs, lifetime employment would lead to a rigid and stagnant work force. Consequently, there is an emphasis on developing generalists who can be rotated among different work places. This also minimizes the emergence of a subculture opposed to the general organizational objectives.[10]

The second requirement is to rank by seniority, as this provides a greater incentive for employees to be dexterous than does a merit system. A member has to remain in the organization if he or she desires to reach a position of responsibility or influence. This seniority is with respect to the number of years spent with the organization and does not necessarily reflect a greater age. Without this protracted process of competitive evaluation, lifetime employment would merely provide a feeling of security to the employees, which may not always benefit the organization.

When these conditions are met, lifetime employment provides a positive contribution to the building of a long-term relationship. Conflicts are better resolved when the mutual exchange of favors can be calculated over a relatively long time period. It serves as a balance to the constant striving between the various groups within the organization. Together with the shared institutional culture, lifetime employment acts as a cohesive force to integrate the conflicting proposals arising from the various QC activities. This force has to be powerful, for the ultimate aim of QCs is to establish an integrated companywide QC movement.

Ethical goals

There is an emphasis on the superordinate, ethical goals of the organization. It is believed that the organization does not exist just to earn profit from society, but to enrich the lives of society's members. Thus it follows that employees do not merely work to receive a justifiable financial remuneration but to make a contribution to this objective.

However, employees must realize that the material reward and the spiritual reward need to be mutually complementary. There is an implicit but strongly held belief that hard work will ultimately be rewarded by a

> *It is believed that the organization does not exist just to earn profit from society, but to enrich the lives of society's members.*

commensurate material gain, even if the pursuit of ethical duties may be the only overtly stated organizational goal. There appears to be a need for balancing this delicate relationship if the spirit of capitalism evident in both the Protestant ethic and the Japanese Buddhist traditions is to develop.[11]

The value of combining an explicit superordinate goal with an implicit feedback of material reward is that it is then possible to create a feeling of common destiny, which can be equally shared by both management and labor. Whereas labor has the task of increasing productivity, management has the responsibility of distributing the increased profit to labor. This mutual trust and confidence has to be constantly maintained by tangible and intangible feedback.

On the one hand, when the company is in financial difficulties it is the first duty of management to take a reduction in remuneration, with the president obliged to take the largest reduction. On the other hand, a special bonus of equal amount may be distributed to all employees on reaching an organizational goal. The savings and expenditure involved may be of token value; however, it substantiates the notion of common destiny among all concerned.

This aspect has particular relevance in the QC context as QC members are first educated to the fact that to increase their salaries, there is a need for all members to improve quality and efficiency.[12] That is, they must enlarge the pie before increasing their share of it. This is one reason why QCs must be transformed to a companywide movement with all members striving; otherwise, the active QC members would eventually become discouraged as their efforts would not be reflected in increased company proceeds. In addition, expansion of the company usually results in an increase of managerial level positions. Thus there is an implicit assumption that there would ultimately be a material reward from their activities in QCs.

QCs as an integral element

From the preceding discussion, several features have become clear. First, as the characteristics of Japanese management are mutually complementary and interdependent, the QC concept has to be totally introduced for it to be effective. Lifetime employment is a feature resulting from and enhancing the cooperative nature of the organization. The explicit superordinate goal and the implicit feedback of material reward are needed to provide the cohesive force for a shared destiny.

Second, Japanese management results in an optimum combination of cooperation and competition, which are provided by reliance on an interdependent group effort. This is the major underlying force in the success of QCs.

Third, it follows that QCs can best be implemented by a comprehensive adoption of Japanese style management. There would be little incentive for workers to participate voluntarily in QCs if they were expressly introduced by management for the purpose of greater productivity. Although there might be a momentary response from the satisfaction of participating in the managerial process, employees would tend to become gradually disillusioned if there did not exist an environment of shared institutional goals and an implicit feedback of rewards.

This conclusion might seem discouraging to those trying to use QCs as a convenient shortcut for increasing productivity. However, it has to be realized that unlike the original introduction of QCs to Japan, where it was a case of applying a scientific methodology, the reintroduction of QCs to the United States requires a change in management perspective.

This task is not as difficult as it seems; Ouchi[13] and Pascale and Athos[14] have illustrated that, given the necessary commitment by management, this transformation could be achieved. Peters and Waterman have gone further, describing the so-called Japanese style of management as not intrinsically Japanese at all, but a style that can be consistently found in successful U.S. companies.[15]

The questions remain of whether this concept of Japanese management could be realized in hospitals in general, and whether the organizational structures of the hospitals would make it more readily achieved in the Japanese or in the U.S. context.

JAPANESE MANAGEMENT AND HOSPITALS

Balance of power

The problems of applying Japanese management techniques to American hospitals have been discussed by Shortell.[16] The principal barrier appears to be that the balance of power within the hospital still tends to lie with the professionals. Not only are they numerically larger than in any other kind of organization, but their interests tend to be individually diversified.

The exocentric forces of professional identity are dynamically opposed to the concept of an intrinsic organizational value system. In this respect, professions may be regarded as resembling a strong labor union that defines the task limits of each occupational group. This sharply contrasts with the Japanese ideal of a homogeneous and malleable work force.

Legal constraints

Professional status not only gives its members corporate and individual leverage regarding the extent of duties, but it is strengthened by statutory regulations. The interpretation of the legal limit in activities tends to be decided more at a national level than at the workshop level. Thus the organization may find that although it would prefer to train and develop the required personnel to meet its objectives, it is barred by legal constraints.

Even in related professions, the usual requirement is to start at the bottom of the professional ladder, at an establishment independent of the hospital. This lengthy and expensive process requires a considerable commitment by the hospital and by the individual if an uncertified but competent and dedicated employee is to be promoted to a managerial position.

These legal restraints also provide the legitimacy for obeying implicitly the orders of the professionals and, in particular, physicians. Not only does the authority of the physician come from superior expertise, but also from the legal process of prime responsibility for clinical care. Thus for a physician to undertake a participative, group approach to clinical care, it is necessary not only for him or her to be committed to democratic principles, but also to be prepared to take any legal responsibility for the resources provided by the group. This may be the basis for the emphasis on individual responsibility in the training of professionals.

While the professional requirements may be a necessity for maintaining quality standards, their entrenchment backed by legal constraints tends to hinder greatly the adoption of Japanese management techniques. The contrast between the professional and the organizational perspectives becomes highlighted where QCs are concerned. QCs implicitly assume that the members are inherently equal in expertise and that the innovative process consists of a slow but continuous drawing out of the potentialities of each participant regardless of qualifications or background.

The qualities most highly regarded come from on-the-job training, and the expertise valued is one that has immediate practical consequence rather than one that comes from formal, structured training. The a priori assumption that professional status indicates greater expertise and authority is dynamically opposed to the democratic principles of QCs. A further problem is that the closely interdependent nature of health care services makes the clear delineation of QC group responsibility difficult.

Thus it is clear that hospitals are not generally a favorable environment for implementing QCs. However, if Japanese management techniques in general, and QCs in particular, are to be regarded as the organizational perspective most suited to the demands of contemporary society, it is necessary to further explore the possibility of adopting Japanese techniques in hospitals even if limited in scope.

BARRIERS TO QC IMPLEMENTATION IN JAPANESE HOSPITALS

As has been emphasized, for QCs to be introduced, there needs to be a commitment to the concept of Japanese management. The top executives of an organization have to convince themselves of its value and to demonstrate to all that they are in fact converted. This primary condition has been generally lacking in Japanese hospitals, mainly because the medical staff occupies an even more dominant position than in most countries.

First, the dominance of the medical staff is apparent from the fact that the CEO of the hospital is a physician who combines the roles of medical director and hospital administrator. This is required by law and is a result of a historical context in which hospitals were opened by physicians and inpatient facilities evolved from the temporary living quarters for outpatients.

Second, in hospitals where a board exists, the CEO of the hospital usually sits as chairperson with most of the senior medical staff as members, so that a triage of power among the board, administrator, and medical staff does not exist. The power of the physician CEO is especially great in the private physician-owned hospitals, which comprise nearly 70 percent of the hospitals in Japan.

Third, most of medical staff members working in the hospitals are full-time salaried employees. This should, in theory, make them more committed to the hospital. However, due to the way the medical staff is organized, this is not generally the situation and merely tends to further concentrate all initiative and decision making in the medical staff.[17]

Most hospitals in Japan are affiliated with a university, and the movement of the medical staff is effectively controlled by the patronage of the university clinical department chairperson. Newly licensed physicians generally enter their chosen speciality department in the same university from which they have graduated. This process is similar to joining a large company; henceforth, their primary allegiance is to the university clinical department.

Thus, although the medical staff are full-time employees of the hospitals, it is difficult to interest them in the long-term goals of the organization. The exception to this situation is when the physician is the hospital owner, or is connected by family to the hospital owner. However, in smaller private hospitals, because the division between ownership and management has not been established, there is a tendency for such physicians to pursue private and short-term benefits.

Until recently, the pressure for running hospitals efficiently had not been as strong as it had been in private companies. The physician-owned hospitals could usually be run with a profit, partly due to the fact that the health insurance system tended to reimburse the services offered by these hospitals in a better way and partly due to more cost-conscious management. In contrast, public hospitals in which high-cost, high-technology medical care tended to be concentrated were usually allowed to run with a deficit, with public authorities paying the balance.

This situation is rapidly changing in the face of growing health care costs, abundant provision of health care, and continuing governmental fiscal crisis. In such circumstances, it is likely that the role of unqualified business managers will grow, and that they, rather than the physician CEO, will initiate the QC concept.

Several hospitals have tentatively tried to start QCs, with the business manager taking the initiative. In other hospitals, nurses have independently attempted to introduce QCs, focusing more on the

> *It is difficult to visualize that business managers or nurses could have the organizational clout to maintain the QC movement in view of the physician-dominated structure of Japanese hospitals.*

quality of nursing care rather than the cost efficiency. Although it may be premature to evaluate these tentative trials, it is difficult to visualize that business managers or nurses could have the organizational clout, that is to say, the prestige and expertise, to maintain the QC movement in view of the physician-dominated structure of Japanese hospitals.

ORGANIZATIONAL AFFINITY FOR QCs IN U.S. HOSPITALS

It has been argued that the organizational structure of the Japanese hospitals that are administered by physicians tends to inhibit the development of QCs despite the basic cultural affinity. On the other hand, it appears that U.S. hospitals are in the opposite position of having greater organizational affinity despite the prevailing individualistic culture.

The first advantage is that the physicians in U.S. hospitals are, at least technically, off the organization chart; they have more of the attributes of the customer. In the United States, lay administrators are CEOs of hospitals that are more similar to the corporate model. Such hospital administrators might tend to have a greater understanding of the cooperative and competitive elements of QCs and have fewer personal conflicts of interests between professional and organizational goals.

The second advantage is the development of multiunit hospitals that take the U.S. hospitals even nearer to the corporate model. The scale of these multiunit hospitals would result in personnel having greater opportunities for upward mobility, even with the constraints of professional and legal restrictions, by providing a wider perspective for self-actualization and development.[18]

It is interesting to reflect that, although there exist in Japan several associations of hospitals such as the Red Cross and the Government Workers' Cooperate, there is little cross flow of personnel. This is partly due to the fact that each hospital tends to be affiliated with a different medical school.

The third advantage is that investor-owned hospitals, which are prohibited in Japan, would also allow the organization to overtly feed back productivity savings to workers, thus providing the implicitly necessary condition for successful maintenance of QCs. However, there is a caveat in this situation: In health care, the efficiency of measures is only justified within the framework of the effectiveness in patient care.

The relative lack of consumer sovereignty arising from imperfect knowledge obliges professionals to see that effectiveness is maintained despite savings. This could become especially problematic in a pre-payment scheme where the physician has a vested interest in making savings, as it could lead to a direct conflict between professional and organizational values.

CONDITIONS FOR A SUCCESSFUL QC IMPLEMENTATION

It has been emphasized that QCs could only be implemented as an integral part of the Japanese style of management. If QCs were attempted in a Theory X type of organization as a panacea for its multiple problems, management would be deluding itself about the expectations, and it would be likely to find that workers would soon perceive the deception. Realizing the principles of Japanese management may be a circuitous route to implementing QCs, but it appears to be the only certain way to success.

Critical tasks

Thus the primary task of top management would be to commit itself to the principles of Japanese management and to provide tangible evidence of this commitment to workers. Goldsmith has outlined a five-point agenda for change. It consists of a defined single recruitment period, investment in orientation, permanent employment for core members, holism and paternalism, and better communication.[19] The provision of these specific targets should make the transformation process operationally more clear for management.

The second task would be in deciding which areas to concentrate on initially in introducing QCs. It can be readily inferred that the most feasible candidates for QCs would be the clerical/ancillary departments, which have nonprofessional personnel. Such services as finance or laundry, being relatively distant from patient care, could develop their QCs without coming in conflict with professional workers.

They would have additional advantages over the other departments. First, there is a close relationship between their activities and those in the non-health care industries, which makes it possible to assimilate and communicate with those industries as to QC activities. Second, their workload tends to be more easily segmented and statistically evaluated. Third, although the effect may be more subtle, there is a greater capacity for self-development of workers in these fields, who have long been suppressed by the professional-dominated culture of the hospital. The success of ServiceMaster Industries may be partly due to this factor.[20]

The next area for introduction of QCs may be the departments comanaged by medical professionals who have the greatest affinity for the methods of QCs. Dietitians, radiographers and laboratory technicians are potential candidates, but much would depend on the caliber of the hospital manager.

The pivotal department for development into hospitalwide QCs would be nursing, as it is not only the largest numerically but also the potential clearinghouse for coordinating proposals generated by all QC activities. Nurses would require a considerable upgrading in their prestige and authority for the coordinating activity to function smoothly.

Total commitment

Whether it is possible to take the final step of involving physicians in QCs is still uncertain. Physicians could argue that they have always been conscious of quality assurance and that their peer group activities antedate those of QCs. However, contemporary QCs emphasize cost containment and effectiveness and as such it would have to wait until this cost orientation could be incorporated into medical ethics. This requires considerable adjustment in the prevailing norm of providing the best possible care to the individual patient regardless of cost.

The role of hospital executives in introducing QCs would be a taxing one, as these executives would have to limit the proposals coming from QCs to those tasks that could be directly implemented by the same QC members. Very often in the initial stages, the QC brainstorming exercises tend to result only in grievances against other groups.

It would be the task of the QC leaders to constantly redirect attention to specific issues that could be constructively dealt with by the members. This would become increasingly difficult as activities were expanded to those more central to the patient, for not only does the interdependence of work tasks become greater, but also the professionals begin to exert a more decisive role.

This highlights the need for an eventual hospitalwide QC movement and a major effort in coordinating QC activities. To realize successful implementation, there is a need for a commitment by all concerned to the organization's values and goals. Whether this commitment could be obtained, particularly from the professionals, is the key question.

REFERENCES

1. Baird, J. "Quality Circles May Substantially Improve Hospital Employees' Morale." *Modern Healthcare* 11 (September 1981): 70–74.
2. Maser, M. "Mount Sinai Invests in Quality Circles." *Health Services Manager* 15 (February 1982): 12–13.
3. QC Circle Headquarters, ed. "Fundamentals of Managing QC Circle Activities" (in Japanese). Tokyo: Japan Association of Science and Technology, 1970, p. 1.
4. QC Circle Headquarters, ed. "Fundamentals of the QC Circle" (in Japanese). Tokyo: Japan Association of Science and Technology, 1970, p. 76.
5. Ibid.
6. MacGregor, D. *The Human Side of Enterprise.* New York: McGraw-Hill, 1960.
7. Williams, J.P. "Can They Learn Japanese in Detroit?" *Journal of Constructive Change* 3 (Fall 1981): 20–22.
8. Hamaguchi, E. "What Is Japanese Collectivism?" In *Japanese Collectivism* (in Japanese), edited by E. Hamaguchi et al. Tokyo: Yuhikaku, 1982, pp. 1–26.
9. Pascale, R.T., and Athos, A.G. *The Art of Japanese Management: Applications for American Executives.* New York: Simon & Schuster, 1981.
10. Pfeffer, J. *Organizations and Organization Theory.* Boston: Pitman, 1982, p. 98.
11. Yamamoto, H. *The Spirit of Japanese Capitalism* (in Japanese). Tokyo: Kobunsha, 1979.
12. QC Circle Headquarters, ed. "Fundamentals of the QC Circle," 27.
13. Ouchi, W.G. *Theory Z: How American Business Can Meet the Japanese Challenge.* Reading, Mass.: Addison-Wesley, 1981.

14. Pascale and Athos, "The Art of Japanese Management."
15. Peters, T.J., and Waterman, R.H. *In Search of Excellence.* New York: Harper & Row, 1982.
16. Shortell, S.M. "Theory Z: Implications and Relevance for Health Care Management." *Health Care Management Review* 7 (Fall, 1982): 7–21.
17. Ikegami, N., and Goldsmith, S.B. "The Japanese Health Service—An Overview," *Journal of Ambulatory Care Management* 5, no. 4 (1982): 78–86.
18. Brown, M. "An American Version of Theory Z." *Health Care Management Review* 7 (Fall 1982): 23–25.
19. Goldsmith, S.B. *Theory Z Hospital Management.* Rockville, Md.: Aspen Systems, 1984.
20. Brown, "An American Version of Theory Z."

Human resource indicators for hospital managers

John D. Aram,
Paul F. Salipante, Jr,
and
James W. Knauf

Hospitals often have large amounts of information available about their human resources, but this information is usually inadequate. The nature of existing human resource information is examined in one large hospital, and the concept of a human resource indicator system is developed. Methods for deriving meaning from the indicators are discussed.

Increases in the power per dollar and the user friendliness of computers are giving rise to an explosion of potentially useful data for managing hospital personnel. Unfortunately, the use of strategic human resource management is now quite limited and significant deficiencies exist in managers' knowledge of how to interpret personnel information. Many years of effort are required to develop a human resource information system that can aid in decision making.[1] As hospital managers attempt to take advantage of new technology, the most likely outcomes are problems of raw data overload. Research indicates that the creation of managerially useful human resources information will be a difficult challenge.

This article addresses that challenge by exploring the data requirements for strategic human resource management in hospitals, focusing on the transformation of personnel data into indicators of human resource management effectiveness. The perspective is that of a large, innovative health care organization, the Cleveland Clinic, where the authors have been involved in an ongoing effort to derive meaning and action implications from the mass of data available. The objective is to identify relevant sources and types of information, to explore their advantages and limitations for hospital operations management, and to provide guidelines for building a system of human resource indicators useful to line managers and staff managers.

John D. Aram, *Ph.D., is Professor of Management Policy in the Department of Managerial Studies, Weatherhead School of Management, Case Western Reserve University, Cleveland, Ohio. His major interests are business and public policy issues.*

Paul F. Salipante, Jr., *Ph.D., is Associate Professor of Industrial Relations in the Department of Managerial Studies, Weatherhead School of Management, Case Western Reserve University, Cleveland, Ohio. He is engaged in research on human resource planning and conflict management.*

James W. Knauf, *Ph.D., is Associate Director of Human Resources, Cleveland Clinic Foundation, Cleveland, Ohio. His interests are human resource processes that foster and maintain healthy working environments.*

The research on which this article is based was supported by the W.M. Keck Foundation Research Scholars Program of the Health Systems Management Center, Case Western Reserve University, Cleveland, Ohio.

STATUS OF PERSONNEL MANAGEMENT IN LARGE HOSPITALS

Personnel management is generally recognized as a key staff function in hospitals. Salaries and wages represent a high proportion of total service costs. Periodically, scarce labor markets for occupations such as nursing make recruitment and retention key organizational tasks. A high proportion of technical, professional, and administrative staff leads to an emphasis on human resource development and conflict management. Because of a variety of factors, the personnel management in large hospitals assumes a particularly significant role.

Paradoxically, hospitals often fail to incorporate human resources or employee relations into their central managerial functions. Organizationally and operationally, hospital personnel activities resemble the normal functions of personnel administration—recruitment, selection, training and development, wage and salary determination, promotion procedures, grievance handling, and so forth. Rarely does the personnel department articulate a human resource philosophy for the hospital; even more infrequently does the department exert sufficient influence to implement a management or organizational philosophy with respect to human resources. Often, the greatest role in planning that the human resource administrator plays is the development of personnel projections and costs.

INFORMATION FOR STRATEGIC HUMAN RESOURCE MANAGEMENT

To develop and maintain a viable human resource strategy and to improve health care delivery, hospital managers need a systematic and ongoing means of understanding the status of human resources, identifying and anticipating early-stage organizational problems, and stressing human resource impacts on service quality and cost. In practice, the aspirations and contributions of personnel administrators often fall short, and they experience difficulty in identifying needed areas of improvement in supervision, communication, group relations, and a host of other behavioral factors. Lacking ongoing knowledge about the organization, hospitals may resort to periodic attitude surveys to learn the perceptions of employees.

Attitude surveys have several drawbacks for hospital managers. First, attitude surveys are carried out infrequently, because of the staff time and expense involved. Second, surveys depend on reports of perceptions and attitudes that may not be fully reliable indicators of the status of the human resource system. Finally, the implications of attitude surveys are often difficult to decipher. Problems or issues are phrased in terms that do not correspond well with line managers' operational responsibilities, so that consequences of survey results for the quality and cost of service are rarely identified.

In short, attitude surveys cannot stand alone; there remains a need for more reliable and timely information. Surprisingly, hospital personnel departments often have a wealth of data available on the nature, change, and impact of the human organization. Data on such factors as voluntary departures, promotions, performance ratings, disciplinary actions, employee grievances, and externally filed discrimination charges are concrete indicators of organizational functioning and are available on a continuing basis.

The negative side of the picture is that the data are rarely transformed into useful information by combining various pieces of data and making comparisons that allow useful conclusions to be drawn. A major reason for this problem is that the data are usually incomplete, unsystematically recorded, fragmented in storage location, and recorded by differing job classifications, occupational groupings, and employee characteristics. In sum, the basis for a managerially relevant, reliable, and continuous system of human resource indicators is present in larger hospitals but is unrealized.

With improvements in data collection and accessibility, human resource data can provide much useful information. Information on a variety of indicators can be simultaneously arrayed and then analyzed in at least the following ways:

- Individual-by-individual comparisons provide a more complete profile of the individual's behavior and identify problem and success cases. In hospitals with large staffs it is surprisingly easy to overlook both types of cases or to rely only on subjective opinions to identify individuals requiring corrective action or having strong potential for promotion.
- Analyzing relationships between indicators and employee characteristics can indicate employee groups requiring changed or additional human resource procedures; for example, high turnover of high-performing, middle-aged employees would raise the possibility of weaknesses in

higher level job ladders or career development practices.
- Summaries of the array of indicators by various organizational units can alert directly involved managers and higher level managers to problems or successes in supervisory practices. They can also alert the organization to problem units and help diagnose the causes of human resource problems in those units. Problems that the summaries reveal to run across many organizational units would indicate a need for hospitalwide revision of particular human resource management procedures.

This type of information, when combined with more qualitative information, allows hospitals to make the best use of their personnel, to catch problems while they are still at an early stage, and to develop and maintain procedures that promote effective use of employees over the long term.

GUIDELINES FOR ANALYZING HUMAN RESOURCE DATA

No single piece of personnel data can be unambiguously interpreted. For example, low turnover in a department may result from high levels of performance, challenge, and satisfaction—a very positive situation—or from supervision that settles for low levels of employee effort and performance. The authors' aim is to produce information upon which managers can act, information that indicates whether human resource problems or successes exist at the level of the individual worker, the work group, the department, or the hospital. The guidelines for transforming data into useful information are:
- Combine several indicators. Look for patterns in the indicators that are representative of effective and ineffective management. Compare indicators across time. For an individual employee or organizational unit, analyze the trend over time and compare the trend with trends for other individuals or units.
- Compare indicators across hospitals. Compare one organizational unit at one point in time with similar units at other area hospitals.
- Do not compare data from only one point in time across units in the same hospital.

These guidelines differ from the form of reports normally produced in hospitals. When personnel information is summarized at all, it is usually presented department by department for a single time period. The consequence is that human resource staff, supervisors, and higher management invariably compare one hospital unit with another, e.g., "turnover is three times as high in medicine as in surgery." The problem is that there are innumerable possible explanations for such differences, and managerial action depends on identifying the operative explanation. Attacking this problem is quite simple if one draws on concepts from the literature on program evaluation.

In the above example, the explanation that a hospital manager should attempt to prove or disprove is that medicine's high turnover is indicative of poor human resource management practices. An alternative explanation is that the inherent nature of work in that department or some other basic, difficult-to-change characteristic of that department is causing high turnover. Which explanation is better can be determined by comparing medicine's current turnover rate with its past rates, medicine's trend in turnover with the trend in the hospital overall, and medicine's rate with the rate in similar units in other hospitals. If the manager finds medicine's rate to be similar to its past rates and those in other hospitals, the manager will have less confidence that the turnover rate indicates a problem requiring intervention. If similar analyses then show surgery's turnover to be higher than its historical rate or much higher currently than that of surgery units in other hospitals in the same labor market, the manager will turn efforts toward surgery rather than toward medicine.

Stated another way, comparisons across time within the hospital ("the turnover trend is up in surgery but down for the hospital overall") and across similar units in other hospitals can indicate areas where managerial action is likely to be fruitful. If other hospitals, or the manager's own hospital at some earlier time, have achieved better results in a particular unit, then that unit should be able to do so at the present time. A study of how the unit was managed in the past or how it is managed in other hospitals may then help managers determine what remedial action is needed. Similarly, comparison of the unit's present performance with its past performance provides its managers with the ability to track the results of their managerial efforts.

The above discussion shows that comparisons for one unit alone across time are inferior to comparisons of trends across units. A deterioration in a particular unit's indicators may be due to events external to the unit and beyond its control. This alternative explanation can be investigated by comparing trends in other units. The point is that both types of comparisons—across time and across units and hospitals—are needed.

But what type of data should be compared? In order to have managerially useful information, several sets of indicators must be examined simultaneously.

In order to have managerially useful information, several sets of indicators must be examined simultaneously.

HUMAN RESOURCE INDICATORS

Four primary indicators of the status of a hospital's human resource system are withdrawal behavior (including voluntary and involuntary turnover), internal grievance actions initiated by employees, supervisory disciplinary actions, and individual and unit performance ratings. Each of these areas of employee or employer action reflects an aspect of the hospital's human resource system, and the four sets of indicators may collectively signal changes taking place in that system. To accomplish the types of analyses discussed, the first step is to establish reliable systems for recording employee actions and categorizing them by meaningful organizational units. The second and more conceptually difficult step is to combine separate indicators to permit interpretation of disparate data. There are various problems encountered in this second step and several approaches to overcoming them. The aim is to interpret indicators in order to guide managerial action.

Withdrawal indicators

Withdrawal rates, i.e., turnover and absenteeism, are significant because of the substantial costs and service interruptions associated with replacing employees. In addition, a large body of research has established that withdrawal behavior is inversely related to employee satisfaction: the lower employee satisfaction is, the higher withdrawal behavior is.[2] Thus, withdrawal data may be used as indicators of employee satisfaction. Employee satisfaction, in turn, can indicate problems or successes in human resource procedures or supervisory behavior.

The key requirement in creating and using indicators of withdrawal to alert the organization to practices requiring change is to separate and measure those withdrawal actions that are voluntary on employees' part and reflect conditions the hospital cannot control from those the hospital can control. To this end, withdrawal may be categorized as

- withdrawal due to illness;
- withdrawal due to employees' situations outside of the hospital;
- withdrawal due to discipline; and
- withdrawal due to perceived deficiencies in the work setting.

While the costs of all types of withdrawal significantly affect hospitals, the third and fourth types have the most potential for helping to identify performance deficiencies and human resource management problems.

In reality, withdrawal data are difficult to separate neatly into the four categories above. The second type of withdrawal is difficult to identify. Employees who quit often do not want to make waves, and they will state that they are leaving for personal, external-to-work reasons rather than report their dissatisfaction with the work situation. Supervisors who must report the reasons for subordinates' quitting also have an incentive to report reasons as external factors instead of internal situations. In practice, then, it usually is better not to separate the second and fourth types of withdrawal on the basis of written statements of the employees' reasons for quitting. The percentage of resignations truly due to external situations should be roughly comparable across time and across similar jobs in different organizations. Hence, combining the second and fourth types of withdrawal should not seriously distort comparisons across time and across organizations and should provide a more accurate indication of withdrawal due to dissatisfaction.

With regard to illness-related withdrawal, research has shown that using the number of different incidents of an employee's absence provides a better measure of voluntary absenteeism (i.e., fourth type of withdrawal) than does using the total number of days absent.[3,4] Such a measure helps remove the effects of long illnesses (first type of withdrawal) from the data.

What about tardiness as an indicator of withdrawal

due to perceived deficiencies? Again, a confounding factor is the individual supervisor's discretion in reporting and disciplining tardiness. For example, if a supervisor's style produces high employee dissatisfaction but the supervisor follows the tardiness rules strictly, employees are not likely to express their dissatisfaction through tardiness.

Transfers out of an organizational unit are a more promising measure of withdrawal due to perceived work setting deficiencies. Conversely, attempts to transfer into a particular organizational unit may indicate that the unit is seen as attractive. For an organization having a job-posting procedure, data on the number of bids received from employees in a particular unit and for certain kinds of jobs are especially useful. Since responses to job postings occur more often than actual transfers, data on them aid statistical stability. When interpreting data on bids or transfers, however, a manager must be aware of the divergence among different jobs and departments in terms of investment characteristics. Some jobs and departments are seen by employees and the organization as training grounds for future advancement. High rates of transfer out of such jobs would not indicate deficiencies. To distinguish such jobs from jobs employees do see as deficient, managers must look at jobs having high transfers in (as opposed to hiring from outside the organization) and moderate or low exit rates (i.e., number of people quitting the organization altogether). Furthermore, one would expect quite high rates of turnover in such jobs, but low rates of other withdrawal behavior—absenteeism and tardiness. Thus, the combination of several measures may be used to identify which jobs and departments in an organization are seen by employees as having investment characteristics; high rates of transfers would not be indicative of job deficiencies—quite the opposite would be true.

Grievance indicators

A grievance procedure is one example of a personnel system common to larger hospitals that generates considerable information about the human resource system—problems experienced by employees and inconsistencies in the administration of hospital personnel policy. In a study of nonunion grievance systems in three large urban hospitals the authors found an average of more than 11 grievances per 1,000 employees per year for two years, indicating a significant incidence of employee action.[5] Moreover, considerable differences were found among the individual hospitals, indicating differences in employee–employer relations and in the effectiveness of the appeals procedures.

None of the personnel staff or managers in the three hospitals systematically used grievance data to evaluate the nature of organizational problems or changes in employee–employer relations. The grievance procedure in one hospital was seen by employee relations staff as a valid source of information about the human resource system of the hospital. Yet, information was not being classified in managerially useful categories such as department or job, was not related to other sources of human resource information, and was not readily amenable to interpretation.

The meaning of a particular grievance rate originating in a hospital unit was not self-evident. A low rate could mean excellent supervisor–employee relations with few employee complaints, or it could mean poor relations with many employee complaints and threats of reprisal toward employees using the grievance procedure to gain recourse for arbitrary supervisory decisions. In other words, grievance data did not offer useful information in the absence of other quantitative or qualitative information.

Various grievance indicators can be measured: rate of grievances, stage of the grievance procedure at which resolution occurred, and percentage of grievances won by employees. Examination of hospitalwide trends in each can point to strengths or weaknesses of the procedure itself. If the grievance procedure is attractive to employees, these indicators can provide insight into the perceived fairness of personnel practices and supervisory behavior.

Interpretation of the frequency of grievances is aided by the ability to compare trends in rates between organizational departments or job groupings and by examination of other sets of indicators. Change in a unit's grievance rate may be related to changes in the unit's level of disciplinary actions and withdrawal. For example, are decreases in grievance rates in a department matched by fewer disciplinary actions taken by supervisors and by lower turnover rates? If so, the department would seem to be improving on a human resource dimension. On the other hand, a decline in grievance rates that is matched by a stable or higher rate of disciplinary actions and by higher departmental turnover may signify a worsening of organizational relations that is being expressed in turnover rather than through internal grievance procedures. These comments suggest the way tentative conclusions can be generated from changes in a combination of indicators included in a human resource database.

Disciplinary actions

Many employee-related actions—grievances, equal employment opportunity (EEO) charges, voluntary turnover—are initiated by employees. Disciplinary actions, on the other hand, are initiated by supervisors and thus represent a complementary side of changes in the human resource system. Virtually all hospitals have written policies describing just causes for disciplinary action and procedural steps of progressive discipline. For minor violations of work rules such as tardiness or abusive language, the progressive discipline typically consists of (1) verbal warning, usually recorded as such; (2) written warning; (3) suspension without pay; and (4) termination. For major violations such as gambling, falsification of records, or use of an intoxicant or narcotic, supervisors may immediately suspend the employee, and then conduct a detailed investigation to confirm or retract this action.

Data generated by disciplinary procedures indicate actions taken by supervisors to address and correct employee deficiencies. Assuming consistency in the administration of policy, changes in a department's disciplinary rates represent the employer's view of improvement or of deterioration in employee work behavior. A decrease in disciplinary rates implies improvement in work behavior in a unit and an increase implies a greater number of problems in employee behavior. Again, assuming consistency in the application of policy, these data can be used to help interpret other human resource information. For example, a decrease in disciplinary actions accompanied by lower absenteeism and turnover and fewer EEO charges would be a positive sign for a unit's personnel system.

Viewed in concert with other indicators, disciplinary actions also give insight into weaknesses in supervision. For example, an increasing rate of disciplinary actions that are not upheld in the grievance procedure directs attention toward supervisory training and development. Similarly, a decrease in disciplinary actions marked by higher absenteeism and turnover and an increase in EEO charges would indicate laxity in supervision and the need for supervisory development to overcome inadequate supervisory control.

Similar to other areas of personnel data, disciplinary action information is rarely easy to retrieve or analyze by job or organizational unit. In addition, substantial variations in recording minor violations are likely to exist as one supervisor records a verbal warning while another supervisor does not. Consequently, data about disciplinary actions may lack reliability. Advantages, however, of this source of information are the relatively high rate of occurrence of disciplinary actions (indicating a significant human resource process), the fact that disciplinary actions are supervisor initiated, and the indication that these data can be integrated with other human resource indicators relatively easily.

Performance indicators

An indicator system should incorporate individual and group performance measures available in the hospital. For example, individual merit ratings are commonly present and may predict some behavioral indicators, such as grievances and turnover, or may coincide with others, such as disciplinary actions. At a group level, performance indicators such as patients per staff or work produced per employee can be integrated with behavioral indicators. Similarly, cost information by unit may be appropriate to track and to compare with other indicators. For example, high turnover and disciplinary actions accompanied by productivity improvements may result from a needed increase in pressure to produce. More generally, it is extremely valuable to establish whether performance indicators are leading, coincident, or following factors in relation to other indicators. Such knowledge will help hospital administrators understand and attend to indicators giving early signals of improvement or deterioration.

Analysis and convergence of indicators

The most important implication of the above discussion is that simultaneous analysis of the four sets of indicators is needed to reveal positive or negative states of particular units or of the hospital as a whole. Analysis over time will reveal that certain combinations of indicators signal particular problems and will help identify styles of supervision and their effectiveness in given units. High turnover accompanied by low rates of discipline and low rates of complaints (grievances) may indicate that informal disciplinary measures are being used, with employees being fearful of retribution if they were to complain. High turnover combined with few attempts at transfers may indicate that the supervisor is hoarding good employees who might otherwise advance within the organization. Low rates of turnover accompanied by high absenteeism and low rates of discipline, promotions, and transfers may indicate slack management and dead-end jobs. Such patterns must be identified and interpreted; using any indicator alone provides, at best, ambiguous information.

Integration of quantitative indicators with qualitative information

As useful as a quantitative system may seem, it is only part of the total picture needed to understand the organization. In addition to the indicators already discussed, qualitative information also must be tapped for a deeper understanding of the state of human resource management.

Analysis of quantitative indicators alone would not permit complete problem identification and diagnosis. No matter how much success is achieved in defining and combining indicators, significant issues of meaning and interpretation will inevitably remain.

The ability to understand changes in a unit's human resource system depends on a related but independent effort to gain information from a personal, qualitative, and verbal process. Qualitative information from particular units would help to assess, qualify, confirm, modify, or reinforce the aggregate indicators. Similarly, an impressionistic and personal view of significant changes in a department or unit may lead managers and staff to look for verification and clarification in the aggregate information. Aggregate information provides a broader, more comparative, historical perspective that cannot be achieved if one relies solely on qualitative information.

REPORTS AND THEIR USES: CURRENT STATUS

Table 1 displays one of the report formats being used by human resource department staff at the Cleveland Clinic. Data on several indicators are presented for each of five years. The measure for any indicator corrects for the overall trend in the organization for that year, and is normalized so that a value of one indicates the same rate for an employee as for the organization overall. This measurement permits easy comparison across different indicators. Reporting to managers is also supplemented by graphs of

TABLE 1

NORMALIZED INDICATOR MEASURES FOR DEPARTMENT A, YEAR END, 1980–1984

Indicators	Measures*				
	1980	1981	1982	1983	1984
Turnover	1.2	1.4	1.4	1.1	1.2
Formal grievances	0.6	0.6	1.4	1.1	0.2
Informal grievances (counseling sessions)	0.2	0.2	0.6	0.8	1.4
Supervisory disciplinary actions	1.7	1.0	1.0	0.8	0.9
Discrimination charges	0.0	0.0	1.6	0.0	0.0
Job bids					
Out	1.6	1.4	1.3	0.8	0.7
In	0.2	0.1	0.3	0.6	1.0

* >1 = higher than hospital as a whole
  1 = equal to hospital
  <1 = lower than hospital as a whole

the same data. The report and graphs allow comparison of a department's trends on several indicators.

By design, the application of this type of human resource information has been limited to date. Middle managers across the organization are provided with aggregated, organizationwide data to heighten general managerial awareness of the human resource indicators available. Reports of departmental data are presented to selected line managers and are used to stimulate their own diagnosis and insights.[6] The human resource department uses the information to identify areas to which its services and interventions should be targeted.

Quite expectedly, the reaction of management to the information has been varied. For some managers who are not data oriented, the reports are seen as just more paper. Some other managers see the reports as highly beneficial, allowing them to make improved decisions relating to employee counseling, performance appraisal, and absenteeism. On balance, the reaction has been positive. Managers have found it useful to examine multiple indicators at one period of time for each employee in a department; such reports have directed attention to particular problem individuals and work groups and have identified inconsistent managerial actions directed toward specific employees.

Another benefit of the information has been dialogue about the meaning of the data. A review of trends over time in the indicators, as in Table 1, has provided insights concerning how major fluctuations in indicators may have been caused by changes in management or by revisions in policies.

• • •

Obtaining useful human resource information is a common difficulty for managers in hospitals. Often, data lacking in context and meaning are distributed to managers. Unfortunately, such practices rarely lead to managerially useful information as few individuals can make sense of mountains of raw and undigested, or only partially digested, reports. As labor-intensive institutions, hospitals need more effective ways of organizing and utilizing human resource information.

Many problems must be solved before human resource information systems fulfill their potential. Numerous technical issues involving the measurement, collection, and interpretation of information pose

As labor-intensive institutions, hospitals need more effective ways of organizing and utilizing human resource information.

large challenges. Means of usefully integrating qualitative as well as quantitative information need to be developed. In addition, a great deal more experience needs to be gained about how hospital managers can learn to use such systems productively.

Unquestionably, the practice of human resource management has a long way to go before attaining the needed level of performance. However, the study presented here suggests that a long-term commitment to developing procedures for using human resource data and a thoughtful process for integrating an indicator system into management decision making can create important benefits for hospitals.

REFERENCES

1. Hilton, B.D. "A Human Resource System that Lives up to Its Name." *Personnel Journal* 58 (1979): 460–65.
2. Srivastva, S., et al. *Job Satisfaction and Productivity.* Kent, Ohio: Kent State University Press, 1976.
3. Hammer, T.H., and Landau, J. "Methodological Issues in the Use of Absence Data." *Journal of Applied Psychology* 66 (1981): 574–81.
4. Huse, E.F., and Taylor, E.K. "Reliability of Absence Measures." *Journal of Applied Psychology* 6 (1962): 159–60.
5. Salipante, P.F., and Aram, J.D. "The Role of Organizational Procedures in the Resolution of Social Conflict." *Human Organization* 43, no. 1 (1984): 9–15.
6. Salipante, P.F., Golden, K.A., and Buck, F.P. "A Partnership for Deriving Meaning from HR Data." *Personnel Administrator* 30, no. 12 (1985): 55–64.

An integrated approach to board development

Robert A. McGowan

Growing demands on hospital boards call for systematic efforts to increase the effectiveness of individual board members. These activities must go beyond the provision of traditional education programs.

Boards of trustees are playing an increasingly important role in determining the current and future directions of today's hospital. The typical hospital board today is confronted with a lengthy list of major decisions that will have a deep impact on the success and survival of the institution in the future. Revising the mission of the organization, assessing the financial and political viability of for-profit ventures, developing aggressive strategies to position the institution's services in order to protect and expand market share, evaluating the pros and cons of entering into new partnerships and alliances with neighboring hospitals, physician groups, and for-profit businesses are but a few of the issues discussed at the average board meeting.

In the past decade there has been increased concern about how to integrate the board, administration, and medical staff in the process of governance. With this issue remaining relatively unresolved in most institutions, health care managers are more concerned about tensions evolving between the goals of the multiple boards within the reorganized holding company, as well as conflicts in their working processes and personalities.

Accompanying these changes is a growing awareness that successful board decision making depends on the performance of individual board members. To participate intelligently in the hospital governance process, board members must be knowledgeable, sophisticated, and politically mature. They must be willing to make hard decisions in an ambiguous environment. They must be strong generalists who understand the many and complex forces affecting their hospital.[1]

The development of individual board members is critical to the success of the governance process. Current board development activities tend to focus on increasing knowledge and skill and get mixed results. There is a need for a more systematic and comprehensive approach. The model described in this article does not offer profound or clever solutions, but it does provide a road map that challenges hospital leadership to diagnose needs more carefully. It also

---

Robert A. McGowan, M.A., is a Partner in The Institute for Organizational Effectiveness, Inc., in Concord, Massachusetts. He has conducted board assessments and board retreats for many hospitals and not-for-profit organizations. He is also a member of the American Association of Health Care Consultants.

emphasizes the importance of selecting a broad range of integrated development strategies.

CLARIFYING THE PURPOSE OF BOARD DEVELOPMENT ACTIVITIES

The goal of all board development activities is ultimately to bring about improved performance on the part of individual board members. Behavioral scientists emphasize that human behavior is the combination of forces within the immediate situation, the environment, and the individual involved.[2] Many development activities oversimplify the process of behavioral change.

Consider for a moment the common types of problems that some hospital leaders identified when asked to describe why there is a need for board development programs: Board members voted against a hospital-based health maintenance organization (HMO) and now a private group has moved in; board members do not understand that the success and survival of the hospital is at stake and they resist change; board members are angry at physicians and put them in an adversarial position; and board members face serious conflicts of interest as the board discusses entry into certain for-profit ventures. These reasons reflect hope that development activities will improve decision making, strengthen relationships, and reduce conflict. These goals cannot be achieved simply by offering traditional educational programs. That is why a systematic approach must be taken.

A MODEL FOR BOARD MEMBER DEVELOPMENT

Figure 1 shows a model that identifies a series of diagnostic questions and a broad range of potential development activities. The first step is to strengthen the foundations for improvement by clearly defining roles and providing performance feedback. Factors that may hinder performance, other than a lack of knowledge and skill, are then addressed, followed by a consideration of educational activities that can impact knowledge and skills. The final step is to identify options available if no change occurs.

The foundations for development

A great deal of ambiguity exists in most hospitals related to the roles and expectations of the board, the medical staff, and administration. It is little wonder that tensions develop as each group, and individuals within them, brings different and frequently conflicting expectations to board meetings.

Organizational theorists emphasize that role clarity is the foundation for development.[3] Within hospital governance systems, bylaws, policies, and procedures are the formal mechanisms that define who owns what turf and who has what level of authority and responsibility. However, as one hospital chief executive officer (CEO) points out, "Their content is frequently the result of a highly political process in which necessary compromises leave a lot of boundaries deliberately ambiguous." Some experts in the field suggest that written job descriptions should be developed for board members.[4] While written documents can provide the skeleton, the ultimate shaping of roles has to be accomplished through careful dialogue and final decisions must be responsive to the unique culture of each institution. Since they are heavily affected by board decisions, members of the medical staff and administration should be involved in this process. This will ensure that the organization is ready to accept and support the roles as defined.

Standards or criteria used for board member selection and evaluation establish the parameters for board member development. Unfortunately they are usually poorly defined. It should not be surprising, therefore, that some board members behave in ways that do not meet the expectations of hospital leadership.

The board must force itself to struggle with the task of developing performance criteria. Ideally, individual performance goals should exist for each board member. Additional criteria can also include the type of knowledge, skill, experience, and personal qualities that a board member should possess. Minimal standards should also be defined related to attendance at board meetings.[5]

The importance of assessment

Performance criteria are of little value if they are not used. They should be clearly communicated. Board members also need to receive honest and straightforward feedback on their performance if they are going to be motivated to change their behavior or participate in development activities. Human development is highly dependent on the receipt of valid and timely feedback that points out the results of behavior.[6] Unfortunately performance feedback for board members is grossly inadequate and frequently nonexistent.

Board Development 103

FIGURE 1
POTENTIAL BOARD DEVELOPMENT ACTIVITIES

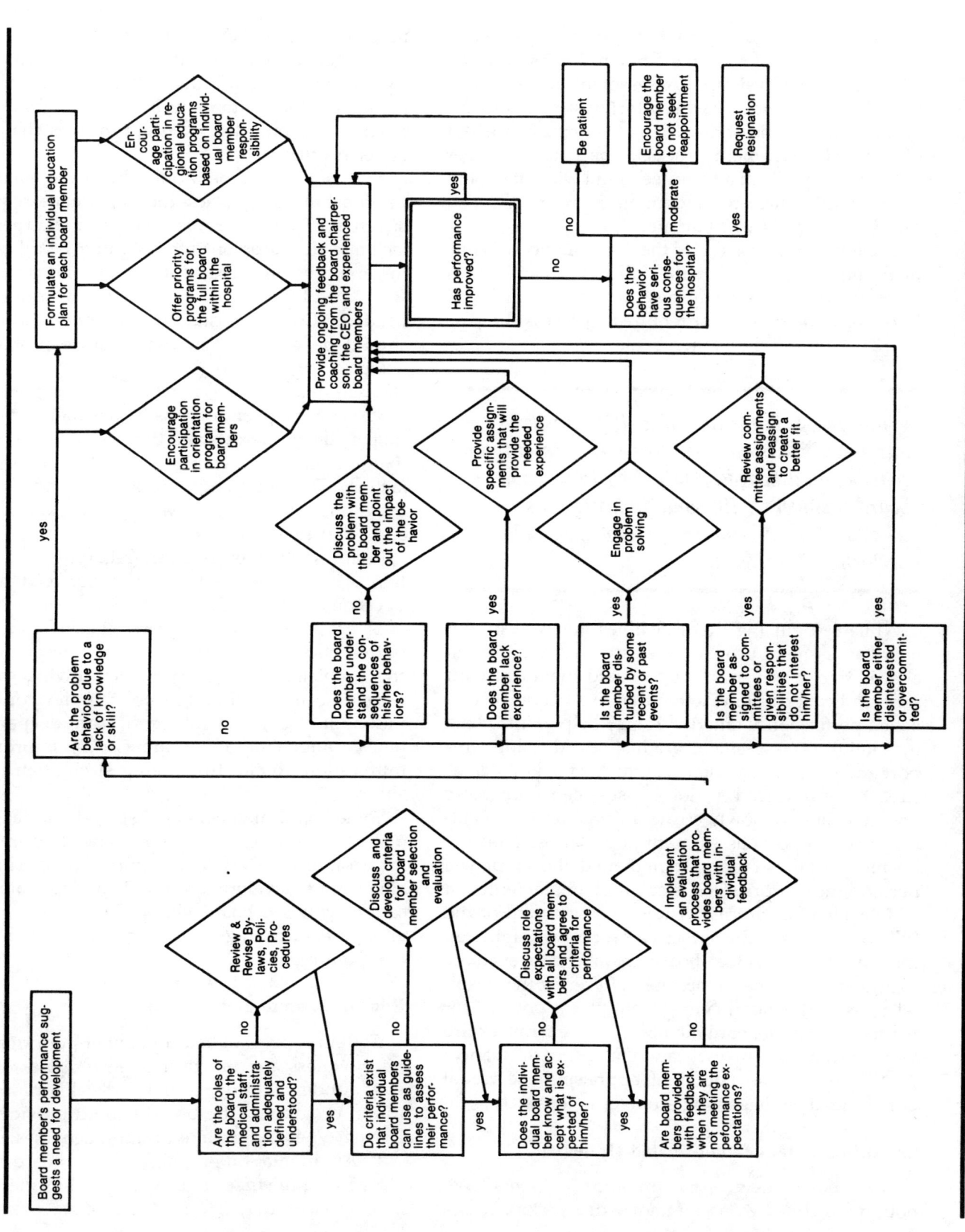

Board members are quick to support the establishment of appraisal systems for employees, for the CEO, and for physicians. Assessment of the board, on the other hand, provokes feelings of defensiveness. In the middle of discussing the appropriateness of a board assessment process, one board member stated angrily, "I am a volunteer and I will not submit my contribution to the judgment of my peers. You can do that when you pay me."

The Joint Commission for the Accreditation of Hospitals now requires that the full board conduct an assessment of its effectiveness on an annual basis. However, boards are not required to assess the performance of individual members.

While recognizing the difficulties involved in any board assessment process, it is essential that hospital boards find a way to provide board members with some indication as to whether they are meeting established performance criteria.

While recognizing the difficulties involved in any board assessment process, it is essential that hospital boards find a way to provide board members with some indication as to whether they are meeting established performance criteria.[7] It may help to introduce the notion of assessment gradually to develop support and minimize resistance. One hospital in Massachusetts introduced various assessment processes one at a time to slowly create an organizational culture that supports assessment at all levels. First the hospital administration strengthened the employee performance appraisal system and implemented a CEO evaluation process. Then an assessment of the full board was conducted and a process was initiated that invited individual board members to evaluate their own performance privately. Recently the hospital revised the nominating committee's policies for reappointment of board members whose terms expire by requiring a careful evaluation of past performance. At this point, the notion of appraising individual board members is gaining stronger acceptance.

Exploring noneducation-related strategies

Many board member performance problems have nothing to do with a lack of knowledge or skill. Sometimes board members simply do not understand the consequences of their behaviors. The board chairperson or the CEO can bring about remarkable changes in many of these situations by talking with these people and coaching them on approaches that will get better results.

New board members often have the ability to do the job but simply lack the experience required to effectively fulfill their role. The board chairperson can facilitate development by assigning board members to certain committees, task forces, and projects that will expand their experience in areas where growth is needed. Coaching should continue and support should be given when performance problems develop.

Occasionally board members are assigned to committees or are given responsibilities that do not match their skills or interests and the performance problems that surface are a symptom of the resulting dissatisfaction. Development in these situations requires providing a better fit between what is asked of them and what they want to do.

Performance problems can frequently be an indication of underlying feelings of anger related to past decisions made by the board. When this is the case, it is essential that the board chairperson, a committee chairperson, or the CEO attempt to resolve the problems. Effective problem solving can make a significant contribution to the development of individual board members. It can deepen everyone's understanding of the issues, improve communication, build trust, and strengthen relationships.

Some board members honestly admit that they are disinterested or overcommitted and they are looking for a comfortable way to get off the board. This generally is not a development problem. A private conversation with the board chairperson or CEO may be enough to encourage the person to resign or not seek reappointment.

Education strategies

Certain aspects of board member performance and effectiveness are directly related to a lack of knowledge, understanding, or skill and call for formal education programs. The board should provide the following types of formal education activities:

1. An orientation program for new board members can familiarize them with the hospital and the services that it offers, introduce them to key

hospital decision makers, increase their awareness of issues and problems the hospital is facing, and clarify the role of the board and the expectations of individual board members. A good orientation will prevent many problems from ever developing.[8]

2. Internal programs for the full board can provide all board members with common knowledge about priority concerns and a common language or framework for decision making. Occasionally members of the medical staff and administration should participate in these internal programs. In addition to expanding knowledge, they offer unique opportunities to improve communication, deepen understanding, and strengthen the working relationships between the three groups.

3. External programs can provide board members with specific in-depth knowledge of important issues and increase board member awareness of what other hospitals are doing. Whenever possible, board members should attend external programs with other members of the board, the medical staff, or administration.

The results from participation in formal education activities will be increased if board members have clearly defined development goals. These goals should flow from performance feedback and from coaching provided by the board chairperson and CEO.

When no change takes place

If application of this development model works and behavior improves, the plan of action should be maintained and ongoing feedback and coaching should be provided. However, if no change occurs and the behaviors seriously hinder the work of the board, the nominating committee should be encouraged not to reappoint the person. If the situation is very serious, it may even be necessary for the board to request that the person resign immediately. These final options have serious political consequences and must be handled diplomatically. However, failure to address serious performance problems affects the credibility of board leadership.

Board development is a complex process. Even the best development efforts will not resolve all problems. However, better diagnosis and more comprehensive integrated development strategies will produce some positive results.

REFERENCES

1. Johnson, E.A., and Johnson, R.L. *Hospitals in Transition.* Rockville, Md.: Aspen Publishers, 1982.
2. Mager, R.F. *Analyzing Performance Problems.* Belmont, Calif.: Feron Publishers, 1970.
3. Melcher, R.D. "Roles and Relationships: Clarifying the Manager's Job." *Personnel* 44, no. 3 (1967): 33–41.
4. Ewell, C.M. "Talking to the Board." *Hospitals* 57, no. 2 (1983): 81.
5. Catholic Hospital Association. *Guidelines on the Responsibilities, Functions, and Selection Criteria for Hospital Boards of Trustees.* rev. ed. St. Louis: CHA, 1974.
6. McCall, M.W. "Leaders and Leadership of Substance and Shadow." In *Perspectives on Behavior in Organizations,* edited by J.R. Hackman, E.E. Lawler, and L.W. Porter. New York: McGraw-Hill, 1977, pp. 375–385.
7. Maryland Hospital Education Institute. *Steps to Self Evaluation of Hospital Board Performance.* Lutherville, Md.: MHEI, 1977.
8. Ewell, "Talking to the Board."

PART III

APPRAISAL, SUGGESTIONS AND INCENTIVES

Performance appraisal systems in rural western hospitals

Thomas C. Timmreck

This article assesses the use of performance appraisal systems in rural western hospitals by discussing and analyzing the findings of 47 health care facilities.

The assessment of employee performance is an essential part of health care management. Even though much has been written and said about performance appraisal, the basics seem to fail to reach many practicing managers including those in small rural hospitals. Improper assessment of employees has reached such a magnitude that it has contributed to job dissatisfaction and is being blamed for contributing to health care employee burnout.[1]

Major hospitals are fortunate to have large management teams and an abundance of human resources to develop and implement good performance appraisal systems. However, small rural hospitals often lack such resources. Small facilities, like larger hospitals, need to have and to implement efficient and effective performance appraisal systems. However, the administrators of many rural hospitals have limited management teams, and when setting up performance appraisal systems, they are forced to rely on past formal education and training seminars, consultants, and literature from professional organizations. Most often the management team of small rural hospitals consists of nurses or technically trained persons who have little training in management and limited in-depth knowledge of correct approaches to performance appraisal systems.

Consultants are expensive; to implement performance appraisal systems, the consultant would be needed for a good deal of time on an ongoing basis. Thus, a major time commitment and great deal of effort are required to establish a well-functioning performance appraisal system in a small facility, and some have done quite well.

This article will present those variables and factors that are important to performance appraisal systems in rural hospitals, provide research findings on performance appraisal systems as found in small rural hospitals in the western United States, and conclude with a discussion of the research findings and the application of proven effective methods and approaches.

ATTRIBUTES OF A GOOD PERFORMANCE APPRAISAL SYSTEM

When defining and identifying performance appraisal uses, goals, and purposes, certain phrases, con-

---

Thomas C. Timmreck, Ph.D. is Associate Professor of Health Care Administration in the Department of Health Science and Human Ecology at California State University, San Bernardino in San Bernardino, California.

cepts, and terms are often used such as identification, evaluation, and development of individual performance.[2] Additional terms are used to pinpoint weaknesses and strengths, such as source for reward, promotion, and change. The performance appraisal system is also used for personal and staff development, job description refinement, planning, organizing, controlling, and directing work performance.[3] The performance appraisal process is complex and has multiple uses.

METHODS AND APPROACHES TO APPRAISING JOB PERFORMANCE

Most performance appraisal systems are implemented to affect the entire organizational structure. According to Price, the system is effectively implemented at the (1) individual level, (2) departmental level, and (3) overall organizational level, with a separate system for upper management.[4] In hospitals a similar but different three-level dichotomy is appropriate—the nonprofessional, professional, and management levels. Evaluation forms and approaches can be designed for use at all three levels. The upper-level managers are best evaluated through instruments that address qualitative factors more than quantitative ones. Some factors that are important to review in top-level administrators include resource utilization, patient care management, financial accountability, effectiveness, efficiency, and coordination of departments and their outcomes. In addition to the above factors, Harvey identifies six additional areas of importance to the top administrator: planning, organizing, quality of medical services, crisis resolution, compliance with regulations, and overall promotion of the hospital.[5,6] The traits used in the two lower levels are determined mostly by job activities, skills needed, and level of work.

TYPES OF EVALUATION APPROACHES

The evaluation of unskilled workers and nonprofessionals should differ from that of professionals. Evaluation of lower-level employees can be accomplished in a variety of ways. Several approaches are commonly utilized for most lower-level and professional employees throughout a health care facility. The most common methods include the graphic rating scale, paired comparison, ranking method, behaviorally anchored rating scales (BARS), forced distribution, and job simulation.[7]

The evaluation of unskilled workers and nonprofessionals should differ from that of professionals.

Graphic rating scale

This scale is used far more than any other approach. The various factors and characteristics on which the worker is to be evaluated are presented on the left side of the form, and then a ranking scale (similar to a Likert scale—least satisfactory to most satisfactory—usually on a five-point scale) is presented on the right side, which assesses the level to which the employee performs each job task of characteristics. By checking the appropriate box for each factor and by adding up the points, the rater can obtain a score for the employee. Effective graphic rating scales use objective descriptors, which describe work behaviors and avoid personality traits and other subjective items. On some of the well-developed forms reviewed in this article, written comments were solicited, sometimes asking for specific input or for general comments.[8-11]

Ranking method

Using this method, the supervisor orders or ranks the workers from best to least best or highest to lowest. The ranking is based on the overall job proficiency or on selected dimensions or characteristics.[12]

Paired comparison

This comparison approach is done by the supervisor comparing each worker with every other worker.[13] This is a more discriminating approach, and, unlike the graphic rating scale, it uses no score. Only one person comes out on top in this method. If factors such as work quantity, work quality, reliability, and effort are used, the subjective nature of this evaluative approach can be somewhat minimized.[14]

Behaviorally anchored rating scales

This is a performance-based approach. It is also more complex to develop, has many pages on which are listed many character traits, and takes longer to administer. However, it is probably the most effective and the fairest of all approaches. The BARS approach is a systematically developed checklist of behaviors, performance activities, or critical incidents. The supervisor, or more often a committee, generates independent dimensions of a given job, and then generates specific behaviors that can be hooked onto (anchored at) different points along a given scale (usually a seven-point scale, but this varies with lists of behaviors). A separate page is developed for each and every job task or characteristic (job knowledge, work habits, and so forth). A list of behaviors, as shown by a short critical incident, is put in question form ("Could be expected to follow correct procedures for writing in the medical record?") for each incident. The critical incident occurs when the worker's behavior results in unusual success or failure in some part of each of the job characteristics or tasks that is identified for each separate form. The form is developed with a vertical scale down the side that rates from high to low. The worker traits or activities are presented in statement form in a high-to-low hierarchical list, so they can be easily anchored onto the scale. The behaviors are placed along the scale according to how the committee agrees they fall. The entire form consists of several pages with seven to ten critical incidents developed for each job task or characteristic but with only one trait or construct addressed per page. The rating process simply involves placing an X or mark on the scale where the worker most closely falls. The supervisor and worker should do this together and should agree on the placement of the mark. The worker's participation in the development of the BARS makes it a much more acceptable process because he or she has a chance to provide input and can participate in the actual evaluation process. The BARS method, though a most acceptable approach, is not often used because it is not easy to grasp and requires considerable time and effort to develop.[15-18]

Job simulation

This method is a unique way of assessing employee work performance. Used more for management personnel, the employee is presented with a structured situation (often away from the job site) that is an example of the work environment. The situation usually involves a series of prearranged problems that the employee has to work through. Then from assessments of the simulation activity, inferences are made about job proficiency.[19] The "In Basket" activity is one job simulation method that is commonly used for managers. The administrative person is given decisions to make or problems to solve, and a team evaluates approaches and results of the process. Clinical "In Baskets" could be used for health care provider personnel.

Of the many approaches available, the most common seen in the health care industry is the graphic rating scale, while the most fair and effective is the BARS system. One reason the graphic rating scale is used most is that it is mass produced by publishers and sold in business supply houses. The problem with this approach is that the publisher pays little consideration to well-established and important management and organizational behavior factors and research findings needed in a performance appraisal system and evaluation form.

PROBLEMS WITH PERFORMANCE APPRAISAL SYSTEMS

There are five major problems with performance appraisal.
1. Supervisors are often poorly motivated to complete the performance appraisal forms accurately. Most often they have not been involved in developing the form or the system, yet are required to use it. Little or no training has been given on how the system works, nor has a great deal of significance been put on why the rating must be accurate.
2. Supervisors are often untrained in filling out the forms or scales properly. Thus supervisors usually end up falling victim to one of the following types of bias:
    - The halo effect (error)—the process of rating a person higher (or lower) on a job trait or task because the supervisor sees the person more favorably or unfavorably on another dimension. One outstanding personality dimension or work effort by the worker outweighs all others, and the supervisor is unreasonably influenced by it.
    - Central tendency error—the supervisor tends to rate the worker in the middle of the range of the scales instead of assessing all dimensions of the person including the high and low areas of performance. Being unwilling to take a stand or afraid of making a true assessment can influence the supervisor to take this position.
    - Leniency error—the supervisor tends to rate the worker in the extreme positive end of the scale and tends not to rate the worker correctly in all parts of the scale. This is similar to the

problem of central tendency with the opposite effect.
- Over critical error—the supervisor or evaluator is unnecessarily petty, critical, or too hard on the worker and thus tends to rate the worker in the extreme negative end of the scale and tends not to rate the worker fairly on all aspects of the scales.
- Personal bias error—the supervisor is influenced by prejudice; is excessively judgmental; or is influenced by beliefs about positions, sexual status, race, and job performance (i.e., women can only be secretaries).

3. When there is no clear system or procedure set forth, only one supervisor is used. The use of multiple raters is not often used; instead, it is usually only the immediate supervisor who fills out a form, often without the employee being present.
4. The characteristics, tasks, and traits used in the forms are often composed of general subjective terms or phrases. This approach lends itself to excessive judgmentalism such as assessing personality traits and mannerisms rather than focusing on work performance. These factors are compounded by poorly constructed forms that encourage use of subjective traits such as personality, attitudes, and how well a person is liked by others, plus supervisor or rater bias and low reliability.
5. The performance appraisal system and approaches are often distrusted and thus, resisted by the employee. It is human nature to not want to be judged and evaluated, and when this feeling is compounded by abuse and misuse of performance appraisal methods, it is little wonder that the employee does not want to participate in the appraisal of his or her performance. In the health care field, the acceptance of performance assessment becomes even more difficult as most workers are highly educated and skilled professionals who view themselves as competent and not in need of appraisal. (To an extent there is some

---

A major management shortcoming is the lack of ongoing and continuous feedback about the performance appraisal system.

---

merit to this position.) Additionally, the system is distrusted because the employee does not know how it will be used and believes it will be used against him or her and for unfair reasons (i.e., to gather evidence or build a file to justify termination).[20-23]

A major management shortcoming is the lack of ongoing and continuous feedback about the performance appraisal system. Often little is done by the supervisor about the performance appraisal process until a memorandum arrives on his or her desk stating that it is time to appraise workers' performance. Performance appraisal, under such circumstances of being ignored and then rushed into, contributes to distrust, as does its infrequent or unpredictable nature. Moreover, frequently the information goes to superiors and is used in salary and promotion determinations without the employee knowing how or on what basis he or she was evaluated. The real reasons for the existence of performance appraisal systems are often missing: to improve work performance; productivity; and personal, professional, and job development.[24-27]

ELEMENTS OF AN EFFECTIVE APPRAISAL SYSTEM

The performance appraisal system, as used in health care settings is most effective and accurate when it possesses the following seven elements:

1. The performance appraisal system contains a high level of trust on the part of both the employee and the superior or supervisor and among coworkers.
2. The performance appraisal system is viewed by the workers as an accurate and fair appraisal of work performance and does not include favoritism, political influence, seniority, and personality factors.
3. The performance appraisal system is perceived by the worker to be specifically linked to salary, rewards, bonuses, promotion, and so forth.
4. The performance appraisal system's extrinsic reward system is relatively noncompetitive; the worker feels there are sufficient rewards available for any employee who receives a high performance evaluation.
5. The performance appraisal form and assessment are made available to the employee and then perceived as fair and unbiased.
6. Employees know they have some control over the

system and results, such as being able to provide input, criticism, or modification of the form or system as needed.
7. The health care facility develops and strongly supports an employee development and training program used to improve deficient areas.[28-31]

Performance appraisal systems are supported best when they are perceived by the employee as a mechanism for personal development and for improving job performance. However, many times the appraisal system used in health care facilities becomes the axe used to eliminate a person, a method for accumulating evidence in the employee's personnel file to legally protect the institution in case the person gets fired (or to gather evidence to justify a termination). Research shows that training can make the performance appraisal system effective and can eliminate many of the problems mentioned above. If performance is truly assessed and subjective personality trait assessment is kept to a minimum, then the performance appraisal process becomes much more fair and acceptable to the health care worker. Furthermore, an effective performance appraisal system can turn around the performance of work groups, causing productivity to increase and mediocrity to vanish. This is an important factor for hospitals in the days of accountability and the push for efficiency due to prospective payment systems.[32-36]

MANAGEMENT CONSIDERATIONS FOR PERFORMANCE APPRAISAL SYSTEMS

It is suggested by K.E. Smith that health care managers must consider several factors when setting forth the criteria for evaluation of the worker's performance. The manager should discern if
- the worker's output can be observed, measured, and evaluated;
- work results can be changed by the individual's effort, motivation, training, or development;
- job performance is defined in terms of behavior and activities that affect results;
- criteria are attainable and realistic for the worker and based on training and experience;
- job activities are consistent with overall department and facility goals; and
- the worker understands work expectations and agrees that the objectives and expectations are fair, realistic, and attainable.[37,38]

The identification and use of objective and measurable criteria are very important. It is realized that this is an ideal to aim for, and that the clear defining of objectives and related work activities can be difficult for some jobs. This need not diminish the effectiveness of the program. It is important that the worker and the direct supervisor understand and accept the criteria as valid.[39]

ACHIEVING A SUCCESSFUL PERFORMANCE APPRAISAL SYSTEM

The hospital administrator and the administrative staff should use the following guidelines to achieve a successful performance appraisal system:
1. Clarify work assignments by assuring a mutual understanding of and agreement on duties and job responsibilities. In other words, have well-developed job descriptions.
2. Clearly set forth the results that are expected. Communicate results to the worker and review objectives on a regular basis.
3. Set specific work standards using objective criteria wherever possible.
4. Be open to problems and shortcomings of the system and be ready to make adjustments to improve the system.
5. Set up a regular schedule for review of progress. Appraisal should not be a yearly function, but an ongoing communication process between supervisor and worker.
6. Use ongoing coaching activities to direct workers toward achieving desired results.
7. Provide leadership through offering personal knowledge and experience.
8. Managers and supervisors should make themselves available as much as possible for questions, discussion, and direction.
9. Provide adequate training and skill development while encouraging workers to develop new and needed skills.
10. Provide appropriate tools, resources, equipment, supplies, and facilities to complete the job.
11. Be open, direct, and supportive.[40]

The success of the small rural hospital administrator directly relates to the effectiveness of the employees. Effective employees can be somewhat controlled by a well-established and effectively run performance appraisal system. Effort spent on developing a well-functioning performance appraisal system can result in improvements in productivity, efficiency, and effectiveness of the entire hospital staff.[41]

RESEARCH QUESTIONS CONCERNING SMALL RURAL HOSPITAL PERFORMANCE APPRAISAL SYSTEMS

What performance appraisal systems are used in small rural western hospitals? What variables are examined? How is the performance appraisal system managed? What aspects of a performance appraisal system are utilized in small rural hospitals? What are the aims of the performance appraisal system and for what purpose do rural hospitals use the findings of the performance appraisals? These and related questions were included in the research of performance appraisal systems found in small rural western hospitals.

RESEARCH METHODS

In order to determine how small rural western hospitals assess work performance of employees and how their performance appraisal systems are managed, questionnaires and letters were sent to over 100 hospitals in the mountain states of the western United States (see Table 1). Forty-seven questionnaires were returned. In the accompanying research letter, a copy of the performance appraisal form was requested. All responding hospitals sent a copy of their performance appraisal form. Using full-screen editing on the CYBER 7 mainframe computer, the data from the questionnaires were entered and then statistically analyzed by the Social and Political Science Statistical (SPSS) program. The results of the questionnaire responses are presented below.

In order to obtain a representative sample of small rural hospitals in the western area of the country, seven states were included in the study (see Table 1). Using the Directory of Health Care Facilities of the American Hospital Association, hospitals were selected by their location (within one of the selected states) and by size.[42] The aim was to select facilities with 50 or fewer beds. One limitation to this sampling method was that some of the listings were outdated; thus, the anticipated size of the facility was not totally accurate. However, this limitation had no major negative impact on the study or the findings.

RESULTS OF THE STUDY

Demographics and informational data

Forty-seven hospitals or hospital–nursing home combinations responded to the questionnaire. Ninety-one percent (43 out of 47 facilities) were located in communities of 20,000 or fewer people. Seventy percent (33 out of 47) of the responding facilities were in communities of 10,000 or fewer people as reported by the questionnaires. Almost one-fourth (11 out of 47) of the facilities were in communities with a population of 3,500 or less (see Table 2).

Thirty of the facilities were listed as hospitals and 17 were presented as being combined facilities (hospital with a nursing home).[43] With an N=47, the smallest hospital had 7 beds and the largest hospital had 100

TABLE 1

HOSPITALS IN RURAL WESTERN UNITED STATES

State	Number of hospitals	Percent
Wyoming	4	8.5%
Utah	5	10.6%
Colorado	12	25.5%
Arizona	6	12.8%
Idaho	8	17.0%
Nevada	5	10.6%
Montana	7	14.9%
	N = 47	100.0%

TABLE 2

SIZE OF COMMUNITY

Population size	Number of hospitals
1–3,500	11
3,600–10,000	22
11,000–15,000	5
16,000–20,000	5
21,000–25,000	3
Over 25,000	1
	Total facilities = 47

TABLE 3

SIZE OF FACILITIES BY ACUTE CARE BEDS

Hospitals N =	47
Largest hospital	100
Smallest hospital	7
Mean size	41.9
Mode size	25
Median size	41.7
Range	93
Variance	450.9
Standard deviation	21.2
Total beds in 47 hospitals	1,970

TABLE 4

SIZE OF FACILITIES BY NURSING HOME BEDS

Total facilities in study N =	47
Facilities with nursing home beds	30
Largest number of beds	120
Smallest number of beds	6
Mean	35.8
Mode	10
Median	25
Range	114
Variance	948
Standard deviation	30.8
Total nursing home beds	610

beds with the average (mean) size of the hospitals being 42 beds. The size of hospital that occurred most often was 25 beds (mode). The middle-sized hospital was 42 beds (median). Beds are defined as acute care (see Table 3).

Even though only 17 facilities were listed as being combined facilities, 30 of the 47 facilities (64%) indicated that they had some nursing home beds. (Many of these possibly were swing beds.) The average number of nursing home beds presented was 36 while the number of nursing home beds listed most often was 10 (mode). The middle number of beds presented was 25 (median). The greatest number of nursing home beds listed was 120, and the least 6 (see Table 4).

Responses to research questions

When asked if the employee was informed of the results of the performance appraisal, 43 of the 47 hospitals (91%) indicated that their employees were informed of the results. When asked if their hospital used a performance appraisal system, 42 out of 46 facilities responding (92%) indicated that they used a performance appraisal system.

When asked if they used a Likert-type rating scale where the worker and his or her various work activities were rated from "excellent performance" to "poor performance," 33 out of 47 facilities (71%) indicated that they did. When asked if the performance appraisal system used in their hospital utilized some form of self-evaluation, 31 out of 45 (69%) indicated that they made no provision for self-evaluation. When the question was approached from a positive context by asking if employees evaluate themselves, only 14 out of 47 facilities (31%) indicated that they did.

It was found that 37% (or 18 out of 47 facilities) used preprinted forms purchased in business supply houses rather than develop forms themselves in-house. When asked if the hospital utilized a separate form for the management staff, only 8 out of 47 (16%) did so. Nine percent (or 4 out 47) indicated that they did not do performance appraisal on all employees. Most of the responding hospitals indicated that they have a formal conference with the employee to discuss the performance appraisal (40 out of 47 or 85%).

RESPONSES TO MANAGEMENT ASPECTS OF PERFORMANCE APPRAISAL

A key research focus was the managerial aspects of performance appraisals in the small rural western hospital. One of the managerial concerns was the goal of the performance appraisal system. Over twenty-five different responses were presented as goals or aims of

One of the managerial concerns was the goal of the performance appraisal system.

conducting a performance appraisal of employees. The goal or aim most frequently cited was "goal setting and improving performance, productivity, and efficiency" (73% or 30 out of 41 facilities responding). Fifty-three percent of the facilities indicated that "reward performance, merit raise, and promotion" were goals of their performance appraisal. These two constructs of (1) improving performance and (2) rewarding performance were the most common goals or aims presented by the responding hospitals. The other common aims or goals of the facilities' performance appraisal systems were to

- determine value of employee—for termination purposes—for placement in personnel files,
- evaluate performance,
- identify workers' strong points—give workers praise—for use as a means for development of the employee,
- determine employee's wage and salary increases,
- encourage cooperation and communication,
- increase quality of employee's work,
- improve quality of patient care,
- plan future inservice training needs and programs,
- enhance job satisfaction,
- understand workers' needs or disagreements about various facets of the job, and
- improve morale.

When the research question, "For what purposes do you use the findings of the performance appraisal?" was presented, 17 different responses were given. These 17 responses were then put into 12 categories. The most frequent use of the findings of the performance appraisal was for wage increases. That is, 17 of the 41 responding hospitals (48%) indicated that wage increase was one of the main purposes of the performance appraisal findings. The second most common purpose was to "determine weak and strong points for period review." (14 out of 41 responding hospitals or 40%).

The following were the remainder of the purposes for which the findings of the performance appraisal were used: promotion/merit raises (32% or 13 out of 41 responses), future reference/documentation (32% or 13 out of 41 responses), set standards to improve performance (32% or 13 out of 41 responses), for performance feedback (29% or 12 out of 41 responses), reward or recognize high levels of performance (17% or 7 out of 41 responses), communicate with employee (15% or 6 out of 41 responses), goal setting (15% or 6 out of 41 responses), for follow-up or to determine training needs (12% or 5 out of 41 responses), for disciplinary reasons (7% or 3 out of 41 responses), and the least mentioned use was to improve quality of care (2.4% or 1 out of 41 responses).

WHO RATES THE EMPLOYEE?

To better understand how the performance appraisal system was managed, three additional research questions were presented: (1) Who rates the employee? (2) How are the employees informed of the performance appraisal results? and (3) How often are performance appraisals done?

The findings of the first question showed five different persons or approaches have been used to rate the employee. The person listed most often was the immediate supervisor (46% or 21 out of 45 responding hospitals). The second most frequently listed rater was the department head (40.4% or 18 out of 45 responding hospitals). The remaining three approaches include being rated by the head nurse, by management, and by a committee.

The second management research question concerning how the employee is informed of the results of the performance appraisal had six types of responses. The most frequent response was, "The supervisor discusses the performance appraisal results with the employee, possibly in an individual conference" (17 out of 47 responding hospitals or 36%). The second most used method to inform the employee of the results was, "The employee is asked to read and/or sign the evaluation" (12 out of 47 responses or 26%). The third most frequent method was, "Both the employee and supervisor fill out the appraisal form together and then discuss the results" (10.6% or 5 hospitals out of 47). The other three responses include "Employee given a copy" (4.3% or 2 out of 47); "Through seeing salary increase from results of performance appraisal" (2.1% or 1 out of 47); "In staff meetings" (2.1% or 1 out of 47).

The third question about appraisal system management concerned how frequently performance appraisals were done. Three different time frames were used in the 47 hospitals in this study. The time frame used most often and by the majority of the hospitals (85.1% or 40 out of 47 hospitals) was a 90-day probationary performance appraisal followed by a yearly appraisal. Twice-a-year appraisals were used by 3 hospitals (6.4% or 3 out of 47 hospitals), and quarterly performance assessments were done by 1 hospital (2.1% out of 47 hospitals).

DISCUSSION OF PERFORMANCE APPRAISAL PRINCIPLES IN SMALL RURAL HOSPITALS

The first variable was hospital size. The largest hospital was 100 beds, but only one hospital was this size. The mode of 25 beds is most representative of the size of the hospitals in this study. It is also significant that in rural areas, many hospitals are not just hospitals but are combined facilities (i.e., a hospital plus a nursing home). The combined facility presents some administrative approaches and staffing patterns that are a bit different from that of a straight acute care facility.[44]

The presence of nursing home care and the type of staff can have an effect on the performance appraisal system and approaches used, because nursing home care uses nursing assistance much more than acute care does. The question then is, how are nursing assistants evaluated? These unskilled workers, for the most part, have limited education and training but yet are involved in many nursing activities. Are they evaluated like kitchen workers or housekeepers or like nurses; or should a special separate form and approach be used for nursing assistants? In the beginning of this article, it was suggested that three levels of evaluation should be conducted: unskilled, professional, and management. These or other similar levels become even more important for the combined facility.

Much encouragement was derived from the research findings in that many facets of a good performance appraisal system were being utilized. A formal conference with the employee is conducted in 85% of the facilities, 91% inform the employee of appraisal results, and 92% have an effective performance appraisal system in place. It was also found that there is room for improvement. Philosophical, political, and managerial changes on some aspects of performance appraisals are needed. Most facilities do not let the employee evaluate his or her performance nor make provision for self-evaluation.

The findings showed that many facilities do not develop their own performance appraisal forms but instead use preprinted forms purchased at business houses. It was also found that in some facilities, one performance appraisal approach or form was used for all levels of employees including management. Also some facilities do not use a rating scale, and some do not even have a performance appraisal system.

Some problem areas were further identified, such as mistakenly believing that performance appraisals will improve morale or enhance job satisfaction. The worker evaluation is more likely to have the opposite effect—contributing to lowered job satisfaction and morale—since most persons find being evaluated anxiety producing because of fear over how results will be used. It was evident that most facilities try to use performance appraisals to improve performance, efficiency, and productivity as well as to set goals for the employee. Other praiseworthy uses of the performance appraisal were the development of the employee or the development of training needs.

Findings of the performance appraisal were used for a variety of purposes, most frequently to justify wage increases. Another major purpose was to determine weak and strong points in a period review. Promotions and merit raises were determined by the findings of the period review and performance appraisals. The findings were also used as a means to set work performance standards as well as to communicate with the employee, to give feedback, and to establish training needs. One of the least often mentioned uses of the findings was goal setting. This item should have been a main purpose for performance appraisal activities and conferences. A small percentage of the facilities also used the findings to reward or recognize high standards or levels of performance. Recognition, promotion, advancement, and better work assignments are positive uses of the performance appraisal findings.[45,46]

The persons who rate the employees in small rural hospitals are similar to those in larger hospitals or other organizations—immediate supervisors and department heads.[47-49] However, it was interesting that a limited few listed a committee or indicated that management was used to rate the employee in the performance appraisal system. How the management did the rating was not explained. To have a committee do the rating is a bit irregular and is considered ineffective and to be avoided for a variety of reasons: concerns for privacy, political problems created among workers, fairness, objectivity, and so forth. Supervisors should have good communication with employees and there should be no surprises in the evaluation process. If the process is handled in a committee, there is limited opportunity to explain the basis for judgments or to allow the supervisor to show concern and commitment directly to the employees.[50-53]

The results and findings of the performance appraisal were presented to the employee in a variety of ways. The supervisor may discuss the results with the employee. This discussion may be done informally with the supervisor or may occur in a formal individual conference. The employee may be asked to read the evaluation, review the form, and sign it or just be asked to sign the form. In a few cases, the employee was given a copy. In only 10% of the facilities in this study did the employee and supervisor both fill out the form and discuss it in a conference. Of the many approaches, this is the one that is most effective and promotes good relations between worker and supervisor. This is especially well received if the employee has the opportunity to do a self-evaluation as part of this process.

Responses from two facilities showed inappropriate means of letting the employee know of the results of the performance appraisal. One facility indicated that the results of the appraisal were presented in staff meetings. This would be fine for providing recognition for positive results. However, questions arise concerning treatment of poor appraisals—are they also presented in staff meetings? If not, does the mere absence of mention present a negative psychological message to the employee and lead to false conclusions by fellow employees? One questionable approach was the salary increase—the employee would learn of the results from the salary increase on the paycheck. If there is no salary increase, does this mean substandard performance, and if so, why? What areas of job performance need improvement?

For the system to be effective, the employees should be included in its development.

One concern most managers have about performance appraisals is frequency. It was found that most facilities use a 90-day probationary evaluation to determine if the worker should be retained and then conduct the performance appraisal annually. This is essentially standard for the industry. Three facilities conducted the appraisal system biannually. This does provide more feedback, allows more frequent communication, and closer assessment. If the performance appraisal system is well-managed and is not too complex, the twice-a-year format may be a very effective system because it shows how important appraisals are to management and the frequency reduces worker anxiety. One facility did the assessment on a quarterly basis. This is too often. The supervisor would begin to feel pressed for time to conduct the appraisal, especially if there was a large number of employees under him or her. The employee would probably feel the hospital is more interested in assessing performance than in getting work done.

There has been a tendency in recent years to want to over-evaluate employees. One area of overkill in the performance appraisal system is frequency. That is, in wanting to do a good job, the process can be done too often. Overkill also involves using a form that is too complex and having an overly complicated assessment system.

OBSERVATIONS AND APPLICATION OF FINDINGS

The performance appraisal system should accurately evaluate employee work and productivity. For the system to be effective, the employees should be included in its development. Fair compensation and career development policies must be included, and the system should encourage workers to perform well and work to their full potential. It must promote constructive communication, comply with state and federal laws and equal opportunity employment directives, and fit within the organizational and managerial methods and philosophy of the organization.[54]

Guidelines for equal employment opportunity law compliance

Many managers believe that the main reason for performance appraisal is to compile evidence in the personnel files of the employees in the event of any legal action or a need to terminate the employee.

With regard to promotion and retention, the *Federal Register* sets forth guidelines that suggest that although the evaluation instrument is not required to be valid, it may require validation if there is an adverse impact on a protected class of worker as defined by racial, sexual, or ethnic background. The selection rate must be fair and balanced. The process and form must also clearly show that it measures only job performance and is not biased against the protected class of worker.[55,56]

Small rural hospital settings

For most workers, the performance appraisal process is fairly threatening. The more subjective the appraisal process is, the more threatening it seems, and the more unfair it becomes. Thus, one challenge to the administrator of the small rural hospital and the management team is to develop an evaluation system that is fair. For the system to be fair, it has to be as objective as possible. The purpose of a performance appraisal system is to develop effective and efficient employees who will be productive and who are assets to the rural hospital's mission.[57]

The era of accountability is upon the rural hospital. Recently, efficiency, effectiveness, and accountability have been put into action due to limitations imposed by the Medicare Prospective Payment System. Effective performance appraisal systems can encourage accountability in the worker and direct him or her toward being a more efficient and effective employee. If the performance appraisal system is handled correctly, personal and professional growth and development experiences can result.[58] A properly developed and administered performance appraisal system can increase accountability, efficiency, and effectiveness while stimulating employee commitment to the job. The patient then benefits by increased quality of care.

Rural hospitals can be major employers in small communities and are a source of a great deal of revenue to the community. Employment in the rural hospital should be more than just a job. Rural hospitals can provide more than employment, compensation, and benefits. Employment in the rural hospital can provide a sense of contributing to the community and to the individuals who reside there.[59]

What the rural hospital workers communicate to members of the town on their off-duty hours can affect the community's perception of the hospital and the quality of care provided there. Consequently, if fair and objective assessment of work performance is done, employees will feel good about the local hospital and will speak highly of it. The opposite is also true: If the evaluation of work is viewed as less than fair and is dominated by subjective judgments, then job dissatisfaction can result, and discontent with the hospital is expressed to family and friends in the community.[60]

The attitude and commitment of the administrator to the performance appraisal process can determine the effectiveness of the system. If the top administrator is committed to the process, the management team and staff will also be committed. However, if a psychological and political message is sent to the management team and staff from the head administrator that it is merely a task to be tolerated, then the staff will produce a less than effective performance appraisal.[61]

In a small hospital, the staff is more intimately acquainted; thus, it is best to use a committee to develop the performance appraisal system so that an objective and fair system results. Moreover, any program that will have an effect on all employees is much more readily accepted and effectively utilized if that program is developed by the employees themselves.[62] In small rural hospitals, the organizational structure is often less formalized, and there is also a tendency to keep the performance appraisal system informal. However, due to the familiarity that develops in small organizational settings, the formalization of the performance appraisal system is necessary. A formalized approach is needed in small rural hospitals due to the familiarity of administration with family and friends of employees as well as the employees themselves. This allows the administration to be objective and fair and to encourage good interpersonal relations among familiar parties.[63]

Each facility should want to develop its own program to meet its own special needs. Management programs developed within a health care facility by the people who work there are more readily accepted. Thus, hospital administrators are encouraged to develop their own system and their own appraisal form, rather than using a canned program and then imposing it upon the employees.[64]

The most important feature of a performance appraisal system for small rural hospitals is that the system, process, and appraisal form should be objective and fair. The second most important feature is the spirit and beliefs behind the performance appraisal. The spirit of a performance appraisal program should be that of personal and professional development. It seems to be human nature to focus on and accentuate the negative traits and shortcomings of individuals and workers. Therefore, the fallibility of the worker has to be somewhat accepted and worked with, while focusing on skill and knowledge development. The performance appraisal has been used by some rural hospitals to point out weaknesses of employees and to document problems, a record to be placed in the employee's personnel file. While the legal ramifications of performance appraisal for personnel managers are recognized, this should not be the focus nor main purpose of the performance appraisal.[65]

The appraisal should be conducted as a formal affair between supervisor and employee. Employees prefer having a special appointment when the form is filled out in a conference with the supervisor. Self-evaluation should be a part of this process. The employee experiences much unnecessary anxiety if the form is filled out, reviewed by the supervisor, and then an appointment is made for a later time. In these circumstances, the time between appointment and review should be kept as short as possible to reduce anticipation anxiety.[66]

Many small rural hospitals do not include the opportunity for the worker to participate in self-appraisal. It is recommended that self-assessment be a part of any good evaluation system. Workers feel much better if they are allowed to express what they have contributed and how they feel about the work environment and the job. The self-evaluation aspect of the system should be recognized and should carry a credible amount of weight in the total appraisal process.[67]

• • •

The research findings showed that in some hospitals one purpose for conducting a performance appraisal was to increase quality of care. The performance appraisal process will not, in itself, increase quality of care. However, the underlying assumption is that quality of care should be the result of increased productivity and better job performance, all of which are the main goals of any performance appraisal system. A good performance appraisal system, which is developed in a thoughtful manner and which focuses on the development of the employee in a positive, humanistic perspective, will reach the ends hoped for and can result in an employee with greater job satisfaction.

REFERENCES

1. Moore, T.F., and Simendinger, E.A. "Organizational Burn-Out: Symptoms and Prevention." *Hospital Forum* 26 (January/February 1983): 7–10.
2. Price, C. *The Management Guide for Developing Group Practice Personnel Policies, Procedures and Employee Handbooks.* Denver: Medical Group Management Association, 1984.
3. Timmreck, T.C. *Dictionary of Health Services Management.* Owings Mills, Md.: National Health, 1987.
4. Price, *The Management Guide for Developing Group Practice.*
5. Hodgetts, R.M., and Cascio, D.M. *Modern Health Care Administration.* New York: Academic Press, 1983.
6. Harvey, J.D. "Evaluating the Performance of the Chief Executive Officer." In *Health Services Management,* 3d ed., edited by A.R. Kovner and D. Neuhauser. Ann Arbor, Mich.: Health Administration Press, 1987, pp. 48–62.
7. Oharari, O., Shaw, E., and Zachary, W. "Industrial/Organizational (Psychology)" *Preparatory Course for the National and State Licensing Examinations in Psychology.* Los Angeles: Association for Advanced Training in the Behavioral Sciences, 1983.
8. Hodgetts and Cascio, *Modern Health Care Administration.*
9. Oharari, Shaw, and Zachary, "Industrial/Organizational (Psychology)."
10. Rakich, J.S., Longest, B.B., and Darr, K. *Managing Health Services Organizations.* New York: W.B. Saunders, 1985.
11. Churden, H.J., and Sherman, A.W., Jr. *Managing Human Resources,* 7th ed. Cincinnati: South-Western, 1984.
12. Oharari, Shaw, and Zachary, "Industrial/Organizational (Psychology)."
13. Ibid.
14. Ibid.
15. Hodgetts and Cascio, *Modern Health Care Administration.*
16. Oharari, Shaw, and Zachary, "Industrial/Organizational (Psychology)."
17. Rakich, Longest, and Darr, *Managing Health Services Organizations.*
18. Churden and Sherman, *Managing Human Resources.*
19. Oharari, Shaw, and Zachary, "Industrial/Organizational (Psychology)."
20. Ibid.
21. Rakich, Longest, and Darr, *Managing Health Services Organizations.*
22. Churden and Sherman, *Managing Human Resources.*
23. Davis, P.A. "Are Your Performance Appraisals Acceptable?" *Health Services Manager* 15 (June 1982): 32–33.
24. Oharari, Shaw, and Zachary, "Industrial/Organizational (Psychology)."
25. Rakich, Longest, and Darr, *Managing Health Services Organizations.*
26. Churden and Sherman, *Managing Human Resources.*
27. Davis, "Are Your Performance Appraisals Acceptable?"
28. Rakich, Longest, and Darr, *Managing Health Services Organizations.*
29. Churden and Sherman, *Managing Human Resources.*
30. Davis, "Are Your Performance Appraisals Acceptable?"
31. Mayer, R.J. "Keys to Effective Appraisal." *Management Review* 69 (June 1980): 60–62.
32. Oharari, Shaw, and Zachary, "Industrial/Organizational (Psychology)."
33. Churden and Sherman, *Managing Human Resources.*

34. Davis, "Are Your Performance Appraisals Acceptable?"
35. Mayer, "Keys to Effective Appraisal."
36. Smith, D.E. "Training Programs for Performance Appraisal: A Review." *The Academy of Management Review* 11 (January 1987): 22–40.
37. Olsen, R.F. *Performance Appraisal: A Guide to Greater Productivity.* New York: Wiley, 1981.
38. Smith, K.E. "Performance Appraisal: A Positive Management Tool." *The College Review* 1 (Autumn 1984): 43–62.
39. Smith, "Performance Appraisal: A Positive Management Tool."
40. Ibid.
41. Ibid.
42. American Hospital Association. *Guide to the Health Care Field.* Chicago: AHA, 1983.
43. Timmreck, T.C., et al. "Acute-Care/Long-Term Care Facility Combinations: New Challenges for the Administrator." *Health Care Management Review* 6, no. 2 (Spring 1981): 63–70.
44. Ibid.
45. Herzberg, F. *Work and the Nature of Man.* New York: Thomas Y. Crowell, 1966.
46. Timmreck, T.C., and Randal, P.J. "Motivation, Management and the Supervisor Nurse." *Supervisor Nurse* 13 (March 1981): 28–30.
47. Wiatrowski, M.D., and Palkon, D.S. "Performance Appraisal Systems in Health Care Administration." *Health Care Management Review* 12, no. 1 (Winter 1987): 71–80.
48. Simpson, D. "Training Program Focuses on Developing Managers' Skills." *Cross-Reference on Human Resources Management* (May/June 1980): 3–5.
49. Golightly, C. "MBO and Performance Appraisal." *Journal of Nursing Administration* 2 (September 1979): 11–20.
50. Forbes, R.J. "Appraiser and Appraised." *Management Today* 33 (January 1980): 33–39.
51. Sandell, R.M. "Building Quality into Employee Performance." *Supervision* 41 (October 1979): 13–14.
52. Mayer, R.J. "Keys to Effective Appraisal." *Management Review* 69 (June 1980): 60–62.
53. Davis,. "Are Your Performance Appraisals Acceptable?"
54. Wiatrowski and Paklon," Performance Appraisal Systems."
55. Ibid.
56. Federal Register 43 (1978): 38308.
57. Timmreck, T.C. "Performance Appraisal: A Necessary Exercise for Rural Western Hospitals?" *Healthcare Forum.* Forthcoming.
58. Ibid.
59. Ibid.
60. Ibid.
61. Ibid.
62. Ibid.
63. Ibid.
64. Ibid.
65. Ibid.
66. Ibid.
67. Ibid.

Performance appraisal systems in health care administration

Michael D. Wiatrowski
and
Dennis S. Palkon

The application of performance appraisal systems to personnel administration represents an important development in health care administration. As health care administrators seek to more effectively manage their resources, performance appraisal systems offer the opportunity to reward and improve productivity by identifying behavioral performance criteria and relating them to promotion, retention, and evaluation systems.

Health care administrators are increasingly being held to higher standards of accountability for the management of their physical, fiscal, and personnel assets.[1-4] Using resources more effectively and efficiently becomes one method of coping in an era of diminishing resources. In the area of health care, personnel administrators must attempt to ensure that their employees are effectively utilized in a manner consistent with the career development of the employee and the needs of the institution.[5-7] Developments in the area of performance appraisal systems (PASs) reconcile both needs and take administrators out of the role of playing God when they must take responsibility for judging the personal worth of their fellow employees.[8]

Personnel administration is a field that has undergone rapid development as new models of management are explored and implemented and the limitations of traditional methods are revealed. The field, particularly in the area of health care, offers the potential to go beyond merely hiring, providing compensation, and administering personnel benefits programs, and has the potential to contribute to the achievement of organizational goals. The development of valid, pragmatic, and reliable PASs ties individual achievement to organizational goals. It is interesting, from a historical perspective, that in the early 1960s some of the first performance appraisal systems were developed in health administration.[9] Yet in recent years, the utilization of these evaluation methods has not kept pace with the progress in the field.[10]

The purpose of this article is to review different methods of employee evaluation and to evaluate their relevance to health administration. It will then suggest that behaviorally anchored rating scales (BARS) come closest to maximizing both individual and organizational effectiveness in health organizations. The article will suggest some criteria necessary for developing a personnel appraisal system; briefly review and criticize commonly used personnel evaluation

---

Michael D. Wiatrowski, *Ph.D., is an Associate Professor in Criminal Justice at Florida Atlantic University in Boca Raton. He has published in the areas of personnel administration, education, and sociology.*

Dennis S. Palkon, *Ph.D., M.P.H., M.S.W., is an Assistant Professor in Health Administration at Florida Atlantic University.*

methods; discuss BARS; and suggest guidelines on how to establish and administer BARS systems.

HISTORICAL EVOLUTION

Personnel evaluation methods assess differences between employees along behavioral and psychological dimensions. Consequently their function is identical to that of psychological tests, and the development of these methods parallels that of psychological tests. The evaluation of employees is not a new development. In the Chinese civil service that existed three thousand years ago, the "Imperial Rater's" task was to evaluate the performance of the royal family.[11]

The first recorded appraisal system in industry was devised by Robert Owens in Scotland around 1800. Owens put a colored block at each worker's place to designate how well the worker had performed the previous day. Different colors indicated higher or lower levels of performance.[12]

In the United States, prototypes of performance appraisal systems were initiated by the federal government and a few city administrators in the late 1800s. Walter Dell Scott began developing the well-known man-to-man rating chart that was used extensively to identify and evaluate military leaders during World War I.[13]

In the 1920s, the graphic rating scale was introduced that rated individuals on personality traits. By the mid-1950s, the orientation of performance appraisal systems was changing and work-oriented qualities were being assessed.[14] In 1970 the United States Supreme Court decision *Griggs v. Duke Power Company* strengthened the need for the construction and validation of appraisal tools that focused on the evaluation of job performance rather than on personality characteristics.[15]

Most recently, Title VII of the 1964 Civil Rights Act, as amended by the Equal Employment Opportunity Act of 1972, specifically states that it is an unlawful practice to discriminate on the basis of race, color, religion, sex, or national origin with regard to any terms, conditions, or privileges of employment. Case law indicates that the courts are including selection and promotion policies under the purview of that act. It is possible that if improperly designed and unvalidated appraisal instruments are used to hire and evaluate employees, these instruments may inadvertently discriminate against an individual or group of individuals; this would bring the practice under the domain of Title VII and its enforcement agency, the Equal Employment Opportunity Commission. Well-developed appraisal systems have demonstrated their ability to protect health care agencies from discrimination litigation.[16]

CRITERIA OF A PERFORMANCE APPRAISAL SYSTEM

The choice of a PAS for a health care system is a complex decision. Health care delivery systems are constantly being influenced by a variety of internal and external environmental factors. Consequently, the evaluation system that is selected must exhibit a number of characteristics. It must be flexible and capable of balancing the technical, social, and environmental aspects of the organization and their interactions with one another. By accomplishing this balance, the goals of the PAS will assist in the achievement of the goals of the organization. With respect to specific personnel administration goals, the appraisal system must accurately assess employee performance to ensure that the following interrelated goals are accomplished: (1) All employees must be included in the system design and implementation process; (2) The system must produce equitable compensation and career development decisions; (3) The system must motivate employees to perform well and develop their capabilities; (4) The system must be consistent with the rational planning of the human resource requirements of the organization; (5) The system must promote constructive communication between managers and employees; and (6) The system must be in compliance with federal equal opportunity employment directives.

Goals of a PAS including all employees in design and implementation

The first goal of a PAS is to have an explicit policy that indicates that all persons who will be affected by an evaluation system should assist in the design of the system by determining what constitutes effective performance. Implementation is facilitated by indicating that all employees are subject to periodic evaluation.[17] It also signifies the agency's concern for implementing a feedback mechanism system that allows employees to know how well they are performing within the organization.

Equitable compensation and career development

Performance appraisal systems relate job performance to compensation decisions. The PAS should therefore be based on measures of performance and not personality measures that may not be related to how well an employee does his or her job. If it is executed properly, employees will believe in the fairness of the appraisal process and will see how they can contribute to the organization.[18] Correspondingly, employees must realize that decisions related to career development, such as termination, training, increased responsibility, or advancement, are dependent upon performance.

Employee motivation

Having implemented an evaluative system that relates compensation to performance, it is important that the employer realize that an evaluation system can be an effective motivating tool. Lawler's expectancy theory states that individuals will attempt to maximize their own welfare and therefore will choose outcomes that are desirable.[19] Systems that relate improved performance to increased compensation are thus potentially powerful motivating tools. Although not all compensations need be monetary, it must be clear that performance is rewarded. It is therefore necessary that the employees believe the performance standards are realistic, that the evaluation procedure is a valid assessment of their performance, and that an ongoing feedback process exists that identifies and communicates behaviors that lead to effective performance.

Organizational and human resource development

One of the major functions of management is to plan for the growth and development of the organization. There is a corresponding requirement to relate that growth to present and future personnel needs. PASs assist managers in organizational development by inventorying those behaviors that are required for the successful accomplishment of the organization's goals. Through information on who is available and what their skills are and who will be needed in the future, performance appraisal systems can promote the development of human resources and coordinate them with the needs of the organization.

If the performance appraisal system is perceived as critical rather than constructive then a powerful motivating tool has been lost.

The development of constructive communication

One of the critical elements in achieving the goals of a PAS is the establishment of effective communications between employees and managers. If the PAS is perceived as critical rather than constructive then a powerful motivating tool has been lost. Consequently, it is the responsibility of those administering the system to create a milieu that results in employees perceiving feedback as constructive. This milieu is created by letting employees know that their performance, not their personality, is being evaluated. If a manager lets the employee know which behaviors are related to more successful evaluations, the employee is then given the opportunity to improve his or her performance. Employees having a role in establishing evaluation standards helps to ensure that employees know what the standards are and that they are reasonable. Employee input is thus an important factor in the standards' legitimization.[20]

Equal Employment Opportunity guidelines

The Equal Employment Opportunity Commission (EEOC) has developed extensive guidelines that regulate selection, retention, and promotion. "Any measure, combination of measures or procedures used as a basis for any employment decision"[21] is a selection procedure. While employers are not required to validate their evaluation instruments, validation becomes necessary when there is an adverse impact upon a protected class as defined by racial, sexual, or ethnic background. In EEOC guidelines, an adverse impact is generally defined by a selection rate that is less than 80 percent of the rate for the most successful group. This is not a fixed standard because in a later section of the same guidelines, employers are warned that adverse impact may be demonstrated by even smaller differences in selection rates. In such a situation, employers may either discontinue the use of the instrument or validate it and show that the system does explain superior job performance.

When PASs are used for promotion purposes, the instrument is being used to predict whether an individual who achieves a high rating in a position will perform equally well at the next level of the ladder. It is therefore not only legally but administratively crucial that the most competent employees are being selected, retained, promoted, and given advanced training. The alternative that exists with defective placements is the loss of confidence by both employees and supervisors in the ability of the system to make effective personnel decisions.

PERFORMANCE APPRAISAL METHODS

A wide variety of performance appraisal methods exist in the field of personnel evaluation. Having defined the goals that PASs should seek to accomplish, a number of widely used evaluation methods will be reviewed, and their shortcomings discussed.

Essay appraisal methods

With the essay appraisal method the rater writes a paragraph or more identifying and assessing the employee's strong and weak points. The data are subjective, and their effective presentation is dependent on the rater's writing skills. While an obvious advantage is that the rater can write an in-depth profile of the employee, there are a number of disadvantages. First, questions regarding the reliability and validity of the method abound. Second, the method is cumbersome and time consuming for many. Third, because of the lack of a clearly identified structure, data about elements of job performance may be neglected while irrelevant information may be included.

Graphic rating scale

The graphic rating scale is the most commonly used method of employee evaluation in the health care field. With the graphic rating scale the employee is compared to a standard. A five- or seven-point scale is presented, and the points on the ordinal scale are labeled "outstanding," "fair," and "unsatisfactory." Despite the apparent simplicity of the method it is difficult to manage for several reasons. First, because of the lack of objective standards, it is possible for two raters to use the same instrument and interpret the scales differently and to give the same person different evaluations. Another serious deficiency is the fact that the points on the scale have no intrinsic meaning and are thus difficult to define. An employee rated as average by one evaluator may be rated superior by another, resulting in a scale with low interrater reliability, which in turn affects the scale's validity. Finally, the ambiguity of these instruments becomes very apparent when it is realized that they provide no feedback for a person who functions at an "average" level and would wish to perform at a superior level. While the cost of implementing this method is low, it fails to meet the basic criteria of a satisfactory evaluation instrument.

Forced-choice rating method

With the forced-choice method, a series of phrases are developed to describe various elements of job behavior or personal qualities. The rater then chooses the phrases that are most and least descriptive of the employee. The rater, however, is kept unaware of the numerical value, resulting in the rater's being unsure that employees are being evaluated in a manner that accurately reflects their performance. The rater is not allowed to modify a phrase nor can the degree to which a phrase applies be specified.[22] Consequently, this rating method fails to provide the specific information needed to counsel employees for improving their performance; it does not aid in human resource planning; and it is of little use in equal employment decision making because it cannot be used to establish the relationship between selection criteria and job performance.

Critical incident method

The critical incident method requires that supervisors keep a record in a "little black book" on each employee in which they record incidents of positive or negative behavior,[23] such as the application of a new therapeutic technique or a misplaced instrument. This method, however, has some serious limitations, such as negative incidents being more noticeable than positive incidents. A delay in administering medication may be recorded while the same person delivering prompt action in an emergency may be taken for granted. The supervisor may put off recording or forget to record incidents, or, on the other hand, may become overly attentive and record everything.[24] This action places the feedback process in an unfavorable light because giving an employee a series of complaints does not provide a format for constructive employee development.

Comparison methods

Comparison methods of evaluation require the rater to rank subordinates along a designated dimension, such as effectiveness or efficiency. Generally, employees are ranked on only one dimension, which may make it difficult to accurately reflect jobs that have multiple skills such as patient communication, administrative procedures, therapeutic methods, and so forth. Thus the validity of comparison methods becomes questionable. Furthermore, the method provides little feedback. If an employee is ranked 6, what must he or she do to become more like employee number 1? Rather than promoting excellence in job performance, peer pressure could lower performance standards and create resentment toward higher ranked employees. The method can best be characterized as ranking employees against each other rather than against an absolute standard. It therefore provides little useful information on how to improve the absolute performance of an entire job category such as operating room technicians or nurses' aides. Consequently, it is deficient in the areas of providing information for compensation and reward systems, counseling and human resource planning, and equal employment opportunities.

Management by objectives

The management by objectives method (MBO) requires the supervisor and the employee to jointly set performance goals, and the employee is rated after a specific time period in terms of how close he or she comes to attaining those goals.[25,26] This method has the advantage of allowing the employee to exert a direct influence on the standards by which he or she is to be appraised. However, any system that relies entirely on goals may not be generalizable to other employees in similar positions and may not fairly reflect an employee's total performance. In addition, its application to health services must be approached with caution.

First, goals may be influenced by variables beyond the employee's control, such as economic conditions or peer performance. Second, by being results oriented, this method excludes any focus on activities that lead to those goals as also being desirable—methods may be as important as ends. Focusing either on activities or results to the exclusion of the other produces undesirable consequences because it causes individuals to emphasize that which is measured to the exclusion of that which is not measured.[27] Moreover, it is difficult to make comparative judgments for compensation purposes by establishing individualized goals.

BEHAVIORALLY ANCHORED RATING SCALES

Behaviorally based performance appraisal systems have been advocated by a number of professionals in the field of health care personnel management.[28–31] In this section the major characteristics of BARS will be discussed and their advantages summarized. Then the steps that are undertaken to develop BARS will be reviewed.

First, BARS emphasize ongoing behaviors rather than traits or attitudes that may or may not affect job performance. A position such as staff nurse is broken up into its component functions. Along a specific dimension of a job, an ordinal scale is created in which behaviors that reflect job performance from poor to excellent are arrayed. This scale then gives the employee specific feedback on the level at which he or she is currently functioning and how specific elements of job behavior might be improved when attempting to improve performance. Tables 1 and 2 illustrate two BARS that have been developed for nurses. These scales are for two dimensions of nursing: patient and family education and the nursing process.

Performance appraisal measurements are designed for a specific job or job classification. The PAS instrument developed for a nurse would be different from that for a medical record specialist because the functions that the instrument is designed to measure are different. In the development of the instruments, two factors interplay: cost and effectiveness. The broader and less job specific the instrument becomes, the lower the developmental costs. This vagueness, however, lowers the usefulness of the instrument for improving the performance of an individual or specific jobs and almost forces the evaluator to return to the assessment of global personality dimensions that may not be related to job performance. Performance appraisal instruments are unique because they evaluate behaviors. Health care personnel can evaluate their performance with organizational expectations and know exactly what they have to do to improve the effectiveness of their work.

Finally, while performance appraisal systems are behaviorally oriented, they do confront the issue of

TABLE 1

RN DIMENSION 1: PATIENT AND FAMILY EDUCATION

The nurse must have the ability to effectively orient the patient and the family to the hospital environment, procedures, and policies, and the ability to effectively educate the patient and family in terms of the patient's problems and capabilities.

Rating	
	Highly effective performance
9	Serves as a resource to other nursing personnel in planning patient education; identifies need for new patient education programs; participates in development or review of patient education programs.
8	Consistently begins patient and family education at time of admission; participates in teaching groups of patients by using established programs; refers patients to community agencies as an additional educational resource.
	Effective performance
7	Consistently provides preoperative teaching based on patient needs; sees that the patient understands all aspects of medication usage, administration, and effects; involves the family in orientation to the hospital; consistently documents outcomes of patient's return to former life style.
6	Provides a set of self-care instructions for patient with a newly applied arm cast; assesses learning needs of patient and family; teaches health care principles based on identified needs coordinated with medical care plan; uses established patient education programs.
	Marginal performance
5	Begins coordinating patient and family education after receiving the physician's written order; fails to discontinue teaching when patient is anxious; rarely refers patient to community agencies as an additional educational program geared to patient's specific needs.
4	Fails to include family in education process; occasionally must be reminded by head nurse or charge nurse to provide patient and family teaching; gives inadequate instruction and supervision to patient in self-administration of insulin; fails to teach significant other how to care for patient when patient is unable to provide self-care.
3	Discusses posthospital care with the patient and family only if requested; avoids patient education when it involves subject matter that is difficult to handle, such as death; sets up the educational television program for patient and family without discussion before or after.
	Ineffective performance
2	Consistently postpones patient education until discharge day; when teaching patient, communicates disapproval of life style that has contributed to illness; does not refer patient to community resources.
1	Provides no patient or family education to deal with an illness; assumes that the patient knows how to change a dressing at home; fails to orient a patient to physical environment; fails to document patient education outcome.

performance. This issue is critical because performance relates the individual's behavior to the achievement of the amount of inputs and to the level of goal achievement. More effective employees should be achieving higher levels of the organization's goals, such as fewer lost records, decreases in patient mortality, and improved inventory control. Thus by linking the evaluation criteria in BARS to organizational goals, better employee performance is translated into an improved organization.

TABLE 2

RN DIMENSION 2: NURSING PROCESS

The nurse must effectively conduct and document the nursing process; assess patient needs; plan, develop, implement, and evaluate the nursing plan; and make nursing care decisions based on sound knowledge and judgment. The nurse must also have the ability to implement and apply this knowledge and theory in the care and safety of patients.

Rating	
	Highly effective performance
9	Meets standards set forth in number 6 below. Functions independently in doing so and is a resource to others. Consistently checks to determine that the care plan is followed through, and is always sensitive to patient's age, sex, and cultural background in the care process; initiates nursing care conferences when needed; documents all patient changes, nurse's actions, and potential problems; begins patient teaching as soon as possible and documents this clearly.
8	Meets standards set forth in number 6. Functions independently in doing so. Identifies and documents patient's understanding of his or her illness and need for patient education; is sensitive to special requirements of patients; documents on the nurse's notes nursing actions, patient progress, changes, and patient teaching.
	Effective performance
7	Meets standards set forth in number 6. Requires minimal guidance in carrying out responsibilities; gathers assessment information to develop a care plan; is aware of cultural differences in reaction to pain; completes nursing care process with little supervision; documents all patient changes.
6	Makes an accurate assessment of the patient's physical and behavioral status; accurately identifies most problems that require nursing intervention; writes a nursing care plan that effectively addresses the nursing problem; implements the nursing care plan; evaluates effectiveness of the nursing intervention and modifies as necessary; requires guidance and seeks it when needed from appropriate resources; determines patient's knowledge base about his or her illness(es); takes cultural differences into account when developing the care plan; documents major patient changes.
	Marginal performance
5	Makes occasionally inaccurate assessments; identifies a few problems inaccurately; occasionally does not address nursing problems in nursing care plan; occasionally does not implement part of the nursing care plan; evaluates effectiveness of nursing intervention but occasionally fails to make modifications; requires guidance but occasionally does not seek it; needs supervision in order to complete the nursing care process; waits for an order to begin coordinating patient education and discharge planning; fails to do nursing assessment within 24 hours of admission.
4	Makes frequently inaccurate assessments; frequently identifies inaccurately problems that require nursing intervention; frequently does not address nursing problems in nursing care plan; frequently does not implement part of the nursing care plan; evaluates effectiveness of nursing intervention but frequently fails to make modifications; requires frequent guidance; frequently fails to seek guidance from appropriate resources; cares for needs of patient but ignores needs of the family during the nursing process; carries out the nursing process yet fails to document it; collects data but fails to arrive at a nursing diagnosis.

continued

Table 2—*continued*

Rating

3	Usually makes inaccurate assessments; usually identifies inaccurately problems that require nursing intervention; usually does not address nursing problems in nursing care plan; usually does not implement part of the nursing care plan; evaluates effectiveness of some nursing interventions and usually fails to make modifications; usually requires guidance; does not seek guidance from appropriate resources; begins implementation of the care plan without assessing patient needs; documents patient's daily activities without relating them to the care plan; needs close supervision in order to complete the nursing process.
	Ineffective performance
2	Fails to do nursing assessments; fails to identify nursing problems that require nursing intervention; does not write a nursing care plan; does not implement a nursing care plan; does not evaluate effectiveness of nursing intervention; is unable to function without near constant guidance; needs constant supervision to complete the care plan and carry out the nursing process; fails to use proper resources to aid in assessing, developing, or evaluating the care plan; conducts an admission assessment interview when patient is in pain; usually provides incomplete documentation of nursing process.
1	Always makes inaccurate assessments; always identifies inaccurately problems that require nursing intervention; never addresses nursing problems in nursing care plan; implements nursing care plan incorrectly so that it harms the patient; evaluates effectiveness of nursing intervention incorrectly and makes the wrong modifications; is totally unable to function without guidance; is unable to complete the care plan without assistance; fails to do initial assessment upon admission; never checks to see if a care plan is being correctly implemented; fails to obtain patient's medication.

The methodology for the development of behaviorally anchored rating scales has been described in the personnel management literature.[32-35] It consists of a number of discrete steps, which are described below.

- First, a behavioral job classification system must be developed. While most large organizations have job descriptions, these may not be behaviorally oriented. The number of employees to be rated should be large and stable, such as X-ray technicians, medical record specialists, medical technologists, billing personnel, nurses, and so forth. A large number allows for the generation of sufficient data to develop the skills and to distribute development costs.
- In the next step, jobs are analyzed and broken down into discrete components or dimensions. The skills needed to perform these functions are then described. The assistance of those performing the functions should be obtained to ensure that the skills are accurately described and to promote constructive communication between those who are developing the instrument and those who are being evaluated.
- Third, supervisors must develop statements from the job description that describe effective performance for the position being examined. These descriptions should reflect behaviors that are observable, and which can be evaluated with a high degree of reliability and validity. Typically a job will have between 10 and 15 major functions.
- In each job function, examples of high, medium, and low job performance will be developed to establish an ordinal dimension to the scale ranging from effective to ineffective. The scale is evaluated by supervisors and if 75 percent of them do not agree with the position of any item on the scale, it should be eliminated from the instrument.

- Values for items on a scale are determined by standard psychometric scaling processes. Elements that have a standard deviation of less than 1.5 are typically used in the scale. When finished, a BARS would consist of 10 to 15 job dimensions, and each dimension would have 7 to 9 examples of job performance, which range from poor to superior or low to high.

IMPLEMENTATION OF PERFORMANCE APPRAISAL SYSTEMS

The implementation of PASs in health care organizations will require an acknowledgement by administrators that current evaluation systems contain weaknesses that are not contributing to the effectiveness of health care organizations. As health care is more closely scrutinized by the public, more effective methods of personnel appraisal will have to be implemented. The reasons for the adoption of PASs can be weighed on a cost–benefit basis. This article has dwelt on the benefits of the more effective employee evaluation systems. The costs of developing PASs are not inconsequential. It may require the full-time attention of a personnel administrator for up to a six-month period to accomplish the job analysis, scale development, and implementation process. These costs, however, must be weighed against systems that do not reward performance and that do not relate performance to the achievement of organizational goals.

• • •

The development of BARS that reflect individual performance and that in turn promote the achievement of organizational goals is a complex task. As resources become constrained, the necessity of evaluation systems may be questioned. BARS are systems that will help ensure that the criteria of goal inclusion, equity, motivation, planning, communication, and legality will be accomplished. The costs associated with the development of these goals are consistent with the mandate to provide the community with an efficient and effective health care system.

REFERENCES

1. Anthony, R.N., and Herzlinger, R.E. *Management Control in Nonprofit Organizations.* Homewood, Ill.: Irwin, 1980.
2. Kaluzny, A.D., et al. *Management of Health Services.* Englewood Cliffs, N.J.: Prentice-Hall, 1982.
3. Turban, E., ed. *Cost Containment in Hospitals.* Germantown, Md.: Aspen Systems, 1980.
4. Spano, R.M. "Performance Appraisal in a Hospital Social Service Department." *Social Work in Health Care* 7 (Winter 1981): 13–37.
5. Kaluzny, *Management of Health Services.*
6. Smith, M. "Documenting Employee Performance." *Supervisory Management* 24 (September 1979): 30–37.
7. Smith, J. "Personnel Management: An Overdue Review." *Health and Social Service Journal* 92 (April 1982): 529–31.
8. McGregor, D. "An Uneasy Look at Performance Appraisal." *Harvard Business Review* 35, no. 3 (1957): 90–94.
9. Smith, P.C., and Kendall, L.M. "Retranslation of Expectations: An Approach to the Construction of Unambiguous Anchors for Rating Scales." *Journal of Applied Psychology* 47, no. 2 (1963): 149–55.
10. Bernardin, H.J. "An Evaluation of the Norfolk General Appraisal System." Norfolk, Va.: Norfolk General Hospital, 1984. Photocopy.
11. Anastasi, A. *Psychological Testing.* 4th ed. New York: Macmillan, 1976.
12. Veninga, R.L. *The Human Side of Health Administration: A Guide For Hospital, Nursing, and Public Health Administrators.* Englewood Cliffs, N.J.: Prentice-Hall, 1982.
13. Haar, L.P., and Hicks, J.R. "Performance Appraisal: Derivation of Effective Assessment Tools." *Journal of Nursing Administration* 6, no. 7 (1976): 21–28.
14. Ibid.
15. Ibid.
16. Regan, W.A. "Thorough Evaluation Reports, Procedures Protect Hospitals from Discrimination Charges." *Hospital Progress* 63 (September 1982): 67–69.
17. Veninga, *The Human Side of Health Administration.*
18. McMaster, J.B. "Designing an Appraisal System that is Fair and Accurate." *Personnel Journal* 58 (January 1979): 38–40.
19. Lawler, E.E. *Pay and Organizational Effectiveness: A Psychological View.* New York: McGraw-Hill, 1971.
20. Nadler, D.A., Hackman, J.R., and Lawler, E.E. *Managing Organizational Behavior.* Boston: Little, Brown, 1971.
21. *Federal Register* 43 (1978): 38308.
22. Oberg, W. "Make Performance Appraisal Relevant." *Harvard Business Review* 50, no. 1 (1972): 64–68.
23. Ibid.
24. Ibid.
25. McConkie, M.L. "A Clarification of the Goal Setting and Appraisal Processes in MBO." *Academy of Management Review* 4 (January 1979): 29–40.
26. Muczyk, J.P. "Dynamics and Hazards of MBO Application." *Personnel Administrator* 24 (May 1979): 51–62.

27. Nadler, Hackman, and Lawler, *Pay and Organizational Effectiveness*.
28. Kearney, W.J. "Behaviorally Anchored Rating Scales—MBO's Missing Ingredient." *Personnel Journal* 58 (January 1979): 20–25.
29. Schneier, C.E., and Beatty, R.W. "Integrating Behaviorally-Based and Effectiveness-Based Methods." *Personnel Administrator* 24 (July 1979): 65–76.
30. Schneier, C.E., and Beatty, R.W. "Developing Behaviorally-Anchored Rating Scales (BARS)." *Personnel Administrator* 24 (August 1979): 59–68.
31. Schneier, C.E., and Beatty, R.W. "Combining BARS and MBO: Using An Appraisal System to Diagnose Performance Problems." *Personnel Administrator* 24 (September 1979): 51–60.
32. Smith and Kendall, "Retranslation of Expectations."
33. Bernardin, H.J., et al. "Behavioral Expectation Scales: Effects of Developmental Procedures and Formats." *Journal of Applied Psychology* 61, no. 1 (1976): 75–79.
34. Campbell, J.P., et al. "The Development and Evaluation of Behaviorally-Based Rating Scales." *Journal of Applied Psychology* 57, no. 1 (1973): 15–22.
35. Harari, O., and Zedeck, S. "Development of Behaviorally Anchored Scales for the Evaluation of Faculty Teaching." *Journal of Applied Psychology* 58, no. 2 (1973): 261–65.

Assessing the value of employee training

Robert Blomberg,
Elizabeth Levy,
and
Ailene Anderson

Scarce economic resources make cost-benefit assessment of employee training programs an important issue. It is helpful to review the role evaluation plays in training and apply cost-benefit assessment to management development programming.

Employee education and training in America is big business. A recent Carnegie study revealed that in 1981-82 American business invested $60 billion on employee training, a sum comparable to the total spent by all four-year universities and colleges in the United States.[1] Forecasts by the National Alliance of Business suggest even greater expenditures will be necessary in the future.[2] The total number of students trained is also nearly comparable. With significant figures like these it is understandable that more and more senior health care executives are becoming vitally interested in the return that dollars invested in employee training are delivering, both in terms of the extent that acquired skills, knowledge, and attitudes are being applied on the job, and the extent that application of training results in increases in effectiveness, quality, and quantity of products or services.[3]

This article describes what training evaluation is, why it is important, and what part evaluation plays in the organizational training process. Through means of an applied example, two of evaluation's toughest challenges are illustrated: measuring the transfer of training to the job and assessing the bottom line dollar value of this transfer.

For purposes of this article, "training" or "corporate education" involves those systematic means used by a business to instill new skills, knowledge, or attitudes in employees, thereby increasing their worth and long-term contribution.[4] The term "cost" refers to that which is given up to obtain something else. When resources are assigned to one good or service, those resources are not free to be assigned or used in an alternative way.[5] The term "benefit" refers to the outcomes from a project, outcomes that can be assessed in monetary terms.[6] Cost-benefit analysis, then, is the means one uses to assess the desirability of alternative investments in monetary terms relative to their respective merits.[7]

Inherent in these definitions are three important points. First is an understanding that corporate education must have economic worth to be considered of value. In a business sense, investment in human capital

---

Robert Blomberg, M.A., is Assistant Director, Training and Development, at Rochester Methodist Hospital.

Elizabeth Levy, M.H.A., is an Administrative Fellow in Hospital Administration at Rochester Methodist Hospital, Rochester, Minnesota.

Ailene Anderson, B.S. Eng., is a Training Consultant in Human Resources at Rochester Methodist Hospital.

seeks a measurable and positive result in enhanced business profits. Denova matter of factly concludes that the only reason for conducting training is an economic one.[8]

Second, investment in human capital must prove economically superior to alternative investments in land or other resources. Merits of investment in human capital will be evaluated with merits of investment in alternative forms of capital. Allocation decisions will be based on expectations of utility, life, and respective contribution margins each competing alternative provides. Investment decisions, therefore, are based on costs relative to benefits, and cost-benefit analysis becomes one mechanism used in guiding the decision-making process.

Third, both benefits and costs occur over time (usually years) as a cost-benefit comparison involves comparing the present values of both at the time the investment is initiated.

CONSUMPTION AND INVESTMENT

Education has consumptive (i.e., joy of learning) and investment (i.e., job skills) value.[9] Benefits of training accrue in different proportions to the persons receiving the training, the corporation sponsoring the training, and the society as a whole. In a distributional and economic sense, costs should be assigned to the individual, the sponsoring organization, or society in proportion to benefits derived.

According to Rogers, education is either generalizable or specific in nature.[10] Generalizable training increases the individual's general worth and is transferable from one organization to another. Learning to read and write or participating in apprenticeship education provide examples of generalizable training. Employers, individuals, and society as a whole all benefit from this type of training. By comparison, specific training increases the worker's value for one specific employer or for one specific job. Orienting a new worker to his or her work place or teaching managers about organizational policies and procedures are examples of specific training.

WHY COST IS AN ISSUE

Hospitals and other health care providers have long lived relatively isolated from the impact of truly free market economics. Transfer payments, entitlement programs, insurance reimbursement, and public demands for quality regardless of price have historically neither rewarded cost conciousness nor fostered competition. Until recently, investments in education and training in health care were rarely questioned. Symbiotic linkages between hospitals, universities, medical schools, and nursing schools were viewed a priori as legitimate, although immune to economic justification through cost-benefit analysis. Hospital environments provided practical work internships, and universities provided inexpensive labor pools and a ready source of graduated students for employment. The government, insurance carriers, and the public at large accepted this arrangement and paid the bill.

Today the environment is substantially different. Government reimbursement schemes, preferred provider organizations, and diagnosis related groups have changed the rules of the game. Much like other businesses, health care providers must maintain accountability for bottom line costs relative to benefits.[11] Generalizable training and education programs are no longer free from the budget cutter's ax. Rather, educators must justify training and development on the same basis as other hospital programs and projects to gain funding, staff, and administrative sanction. This justification is challenging for face valid types of specific training, such as teaching nurses how to prevent infections or how to do heavy lifting, but becomes even more difficult when dealing with topics of less observable or operational value, such as teaching nurses how to manage or how to conduct employee performance appraisals.

What then can be done to demonstrate whether the training an organization is providing is contributing measurably to organizational objectives and doing so in a cost justifiable way? A part of the complex answer to this question lies in the systematic application of sound management practices related to program evaluation. A partial understanding of program evaluation is facilitated through the development of both a conceptual picture of its role in the process and through the use of an applied example.

EVALUATING TRAINING IN TERMS OF COSTS AND BENEFITS

Literature on training and education describes two types of evaluation, interchangeably called formative and summative, conduct and outcome, or process and product. Formative evaluation typically audits administrative and process elements of a program, such as needs assessment technique, program materials, instructor qualifications, and teaching methods.[12,13] A successful summative evaluation determines the extent to which the program has achieved its objectives and the extent to which the accomplishments of these objectives can be attributed to the program.

In relation to training, summative program evaluation typically measures program effectiveness along four dimensions: (1) Did participants like the learning experience (reaction)? (2) Did participants learn what the instructor wanted them to learn (learning)? (3) Did participants transfer learning to the job—did they actually change behavior and apply their newly acquired skills on the job (transfer)? (4) Did participants demonstrate results of their training on the job (effectiveness, that is, did training increase efficiency, save time, increase quality, and so on)?

Training and education programs are no longer free from the budget cutter's ax.

Evaluative methodologies should be judged by three criteria: rigor, relevance, and economy. Rigor refers to the reliability, validity, and precision of measurement. Relevance connotes a link to organizational goals. Economy analyzes the trade-off between costs and benefits.

No single evaluative method is uniformly superior to others on all measures. Clearly, different evaluative tools (see box, "Ten Common Evaluative Measures") must be jointly and situationally applied. Maximizing rigor, for example, through an experimental and control group design, may be expected in a university, but may be politically or economically impossible in business. In business, maximizing economy by using reaction sheets alone may lack both the validity and relevance necessary to sell the program's continuation. In fact, the selection of appropriate evaluation techniques for a given course or program is a professional judgment call based on the trainer's situation and abilities.[14,15] Acceptable degrees of validity and reliability can be obtained by using one rigorous method, such as experimental and control group studies, or by combining several sequential or concurrent measures, such as reaction sheets, testing, measures of transfer including work samples, and measures of results including interviews and post-post research.[16,17]

One useful example of a health care organization using evaluation techniques is Rochester Methodist Hospital (RMH) in Rochester, Minnesota.

Ten Common Evaluative Measures

1. Reaction questionnaires (e.g., asking, "Did participants enjoy the class?")
2. Tests on content (e.g., short-answer or multiple-choice verbal or written questions)
3. Interviews with key persons on application of training (e.g., asking participants, "What did you use on the job from the class?")
4. Measurement of decreases in costs (e.g., absenteeism has decreased; savings in labor cost have resulted)
5. Measurement of increases in quality (e.g., employee performance appraisals are of higher quality)
6. Before-and-after research (e.g., supervisor X had 10 grievances filed against him last year; he has had 2 this year after labor relations training)
7. Direct observation (e.g., observing a supervisor orient and train a new employee)
8. Work sampling (e.g., reading reports written before and after skills training on writing)
9. Post-post research (e.g., reviewing the quality or quantity of work done at intervals after training)
10. Control and experimental group research (e.g., half the supervisory group receives X training, while the other half does not. What differences appear in production, quality, effectiveness, and so on between the two groups?)

Reprinted with permission from "Accountability Based Management Development," *Mobius* 7, no. 1.

EXAMPLE OF TRAINING PROGRAM AND EVALUATION

Program description

RMH is a 700-bed acute care facility affiliated with the Mayo Clinic and Saint Marys Hospital to form the Mayo Medical Center.

The hospital initiated an intensive management development program in 1982 to achieve the following goals:

- To enculturate managers with those basic values and goals essential to success.
- To build a cohesive management team capable of winning the commitment, energy, and ingenuity of every employee.
- To teach sound management skills and proven techniques.

FIGURE 1

PROGRAM EFFECTIVENESS SURVEY

DIRECTIONS: For each statement listed below, circle the response that accurately represents your opinion. In the left hand column, circle the response as it applies to you personally. In the right hand column, circle the response as it applies to supervisory/managerial employees reporting to you. If a question does not apply, leave it blank.

Answer Questions In This Column As They Apply To You Personally (TO A GREAT EXTENT … NOT AT ALL)		Answer Questions In This Column As They Apply To Management Persons Reporting To You (TO A GREAT EXTENT … NOT AT ALL)
1 2 3 4 5	1) The relevance of topics covered to the job done	1 2 3 4 5
1 2 3 4 5	2) The compatibility of the Management Development Program with corporate goals	1 2 3 4 5
1 2 3 4 5	3) The reasonableness of time demanded for training and development	1 2 3 4 5
1 2 3 4 5	4) The extent that management quality has improved as a result of this program	1 2 3 4 5
1 2 3 4 5	5) The extent that the corporation received a return on investment from management development	1 2 3 4 5
1 2 3 4 5	6) The extent to which learning was transferred to the job	1 2 3 4 5
1 2 3 4 5	7) The extent to which mandatory enrollment is necessary	1 2 3 4 5
1 2 3 4 5	8) The extent that class scheduling was compatible with work schedules	1 2 3 4 5
1 2 3 4 5	9) The extent to which class scheduling allowed you to take classes you wanted to take	1 2 3 4 5
1 2 3 4 5	10) The extent to which classroom facilities were of high quality	1 2 3 4 5
1 2 3 4 5	11) The extent to which managers were encouraged to apply classroom learning to their jobs	1 2 3 4 5

The Value of Employee Training

1	2	3	4	5		12)	The extent to which employees were encouraged to participate	1	2	3	4	5
1	2	3	4	5		13)	The extent to which you support the continuation of this program	1	2	3	4	5
1	2	3	4	5		14)	The extent to which clarity in the role and expectations of management has been gained	1	2	3	4	5
1	2	3	4	5		15)	The extent to which management morale has improved as a result of these programs	1	2	3	4	5
1	2	3	4	5		16)	The extent to which interdepartmental communication and understanding have been gained	1	2	3	4	5
1	2	3	4	5		17)	The extent to which union and management relations have been improved as a result of these programs	1	2	3	4	5

Please answer the following question in narrative form;

1) What specific comments and suggestions, if any, do you have for modification of this program? Consider the performance of instructors, the content of the program, time demands, etc.

_____
_____
_____

2) <u>Demographics</u>

What department do you work in?

___ Administration
___ Admissions/Business Serv.
___ Auxiliary
___ Chaplain Services
___ Dietetics
___ Finance
___ Housekeeping
___ IMS
___ Materiel Management
___ Nursing/Unit Mgmt./Risk Mgmt.
___ Personnel/Health Svc./Learning Res.
___ Pharmacy/Central Supply
___ Plant Services
___ Public Relations
___ Surgery
___ Charter House

3) <u>Your Position</u>

___ Administrator, Assoc. or Asst.
___ Director, Assoc. or Asst.
___ Head Nurse
___ Manager or Supervisor
___ Other

4) Your Involvement In The Program

___ 0-1 year
___ 1-2 years
___ 2-3 years

Since the program's inception, over 235 managers have participated in 40 hours of management training annually. During their first year in the program, managers complete the following six required core courses:
1. The Leadership Role of the Rochester Methodist Hospital Manager;
2. Rochester Methodist Hospital Health Services, Inc.—Who We Are, What We Do, Where We Are Going;
3. How to Conduct Effective Performance Appraisals;
4. How to Discipline and Discharge Employees;
5. What You Need to Know About Personnel Policies and Controlling Absenteeism; and
6. How to Build Positive Human Relations Skills.

Affiliation agreements with two local colleges enable RMH to grant undergraduate or graduate college credits to interested participants.

After completing the required classes, managers select from four categories of elective courses: (1) topical seminars and workshops, such as "How to Build an Effective Work Team" or "Writing Business Reports"; (2) management forums, a lecture series featuring chief executive officers from corporations across the United States; (3) short courses—half-day sessions on specific topics such as materials management and microcomputing software; and (4) roundtable discussion sessions on topics with RMH administrators, such as health care finance or ethics.

From a formative perspective the program appears well conceived, well organized, and well run. By most measures it sets an industry standard for management training. But does the program make a difference in a summative sense? That is, does the program change the behavior of managers on the job, and what related costs and benefits are revealed? A subjective yet comprehensive summative program evaluation was conducted to help answer these questions.

Method of evaluation

At RMH, two types of subjective methodologies assessed program effectiveness and worth. In 1986 an effectiveness survey elicited ratings of various items on a scale of 1 to 5, with 1 being the highest rating and 5 being the lowest. The entire population of program participants at that time (N = 235) was approached. The return rate was 135 (57 percent) in 1983, and 180 (77 percent) in 1986. The survey is shown in Figure 1. Results are shown in Table 1. An identical survey had been distributed to participants in March 1984, one year

TABLE 1

RESULTS OF PROGRAM EFFECTIVENESS SURVEY

Item no.	Average rating for respondent 1984	1986	Average rating for management persons reporting to respondent 1984	1986
1	2.0	1.8	1.9	1.9
2	1.7	1.7	1.7	1.8
3	2.0	2.0	2.0	1.9
4	2.3	2.6	2.5	2.4
5	2.2	2.0	2.3	2.1
6	2.2	2.2	2.3	2.3
7	2.6	2.7	2.5	2.5
8	2.3	2.5	2.3	2.5
9	2.1	2.0	2.1	2.1
10	2.1	1.9	2.6	2.1
11	2.1	2.0	2.2	1.8
12	1.6	1.9	1.7	1.7
13	1.5	1.4	1.4	1.4
14	2.1	2.0	2.2	2.1
15	2.3	2.2	2.2	2.0
16	2.1	2.0	2.1	2.0
17	2.9	2.9	2.7	2.6

after the hospital initiated the management development program. Responses to the most recent survey were compared to results obtained from the 1984 survey, statistical means were calculated, and a variety of demographic comparisons determined the amount of change that had occurred between 1984 and 1986. Next, statistical means were run by department (e.g., nursing, dietetics), by participant level (e.g., first-level, middle-level, or administrative manager), and by years of involvement in the program.

The second evaluative methodology was an open-ended questionnaire delivered through personal interview. Evaluators randomly selected 60 managers (a 30 percent sample) representative of all hospital departments to answer 12 specific questions, which are listed in the box, "Open-ended Questions."

Evaluation can be an extremely sensitive issue. To ensure objectivity in gathering critical data and to ensure the greatest candor among survey and interview participants, an administrative resident from the Uni-

Open-ended Questions

1. Of the courses that you have completed, which have proven to be the most helpful? Least helpful? Why?
2. Of the two types of instructors, in-house trainers or managers and outside consultants, which have proven to be most beneficial?
3. Please provide at least one example of how you have applied learning from a management development course to the job.
4. Have you changed your management style as a result of the Management Development Program?
5. What are the results of these changes?
6. What circumstances in the environment support or inhibit the use of ideas, techniques, and skills you have learned in the Management Development Program?
7. If you were to assign a dollar value or worth to the application of the management development you receive in your year at RMH, what would the dollar value be? How did you arrive at that figure?
8. When people take time from their job to attend management development classes, is there a change in productivity within the area? What is this change?
9. Do you think the Management Development Program is fulfilling its three fundamental goals?
10. Have you attended any of the management forums? If so, please specifically state the favorable or unfavorable impressions you received from these sessions.
11. Please provide any ideas for future courses, topics, and speakers that you think would benefit the program.
12. What overall effect do you think the Management Development Program has had on RMH?

Reprinted with permission from "Accountability Based Management Development," *Mobius* 7, no. 1.

versity of Minnesota and a personnel intern conducted the evaluation. Alternatively, this evaluation could have been purchased on a fee-for-service basis from a consulting firm, done by peers on a trade or gratis basis, or done by internal training personnel.[18,19] In the case at hand, cost and objectivity criteria were optimized through the use of unbiased resident and intern evaluators.

In sum, the first data-gathering technique secured paper and pencil scale type response data from program participants. The second methodology elicited narrative qualitative data, including a wealth of testimonial detail on learning application, behavior change (transfer), and cost-effectiveness.

Results of evaluation

Evaluations like this usually elicit high marks for programs during their first two years of operation. Typically, perceptions and reports on program effectiveness dip during years 3 to 5, and then enjoy a sustained rebound beyond the fifth year. In aggregate, the effectiveness ratings of RMH's program have not followed the norm (typical decline). However, specific segmented groups (by department, for example, or by length of time in the program) do demonstrate greater variance in effectiveness perceptions. Problem areas, by absolute level of response, were identified in Plant Services and Pharmacy, for example, and special efforts will be undertaken to enhance program options in those departments, especially for those persons with three or more years of course involvement. Other targeted areas for program improvement include efficiently scheduling classes, encouraging employees to participate, and reinforcing learning transfer. The responsibility to respond to these and other related issues, including a reexamination of staffing ratios, rests with hospital administrators. Without the evidence gathered through the survey, these problems would probably have remained hidden, or at best would have been seen as intuitive and emotional rather than quantitative and descriptive in nature.

The answers to the questions in the open-ended part of the survey provided narrative data regarding the learning transfer and dollar benefits, as well as a good deal of formative data directed toward program improvement. Seven summative conclusions resulted from the interview data:

1. Management development effects significant learning, and learning is being positively transferred to jobs by RMH managers. All managers who completed the survey and all who were interviewed responded positively regarding the application of classroom learning to their jobs in the hospital.
2. Management development positively contributes to the hospital's bottom line. Subjective commentary by program participants documents savings in excess of $213,000. The authors are currently in

the process of gathering objective cost-benefit data, and evidence to date suggests that subjective benefit estimates provided by participants may be low.
3. Management development contributes significantly and measurably to team cohesiveness—interdepartmentally, interprofessionally, and interpersonally.
4. Management development promotes progress, as evidenced by increased employee morale, heightened clarity in the role and expectations of managers, and increased quality of organizational communications.
5. Management development gives management team members feelings of being appreciated and recognized by senior management through administrative support of and commitment to managerial training. Managers consistently commented on administration's demonstrated confidence in management's ability to hear a variety of managerial viewpoints and distill what is applicable to RMH.
6. Management development does not interfere with other activities. A consensus (91 percent) reflects no negative effect on organizational productivity despite an organizational commitment of about 8,800 personnel hours per year to management development (37.5 hours for each of 235 managers).
7. Management development must be encouraged. Administration's support of managerial attendance in programs and application of management training on the job must grow or program quality and potential benefit may backslide.

• • •

What are employee development programs worth? Do these programs affect employee behavior and performance? What types of training secure a return for invested dollars? More and more future-minded, cost-oriented health care managers are beginning to pose these questions; few are receiving good answers. It is surprising how little is written in current literature on the cost-benefit assessment of private sector training. This appears to be an area in great need of quantitative study and research, especially since evidence indicates that rigorous but practical evaluation techniques can be economically applied to help trainers, program participants, and health care executives ensure the greatest return on investments in training and development.

REFERENCES

1. Eurich, N.P., and Boyer, E.L. *Corporate Classrooms*. Princeton, N.J.: Carnegie Corporation for Advancement of Teaching, 1985.
2. "More Employee Training Needed." *Bulletin on Training* 11, no. 5 (1986): 1.
3. Goldstein, I.L. *Training: Program Development and Evaluation*. Belmont, Calif.: Wadsworth, 1974.
4. Bass, B., and Vaughan, J. *Training in Industry: The Management of Learning*. Belmont, Calif.: Wadsworth, 1969.
5. Levin, H.M. *Cost Effectiveness, A Primer: New Perspectives in Evaluation*. Vol. 4. Beverly Hills, Calif.: Sage Publications, 1983.
6. Ibid.
7. Cohn, E. *The Economics of Education*. Cambridge, Mass.: Ballinger, 1979.
8. Denora, C.C. *Test Construction for Training Evaluation*. New York: Van Nostrand Reinhold, 1979.
9. Lewis, D. "An Introduction to the Economics of Education." In *Vocational Education and the Nation's Economy*, edited by W.G. Meyer. Washington, D.C.: The American Vocational Association, 1977, pp. 43-58.
10. Rogers, D., and Ruchlin, H. *Economics and Education: Principles and Applications*. New York: Free Press, 1971.
11. Larsen, R.E. "The Value in Evaluation." *Training* (January 1985): 92.
12. Brinkerhoff, R.O., et al. *Program Evaluation*. Boston: Kluwer-Nijhoff Publishing, 1983.
13. *Standards for Evaluations of Educational Programs, Projects, and Materials*. New York: McGraw-Hill, 1981.
14. Komras, H. "Evaluating Your Training Programs." *Training and Development Journal* 39, no. 9 (1985): 87-88.
15. Swierczek, F.W., and Carmichael, L. "The Quantity and Quality of Evaluating Training." *Training and Development Journal* 39, no. 1 (1985): 95-99.
16. Bennett, B., and Griswold, D.F. "Proving Our Worth: The Training Value Model." *Training and Development Journal* 38, no. 10 (1984): 81-83.
17. Pyle, B. "Evaluating White-Collar Technical Training Requirements." *Training and Development Journal* 39, no. 8 (1985): 56-58.
18. Rothwell, W.J. "The Case for External Peer Review." *Training and Development Journal* 39, no. 6 (1985): 78-79.
19. Jung, S.M., and Hamilton, J.A. "Third Party Validation of Training Effectiveness." *Training and Development Journal* 37, no. 3 (1983): 6-8.

Financial incentives for middle managers: pilot program in an inner city, municipal teaching hospital

Linda Shyavitz,
David Rosenbloom
and
Lynn Conover

Can incentive programs work in not-for-profit settings? The experience of a large public hospital says yes. They can motivate middle managers to effect change that will improve the financial performance of the hospital.

Efforts to fix or generally improve the health care delivery system appear to be ongoing in the health care industry. Regulation per se seems out of vogue at the moment, but statutory, regulatory and private initiatives that "structure the right incentives" are thought to hold great promise. Medicare prospective payment, new employer health benefits programs and some state all-payer systems have been designed in ways to provide hospitals incentives to deliver care as cost-effectively as possible.

The trick, however, is to translate an incentive into the work environment of large not-for-profit or public institutions. These institutions and the people within them do not always accept the logic in the policy papers that reshape the industry. As always, progress from policy formulation to systems change can be painfully slow and haphazard, or stymied altogether.

This article describes and evaluates an attempt to build an incentive system into the operating level of a large public hospital. At the time the program was introduced, in July 1981, the financial principles under which this hospital operated were different than they are now. The primary financial objective then was to maximize revenue and tailor expenditures to fit revenue potential. Today hospitals must manage unit costs to ensure profitability given both third party payer restrictions and the impact of price competition on a hospital's market share and revenue.

Despite the significance of these reimbursement changes, the experience this article describes is still relevant for hospitals considering introducing financial incentive programs for their middle managers to facilitate changes essential to improving or maintaining the financial performance of the hospital. The issue is the same: the hospital has incentives to per-

---

Linda Shyavitz, *M.S., is external affairs administrator of University Hospital, Boston, Massachusetts, and adjunct associate professor in the Health Care Management Program at Boston University. She has served as a consultant to numerous health care organizations, primarily in strategic and marketing planning.*

David Rosenbloom, *Ph.D., was formerly the commissioner of the City of Boston, Department of Health and Hospitals. He is currently vice president of Health Data Institute, Newton, Massachusetts.*

Lynn Conover, *M.B.A., is a senior planner at the Central Massachusetts Health Systems Agency, Worcester, Massachusetts.*

form in a certain way. Can the employees—particularly of not-for-profit and public hospitals—be persuaded to manage their behavior to conform to those new performance standards?

One of the reasons individuals select not-for-profit or public institutions is that service, not profitability, is the primary institutional objective. Achieving profitability used to be easier than it is under these new reimbursement initiatives. Hospitals may not have a lot of time to learn to adjust. Can financial incentive programs for middle managers help accelerate the process?

LITERATURE REVIEW

The literature on the use of employee incentive systems in health care organizations is sparse. Only ten identified plans appear to have been reviewed. (Another plan that gives staff a "finder's fee" is mentioned in "The Newest Idea."[1]) There appear to have been no reviews of public agency employee incentive programs.

The hospitals in which incentive plans have been reported include:

- Baptist Hospital, Pensacola, Florida[2]
- Memorial Hospital, Long Beach, California[3]
- Maine Coast Memorial Hospital, Ellsworth, Maine[4]
- Holzer Medical Center, Gallipolis, Ohio[5]
- St. Joseph's Hospital, Burbank, California[6]
- St. Luke's Hospital, St. Louis, Missouri[7]
- Hermann Hospital, Houston, Texas[8]
- Hospital of the United Methodist Church, Oak Ridge, Tennessee[9]
- Jewish Hospital of St. Louis, Missouri[10]
- Name unknown.[11]

All of these plans focused on improving operating efficiency and reducing costs that employees could control (such as labor and supplies). Nine of the ten were successful in reducing costs as a result of their incentive plans. None of the evaluations of these plans, however, were comprehensive or rigorous.

Additional positive consequences of these plans occurred regardless of whether they were intended or not. Table 1 lists those that were reported and the hospitals in which these additional benefits occurred. Many of these benefits—increased productivity, decreased overtime, decreased work hours, improved management control—have particular appeal given current hospital management challenges.

TABLE 1

HOSPITALS IN WHICH BENEFITS OCCURRED

Benefit	Institutions where occurred
1. Decreased turnover	Memorial Hospital, Long Beach; Hermann Hospital
2. Increased productivity	Memorial Hospital, Long Beach; Maine Coast Memorial Hospital; Baptist Hospital, Pensacola
3. Decreased overtime and decreased sick time	Memorial Hospital, Long Beach; Maine Coast Memorial Hospital; St. Joseph's Hospital, Burbank; Hermann Hospital
4. Decreased work hours	Baptist Hospital, Pensacola
5. Increased communication with employees	Memorial Hospital, Long Beach; Maine Coast Memorial Hospital; Baptist Hospital, Pensacola; Holzer Medical Center
6. Improved management control systems	Memorial Hospital, Long Beach; Baptist Hospital, Pensacola; Holzer Medical Center
7. Increased cash	Memorial Hospital, Long Beach
8. Above-area average employee compensation	Maine Coast Memorial Hospital; Baptist Hospital, Pensacola

The limited literature implies limited use of incentive plans in hospitals. There are several reasons reported to explain why incentive plans have not been used frequently in not-for-profit and public hospitals. First, the IRS historically objected to nonprofit organizations distributing "profits" to employees. It was perceived as diverting assets away from the organization. Baptist Hospital, Pensacola, spent two years negotiating with the IRS. The issue was resolved when their plan was called a "savings" sharing plan.[12] While Baptist was the only hospital reporting this problem, Cleverley advises that plans be cleared with the IRS.[13] Hellwig reported that only 8.4 percent of nonprofit organizations that he studied offered incentive plans, and that none of these had been cleared with the IRS.[14] Recently, IRS General Counsel report-

> *Recently, IRS General Counsel reportedly advised that tax-exempt organizations be allowed to adopt employee incentive compensation plans provided they are properly adopted and administered.*

edly advised that tax-exempt organizations be allowed to adopt employee incentive compensation plans provided they are properly adopted and administered.

A second reason that incentive plans have been infrequently used in health care organizations is that the health care product is not easy to define and the relationships among individual performance, process and outcome are not absolute.[15] The relationship may be clearer in other industries that use incentive systems extensively, e.g., sales organizations. The shift to unit costs as the focus for hospital financial performance, however, creates opportunities for assessing performance objectively. Once hospitals are able to accurately define their unit costs, appropriate intra- and interhospital comparisons can be the bases for measuring performance.

Third, if professional identification is greater than institutional loyalty, it is difficult to gain cooperation from health professionals in an incentive plan.[16-18] Moreover, if the goal of an incentive program is improved institutional financial performance, professionals may be insufficiently motivated by such a goal. Yet the challenge of Medicare prospective payment and many all-payer systems is to find ways to motivate professionals to deliver high-quality care with appropriate institutional financial considerations in mind.

MANAGEMENT INCENTIVE PROGRAM

In July 1981 Boston City Hospital (BCH), a nationally renowned municipal, general, acute care hospital in Boston, Massachusetts, launched a management incentive pilot program in its Ambulatory Care Center (ACC). Its primary purpose was to motivate clinical and nonclinical managers to generate increased revenue for BCH or reduce expense. Additional objectives were specified and are reported below. The pilot program offered a salary bonus to middle-level medical, nursing and nonclinical managers who made a defined, measurable contribution to improving the financial status of the hospital.

The program made cash incentive payments available to the management teams of four clinical areas within the ACC if these teams exceeded revenue or utilization targets negotiated with them before the program began. In fact, another option was to reduce expense, but none of the participating teams chose that challenge. That is because the teams had already absorbed expense reductions of 25 percent on average due to municipal tax reductions resulting from a state property tax initiative.

Features of the program included the following:

1. Four out of 15 management teams in the ACC were chosen to participate in this pilot program. Two of the four participating teams—Orthopedics Clinic Team and Cardio-Pulmonary Clinic Team—consisted of three managers: clinic medical director, clinic nurse manager and clinic manager. Two of the four teams—Eye Clinic Team and Women's Center (Ob/Gyn) Team—consisted of two managers: clinic medical director and clinic manager in the case of the Eye Clinic, and clinic medical director and clinic nurse manager in the case of the Women's Center. (For two years prior to this pilot, all ACC clinics had been managed by such teams.)

2. Prior to implementation of the program, each pilot team selected a utilization or revenue target above which they could earn incentive payments. As an example, following is the formula used for a team that chose revenue targets:

$$\frac{\text{Revenue in Excess of 3\% Increase (from prior year) Corrected for Price Increase}}{2\ (50\%\ \text{of Excess to Revert to BCH})}$$
$$= \text{Management Team Incentive Payment}$$

Costs could not rise above what were budgeted given strict budgetary controls.

3. Management Team Incentive Payments were divided equally among the members of the management team, i.e., physicians, nurses and nonclinical managers were treated equally.

$$\frac{\text{Management Team Incentive Payment}}{\text{Number of Management Team Members}}$$
$$= \text{Individual Manager Payments}$$

4. Management Team Incentive Payments were capped at $15,000 for the year. Regardless of performance, no team could earn more than $15,000.
5. Management team members could choose to share all or part of their Incentive Payments with their staff or could retain the payments.
6. Performance was measured quarterly, by ACC senior management, using previously agreed upon utilization and/or financial data bases. Payments, if earned, were to be paid quarterly.
7. There were no penalties for failure to reach the incentive thresholds.

The pilot program was in operation for one year, BCH's 1982 fiscal year, July 1, 1981, through June 30, 1982.

In August 1982 an evaluation program was planned and developed to systematically assess the success of this pilot program.

METHODOLOGY

The evaluation program was to measure the success of the plan in achieving its original four objectives. Those objectives were to:

1. generate increased revenue for BCH;
2. establish that providing personal incentives for clinical and nonclinical leadership in a public hospital would lead to increased hospital revenues;
3. reinforce the collegial interdependence of medical, nursing and nonclinical managers of the clinical units by making them equal partners in the venture;
4. provide a form of salary bonus to medical, nursing and nonclinical managers who made a measurable contribution to improving the financial status of the hospital.

Success in attaining the first two objectives was quantitatively measured. The FY 1982 performance of the four pilot ACC teams was compared with their FY 1981 performance and, in turn, further compared with the performance of the 11 remaining teams. Specifically, both utilization changes (from FY 1981 to FY 1982) and revenue changes (from FY 1981 to FY 1982) were measured and compared.

The comparative revenue and estimated income analyses were conducted using the hospital's FY 1981 and FY 1982 automated accounts receivable financial supporting summary. Variations in clinic revenue were to be compared after adjustments were made to correct for price increases in FY 1982. However, the quality of that data base was marginal, and the analyses were accordingly compromised. Among the limitations in the accounts receivable data base were the following:

1. The revenue recorded is a function of payer class and is posted prior to payer verification. Given the proved high rates of error in patient declarations of payment source, the actual revenue per clinic has been found to be significantly different from the initially assumed revenue which is what is recorded on the automated data base.
2. The collection rates used to estimate income, as opposed to revenue, have not been systematically verified.
3. There is not a high value placed on correctly allocating revenue and estimated income to the correct clinical cost center.

Utilization increases, as measured by clinic visits, were one of the acceptable measurement criteria a team participating in the plan could select. This assumed that utilization increases could be expected to be even across payer classes, i.e., covered versus indigent patients, and accordingly increased utilization would yield increased revenue. Accordingly, the utilization changes from FY 1981 to FY 1982 of the four pilot teams, as measured by visits, were compared with each other and, in turn, with the utilization changes of the 11 remaining teams. ACC utilization reports, which were prepared monthly by ACC senior management from reports generated by the clinic managers, were the data sources used for these analyses. Given that there had been an intensive two-year management program to improve the quality of reporting of utilization statistics, with frequent audits for accuracy and completeness, there was a high degree of confidence in the data base.

Success in attaining the third and fourth objectives was qualitatively measured. An organized series of interviews was conducted with members of the teams participating in the pilot, as well as the senior ACC management. Even team members who had left employment since the pilot year were interviewed. Only one clinic manager who had left employment was not interviewed.

The same interview format and interview questions

were used for all interviews with pilot team members. Slightly varied questions were used for senior ACC management. The team members' interview format included questions related to (1) mechanics of the plan, (2) group dynamics, (3) performance under the plan and (4) evaluation of the plan.

Each team member was interviewed separately, although the ACC senior management was interviewed using a group interview format. The team member interviews were recorded and transcribed and subjected to rigorous content analysis. The analytical design captured indications of success or failure in attaining the aforementioned objectives.

Some of the team member interviews were conducted by a former director of the ACC who was not only well known by the team members but also known to have been the architect of the incentive program, although she had left BCH prior to implementation of the plan. The results of interviews conducted by this interviewer could be highly suspect, given the bias she would have been likely to introduce into the interview situation. Concern regarding this was alleviated by the fact that half of the interviews were conducted by a graduate student unknown to the team members. The content of these interviews was not different from that of the interviews conducted by the former director.

Three of the four clinic management teams earned incentive payments.

FINDINGS

Three of the four clinic management teams earned incentive payments. The Women's Center management team earned the maximum amount, $15,000, in the second and third quarters. This team would have continued to earn payments in the fourth quarter if they had not hit the cap. The Women's Center had chosen increased visits as its target, i.e., visits in excess of a 3 percent increase from the prior year. The Orthopedics Clinic earned $11,166 in the first two quarters. Team members failed to earn a payment in the third quarter, and the fourth quarter has yet to be resolved. The Orthopedics Clinic had chosen increased revenue as its target, i.e., revenue in excess of a 3 percent increase from the prior year. The team earned incentive payments by ensuring capture of charges related to crutches, splints and the like. Previously, the clinic had been lax in capturing charges for these items. The Cardiac Team earned $2,200 in the first quarter. Its members had chosen increased visits as its target. The Eye Clinic failed to earn any incentive payments. The Eye Clinic team also had selected increased visits as its target.

Not surprisingly, interview data strongly indicated that the Women's Center, Orthopedics and Cardiac teams had more clearly developed plans for maximizing utilization or revenue than the Eye Clinic team, which adopted a more passive approach. On the other hand, the Women's Center team was not more focused or aggressive in pursuing its strategy than the Cardiac Team. Given that the Women's Center had had a relatively poor year in FY 1981, the comparison year, and the Cardiac Clinic a good one, it may have been easier for the Women's Center team to consistently exceed the target number of visits and more difficult for the Cardiac team to do the same.

A conclusive evaluation of the ACC incentive plan's success in meeting its first objective—to generate increased revenue for the ACC—was not possible. It is clear that the Orthopedics team, which chose increased revenue as its target, did successfully satisfy the objective. With regard to the other teams that earned incentive payments, however, the financial management information system did not capture or record revenues related to clinic visits with sufficient precision or accuracy to determine if utilization increases yielded revenue increases. While common sense suggests that they did, it is possible, although not probable, that all the increased visits in the Women's Center and Cardiac Clinic were self-pay. At BCH, virtually 100 percent of outpatient self-pay revenue is written off as bad debt or free care. Therefore, while it is probable that the first objective was generally satisfied, it cannot be said with certainty.

Given the success of three of the four incentive teams, it would appear that the second objective was also satisfied, i.e., to establish that providing personal incentives for clinical and nonclinical leadership in a public hospital would lead to increased hospital revenues. Conclusively establishing success in satisfying this objective is compromised for the same reason described above.

Causality, however, was not proved. That is, assuming now that the plan did yield increased revenues, data analysis did not establish that the incentive payments were the cause of the increased revenue. In an effort to indirectly assess causality, the performance of the three teams that earned incentive payments was compared with that of the remaining 12 clinic management teams and outpatient areas outside the ACC management team structure. In each quarter of FY 1982, there were two to three clinic teams that experienced utilization increases over a comparable period in FY 1981 that were comparable to the Women's Center and Cardiac Clinics. While identifiable reasons, other than management initiative, are possible for many of the increases experienced by other teams, this cannot be proved. For example, while it is probable that Dental Clinic increases were a consequence of the Dental Clinic simply being the passive beneficiary of the City of Boston Department of Health and Hospitals closing its community dental clinics, this cannot be proved. Furthermore, not all the incidences of positive performance by other teams can be explained away. Regarding comparative revenue performance, the Women's Center and Orthopedics Clinic appear to have outperformed all the other areas, yet these revenue data are highly suspect for the reasons described previously. Future incentive programs should have better systems for assessing causality.

Interview data strongly suggest that the incentive plan was successful in meeting its third objective—to reinforce the collegial interdependence of medical, nursing and nonclinical managers of the clinical units by making them equal partners in the venture. All members of the four teams thought the incentive plan was a good idea, and all thought that it positively contributed to strengthening the management team concept. The equal participation of the medical director, clinic nurse manager and clinic manager—managers with widely varying status, income, ages and education—yielded interesting consequences. Some teams, primarily those with less pronounced status differences among the management team members, integrated the concept of equity more smoothly than those teams with more pronounced status differences. Yet, even the members of the team that had to work through major conflict over the equity issue, when it came time to divide the incentive payment, thought that the plan had, in fact, strengthened the collegial interdependence of the team members.

The incentive plan had mixed success in meeting its fourth objective—to provide a form of salary bonus to medical, nursing and nonclinical managers who made a measurable contribution to improving the financial status of the hospital. Assuming for the sake of this discussion that the enhanced revenue performance of all three teams that earned payments was firmly established, it is, of course, true that contributions to the financial status of the hospital by three of the four teams resulted in incentive payments to those three teams. Each of the teams "spent" those payments differently, however. No clinic team spent its payment exclusively on bonuses for the clinic management team members.

The Cardiac team and Orthopedics team distributed some of their team payments as bonuses to the clinic nurse manager and clinic manager. These managers appreciated these bonuses and clearly identified them as rewards for performance. The medical directors' portions reverted to their respective departmental discretionary funds for educational purposes or other special ACC clinic programs. Departmental policies or medical director philosophy precluded these two medical directors from accepting the incentive payment as a personal bonus. The Women's Center managers took none of the incentive payments as personal bonuses. In the case of both the medical director and nurse manager, personal philosophy precluded their accepting the incentive payments as personal bonuses, although they did not oppose the idea in principle.

All three of the teams chose to distribute portions of the incentive payment as personal bonuses to personnel working under their supervision in the clinics. The teams differed in the size of the bonuses awarded to their employees. The managers reported that employees clearly related the bonuses received to performance and were pleased to have a reward related to performance. Accordingly, the incentive plan appears to have had unplanned for, although not unanticipated, positive impact on the morale and subsequent performance of rank and file staff in the clinics.

DISCUSSION

Rewarding managers and employees for larger than predicted increases in volume and revenues is a relatively common practice in private enterprise. BCH leadership had considerable doubt that the tech-

$$\frac{\text{Clinic A Profit (Income Minus Direct and Indirect Expense) in Excess of 3\% Increase from Prior Year}}{2 \ (50\% \ \text{of Excess to Revert to Hospital})} = \text{Management Team Incentive Payments}$$

nique was transferable. This pilot program shows that even in as complicated an environment as a public, inner-city, highly unionized teaching hospital, it is possible to establish and run an incentive program with financial rewards for performance. In this sense, the existence of the incentive program is more important than being able to prove that it did or did not cause the assumed revenue increases that occurred in three of the four clinics.

Most important, the program showed that it is possible to meaningfully involve middle managers, particularly clinical middle managers, in the pursuit of overall institutional goals, even financial goals. However, the extent of that involvement appears, in part, to be a direct consequence of the clarity and consistency by which the goals are expressed and reinforced. For example, while the hospital was providing personal financial rewards for increasing volume, some of the physician managers were not allowed to keep their bonuses. They were required by the rules of their medical departments to turn the money into a central fund controlled by the department chairman. Not unexpectedly, this diluted the impact of the incentive for those who were affected. There are many drumbeats to which the professionals can march. They hear those played by their peers and department chairmen most clearly. These clinical leaders must be committed to the behavior changes that incentive plans can facilitate and support. Current reimbursement changes may require many hospitals to introduce significant clinical practice and management changes. This program suggests that financial incentives could facilitate these changes.

On the other hand, the actual behavior of some of the clinicians who earned cash rewards suggests possible limitations or variations to incentive programs in not-for-profit organizations. Three of these clinicians were sufficiently ambivalent about their professional and managerial identities that they were uncomfortable accepting any of the incentive payments for themselves. They were highly motivated in this situation, but to help others. In this case the beneficiaries were their employees rather than patients. That experience may have meaning for other hospital incentive programs or productivity initiatives. Or, the ambivalence may disappear if bonuses become more personally compelling to middle managers in the future as hospital expense increases slow down.

While this program focused on revenue increases per se, the program is wholly adaptable to current reimbursement principles. That is, the program could be run with a focus on clinic or nursing unit profitability. The formula provided earlier could be easily converted as shown in the boxed insert.

Of course, both the threshold that increased profitability must exceed and the proportion of the excess available for incentive payments should be established according to the particular financial circumstances of the hospital implementing the plan. A minor modification of this approach would be to establish profitability targets and allow managers to share in any excess above those targets.

The other features (the objectives described above) remain applicable.

1. A pilot approach initially involving only selected management teams makes sense if one wishes to experiment with the concept or particular plan features and if prompt implementation is a goal. Even a pilot program will require considerable education, discussion and negotiation with clinical leaders and other senior managers but less than a hospital-wide initiative. Furthermore, a pilot program implies a tentative commitment, and that may be easier to win. The BCH ACC pilot program took eight months to plan and negotiate. The equal participation of all team members alone took four months to win approval.

2. Equal participation by all managers on a unit or in a clinic is critical. As each can significantly impact the profitability of the unit or clinic, they are equal from this particular management perspective. Existing income and other status distinctions already recognize their differences.

> *Incentive payments should be capped to ensure that excessive payments are not made to managers.*

3. Incentive payments should be capped to ensure that excessive payments are not made to managers. The potential for substantial economic gain could cloud some people's judgment. On the other hand, the cap should not be so low that managers are not motivated.
4. Participating management teams should be required to establish at the outset how they will spend any payments earned, e.g., what proportion will be used for manager bonuses and what proportion distributed to staff. It is better to make those decisions before real dollars are at stake, when the decisions can be more cerebral and less emotional.
5. Performance should be measured regularly and according to previously accepted standards and previously agreed upon data bases. Of course, the hospital must have acceptable systems for measuring profitability by unit or clinic.
6. There should be no penalties for failure to reach the incentive plan formula threshold. Senior management has other recourse for poor performers. The incentive plan should have the potential for gain, not loss, to win the support and commitment of senior as well as middle managers. The hospital has nothing to lose by structuring it this way.

This pilot program, although inconclusive, suggests that the introduction of incentive programs more traditionally associated with the for-profit sector can work in not-for-profit settings, if structured properly. Given the current challenges facing hospital management, the introduction of such incentive systems may help provide the motivation necessary to succeed at reaching essential goals.

REFERENCES

1. "The Newest Idea: Give Employees a Share of What They Save." *Cost Containment* 1, no. 6 (1979): 3.
2. Jehring, J.J. "The Use of Subsystem Incentives in Hospitals: A Case Study of the Incentive Program at Baptist Hospital of Pensacola." Center for Study of Productivity Motivation, Graduate School of Business, University of Wisconsin, 1968.
3. Jehring, J.J. "Increasing Productivity in Hospitals: A Case Study of the Incentive Program at the Memorial Hospital of Long Beach." Center for Study of Productivity Motivation, Graduate School of Business, University of Wisconsin, 1966.
4. Bush, D.V. "Increasing Productivity through Incentives and Motivation." Chicago: Center for Hospital Management Engineering, 1979.
5. "Holzer Medical Center's Incentive Plan In Action." *Cost Containment* 1, no. 6 (1979): 2.
6. "The Newest Idea."
7. Ibid., 4.
8. Ibid., 4.
9. Williams, F. "Employee Incentive Systems." In *Handbook for Health Care Accounting*, edited by W.O. Cleverley. Rockville, Md.: Aspen Systems Corp., 1982, pp. 395–408.
10. Ibid.
11. Ibid.
12. Ibid.
13. Ibid.
14. Hellwig, H. "Base Salary and Incentive Compensation Practices." *Compensation Review* 10, no. 4 (1978): 44.
15. Williams, "Employee Incentive Systems," 398.
16. Ibid.
17. Cleverley, W.O., and Mullen, R.P. "Management Incentive Systems and Economic Performance in Health Care Organizations." *Health Care Management Review* 7, no. 1 (1982): 7–12.
18. Gustafson, D. *Employee Incentive System for Hospitals.* Health Services and Mental Health Administration, Hospital Branch, USDHEW, 1972, pp. 5–6.

An incentive program to increase revenue in a public hospital

Linda S. Chan,
Park W. Wagers,
Ramona Hernandez,
and
Sol Bernstein

Using financial reward systems to enhance revenue generation or promote cost savings has been more difficult in public than in private hospitals. The program at the Los Angeles County–University of Southern California Medical Center has demonstrated, however, that it can be done.

Public hospitals have been undergoing drastic reform in recent years. The economic and political climate in the 1980s called for increasing accountability.[1] Programs for cost containment and revenue generation continue to be essential to the survival of public hospitals. Many reviews and guidelines have been published on using management programs.[2-11] Many hospitals have used utilization management as a cost-containment tool to control costs by influencing physicians' decisions on the use of health services.[3,4,6,11-14]

The factors influencing physicians' roles in controlling health care cost include physicians' lack of knowledge about costs and charges, lack of incentives for efficiency, inexperience in cost management, malpractice fears, individual training, practice style, and personality.[15-17] Methods and approaches to increasing physicians' awareness include the following:

- developing profiles of current physician practice patterns and distributing these profiles, with interpretation, to the physicians;
- educational programs;
- information systems for monitoring physicians' decisions and efforts to control cost;
- peer review and performance feedback;
- increasing physician involvement in setting department priorities and budgets;
- developing and using guidelines, standards, and protocols for tests, treatments, medications, procedures, diagnostic workups, and lengths of stay;
- concurrent utilization review;
- developing indicators for severity of illness and adverse patient outcome; and
- rewards and penalties.[4,7,12,15,17-20]

Linda S. Chan, *Ph.D., is Director of Research and Evaluation at the Los Angeles County–University of Southern California Medical Center. She is also Associate Professor of Research Pediatrics at the University of Southern California School of Medicine.*

Park W. Wagers, *M.D., is Medical Director of Utilization Review at the Los Angeles County–University of Southern California Medical Center. He is also Professor of Clinical Medicine at the University of Southern California School of Medicine.*

Ramona Hernandez, *B.A., is Administrator of Utilization Review and Quality Assurance at the Los Angeles County–University of Southern California Medical Center.*

Sol Bernstein, *M.D., is Chief of Staff of the Los Angeles County–University of Southern California Medical Center. He is also Associate Dean and Associate Professor at the University of Southern California School of Medicine.*

Some researchers argue that financial incentive plans are the most effective utilization management methods for cost containment.[2-6] In a review of medical costs, educational programs, and financial incentive plans for physicians, Egdahl and Taft conclude that, although they raise ethical questions, financial incentive systems are the most effective in encouraging physicians to hold down costs: Educational programs are too short-lived in their effect.[5]

The use of financial incentives for managers and employees is uncommon in the hospital industry, especially in public settings.[14] In their report about the financial reimbursement program in Boston City Hospital, Shyavitz et al. identified only 10 such plans in the United States.[2]

The Los Angeles County–University of Southern California (LAC+USC) Medical Center, a large public teaching general hospital, implemented several innovative programs for cost containment and revenue generation as a result of the financial pressure on public hospitals beginning in the early 1980s.[21] As a continuous effort to maximize reimbursement from third party payers, the Utilization Review Division of LAC+USC Medical Center launched a concerted effort to maximize reimbursement from Medi-Cal (California's Medicaid program). This included the introduction of a pilot financial incentive program. A year's implementation has shown evidence of success.

Comparison of data before and after the program revealed a statistically significant improvement in the reduction of the number of days denied reimbursement and an increase in the number of days approved for reimbursement in 20 of the 26 clinical departments. The pilot financial incentive program has generated a high level of satisfaction in clinical departments and has gained the approval of the Los Angeles County administration for continuation. The rewards generated by the program have supported items related to direct patient care and medical education that otherwise might not have been available through the usual county budgeting process.

BACKGROUND

The LAC+USC Medical Center is the largest of six county hospitals in Los Angeles County. In fiscal year 1986–87, 78,736 admissions, 16,611 births, and 557,585 outpatient and emergency department visits were recorded. This medical center serves a predominantly indigent population consisting largely of Medi-Cal eligibles and self-pay or uninsured patients. Despite the large number of poor and uninsured patients, the medical center's reimbursement from Medi-Cal has been disproportionately low as compared to actual cost of providing care to these patients. The percentage of hospital days denied for reimbursement by Medi-Cal was 19% in fiscal year 1985–86 and amounted to a loss of $28 million to the medical center.

THE AGGRESSIVE UTILIZATION REVIEW PROCEDURES

The Utilization Review Division of LAC+USC Medical Center has been monitoring days of denied reimbursement by third party payers. An acute hospital inpatient day approved for reimbursement by Medi-Cal is referred to as an *approved day*, and an acute hospital inpatient day that is denied for reimbursement is referred to as a *denied day*. Denial is based on the judgment of Medi-Cal physicians that the patient's care does not necessitate an acute level of hospitalization. The reasons for denial of reimbursement from Medi-Cal have been categorized as related to inappropriate or nonemergent admission, delays in performing tests and procedures, nonacute care, delayed discharge for various placement and transfer reasons, service not covered, and documentation problems, for example.

In November 1983, Medi-Cal removed from the county hospitals a waiver that exempted the medical center from prior treatment authorization requests. In order to meet the treatment authorization request requirements and to improve the approval rate of reimbursement for inpatient days, the Utilization Review Division launched an intensive monitoring effort for reimbursements from all third party payers. This effort consisted of the following elements:

- implementation of 100% concurrent review of all admissions;
- assignment of a full-time senior physician as the director of utilization review to oversee the entire operation;
- direct discussions of denial criteria and specific cases between faculty physicians and reviewing physician representatives of third party payers;
- discussion and review of controversial policy areas and specific denied cases among involved physicians; and
- regular physician performance feedback and problem review with department chairpersons.

A computerized information system was implemented to assist with the effort by providing prompt and reliable information for review and evaluation.

> *The percentage of hospital days denied for reimbursement by Medi-Cal was 19% in fiscal year 1985–86 and amounted to a loss of $28 million to the medical center.*

The effort emphasized the involvement of all medical staff in reviewing cases, identifying problems, and developing solutions. The senior physician assigned full-time to oversee the operation met individually with each department chief to review problems and procedures in detail. These changes provided an educational process that enabled creation of guidelines and development of programs and systems designed to correct many of the deficiencies, some of which had existed for a long time.

The aggressive utilization review effort gave rise to the introduction of a financial incentive plan to reward positive efforts put forth by individual clinical departments to reduce reimbursable denied days and increase approved days. During fiscal year 1986–87 the pilot financial incentive program was tested.

THE MANAGEMENT INFORMATION SYSTEM

A key turning point in the utilization review effort was the implementation of a computerized management information system to record and analyze all utilization review activities. A minicomputer, a DEC PDP 11-84 with 1 megabyte of RAM (random access memory) and 500 megabytes of disk drive, was dedicated entirely to on-line interactive data recording, editing, and reporting of all utilization review cases. The minicomputer supported 12 terminals for data processing and reporting. It is currently storing 250,000 records over a 2.5-year period. The software programs were developed in-house by a physician–computer specialist. The database was then sent to the medical center's mainframe IBM computer for further statistical analysis.

The primary data source was the Utilization Review Disposition Form, from which data on clinical department, month and year, number of patients discharged per month, number of approved days per month, and number of denied days per month were retrieved. Statistical analyses were conducted by the Research and Evaluation Unit of LAC+USC Medical Center, and Statistical Analysis System (SAS) programs were used.[22]

Three management reports with appropriate statistical analyses were generated for monitoring and evaluation purposes. The first was an annual comparison of the percentage of denied days and odds ratios of approved days to denied days between two fiscal years by departments; the second was a quarterly comparison of percentage of denied days and odds ratios; and the third was a trend analysis within and between fiscal years.

The information reports were studied, analyzed, and interpreted quarterly by the Utilization Review Division staff and shared with the clinical departments.

THE FINANCIAL INCENTIVE PROGRAM

LAC+USC Medical Center also developed an incentive program termed the Utilization Review Incentive Plan to positively reward constructive endeavors that generate revenue or reduce the number of denied days. It specifically rewards departments that continue to improve the ratio of approved days to denied days of the current year over the prior year.

For the pilot program in fiscal year 1986–87, only clinical departments providing direct inpatient care were included. Not included were the supportive clinical departments such as laboratory, radiology, and anesthesiology, which required different measures of improvement of reimbursement.

The basic data elements needed to determine the amount of incentive funds each department received are the number of approved days and the number of denied days for each of the two fiscal years for each clinical department and the total amount of allocated incentive funds.

The first step in the plan was to determine which departments showed a significant reduction in the percentage of denied days or a significant increase in the percentage of approved days from one fiscal year to another. A chi-square was used to test the difference in the percentage of approved days or denied days between the two years. A p value of less than .05 was used as the criterion of significance. Clinical departments with significant reductions in the percentage of denied days were eligible for reward.

The second step in the plan was to distribute the total amount of incentive funds dedicated by hospital administration to this program among the departments that showed significant improvement. The distribution of the funds among these departments was based on the weighted number of approved days, the weights being the amount of improvement in the ratio of approved days to denied days from one fiscal year to another.

The share of the reward that each clinical department received was determined through the following six steps: (1) Calculate the ratio of approved days to denied days for each of the two fiscal years for each significantly improved clinical department. (2) Calculate the odds ratio for each department, which was defined as the ratio of approved days to denied days of the later fiscal year to the prior fiscal year. (3) Calculate the number of rewardable units for the department by multiplying the number of approved days by the odds ratio. (4) Derive the total number of rewardable units by summing the number of rewardable units of all significantly improved clinical departments. (5) Calculate the percentage of rewardable units of each department. (6) Calculate the share of the total amount of incentive funds by applying the percentage of rewardable units of each department to the total amount of funds.

Because of the monetary nature of the award and potential legal issues of the incentive plan, categorized limitations were imposed on the expenditure funds.

The distribution of the incentive funds can be approached in many different ways, depending on the weighting system and the method for deriving the number of rewardable units. The medical center used the odds ratio—rather than the difference in the percentage of denied or approved days between the two fiscal years—as a weight to be applied to the number of approved days. (See the appendix for formulas and procedures for these calculations.)

Because of the monetary nature of the award and potential legal issues of the incentive plan,[23,24] categorized limitations were imposed on the expenditure funds. First, each department's expenditure of the reward money had to be a conjoint decision of that department's medical staff, nursing staff, and administration. These decisions were subject to review and approval by the medical director, nursing director, and executive director of the medical center. Second, because the reward was a single annual allocation, hiring of personnel whose salary would have to extend beyond the single expenditure was not authorized. Third, to eliminate any issue of conflict of interest involving individuals who might profit from utilization-review cost containment with compromise of patient care, the money for this reward could not be given to individuals: Each service was instructed to give priority to (1) equipment, (2) supplies, and (3) professional education endeavors.

EVALUATION OF THE EFFECTIVENESS OF THE PILOT PROGRAM

In fiscal year 1985–86, the year before the implementation of the pilot financial incentive program, most of the Utilization Review Division's aggressive monitoring procedures were in place. The only major change in fiscal year 1986–87 was the implementation of the pilot financial incentive program. The effectiveness of the pilot program could be evaluated, therefore, by studying the differences in reimbursement for inpatient days between fiscal years 1985–86 and 1986–87.

Comparison of percentages of denied days by services

The medical center as a whole produced a 4.1% drop in the percentage of denied days (from 18.1% in fiscal year 1985–86 to 14.0% in fiscal year 1986–87). The number of denied days dropped by 4,027 (from 45,248 to 41,221) and the number of approved days increased by 48,923 (from 204,707 to 253,630). Table 1 presents the annual comparison of percentage of denied days by departments between the two fiscal years. Note that 20 of 26 clinical departments improved significantly (p is less than .05).

Odds ratio analysis

Table 2 presents the ratio of approved days to denied days for each of the two fiscal years and their odds ratios and 95% confidence intervals by clinical departments. The odds ratio was the ratio of approved days to denied days of fiscal year 1986–87 to fiscal year 1985–86. Therefore, an odds ratio greater than 1 can be interpreted as an increase in approved days or a decrease in denied days. On the other hand, an odds ratio less than 1 indicates a decrease in approved days or an increase in denied days. A ratio of 1 means no difference between the two years.

The 95% confidence interval for the odds ratio in Table 2 provides an estimate of the variation of the odds ratios. In the calculation of the odds ratio and its standard error, a 0.5 adjustment factor was used to avoid having the number of denied days equal zero. For a discussion of the odds ratio analysis the reader is referred to Fleiss.[25]

TABLE 1

COMPARISON OF PERCENTAGE OF DENIED DAYS IN 1985–86 AND 1986–87

Services	Fiscal year 1985–86	Fiscal year 1986–87	Difference
Medical			
1. Evaluation ward	49.9	41.6	–8.4*
2. Renal	16.7	7.6	–9.0*
3. Medical admitting	16.8	9.1	–7.7*
4. Medical 6	20.4	20.5	+0.1
5. Medical 7	24.6	20.5	–4.1*
6. Medical 8	22.0	16.9	–5.1*
7. Cardiology	26.0	15.9	–10.1*
8. Diabetes	13.2	10.2	–3.0*
9. Chest medicine	12.4	15.2	+2.9†
10. Rheumatology and hematology	13.0	10.3	–2.7*
11. Oncology	17.3	16.9	–0.4
Medical services	20.4	16.8	–3.7*
Surgical			
12. Orthopedics	27.1	21.4	–5.8*
13. Otolaryngology	22.2	12.1	–10.2*
14. Urology	19.0	17.0	–2.0‡
15. Neurosurgery	29.8	19.9	–9.9*
16. Neuromedicine	35.6	18.2	–17.4*
17. General surgery	14.0	8.6	–5.4*
18. Ophthalmology	19.4	13.1	–6.3*
19. Thoracic surgery	11.9	15.1	+3.2†
20. Burn	6.9	3.7	–3.2*
21. Renal transplant	13.2	4.8	–8.4*
Surgical services	21.8	14.9	–6.9*
General hospital	21.0	15.9	–5.1*
Women's hospital			
22. Gynecology	12.6	10.2	–2.3*
23. Gynecology–oncology	19.4	18.3	–1.2
24. Sick newborn	1.7	1.4	–0.3
Women's hospital§	7.8	7.4	–0.4
Pediatric pavilion			
25. Pediatrics	5.5	2.7	–2.8*
26. Communicable diseases	17.5	14.0	–3.5*
Pediatric pavilion	8.0	5.0	–3.0*
LAC+USC Medical Center	18.1	14.0	–4.1*

* Significant decline at $p = .001$ level.
† Significant increase at $p = .001$ level.
‡ Significant decline at $p = .01$ level.
§ Obstetric service was not ready for the study at this time.

TABLE 2

RATIOS OF APPROVED:DENIED DAYS FOR TWO FISCAL YEARS

Service	Fiscal year 1985–86	Fiscal year 1986-87	Odds ratio	95% confidence intervals
Medical				
1. Evaluation ward	1.002	1.406	1.40	1.25–1.56
2. Renal	4.998	12.103	2.42	2.03–2.81
3. Medical admitting	4.955	10.024	2.02	1.81–2.24
4. Medical 6	3.904	3.873	0.99	0.95–1.03
5. Medical 7	3.071	3.883	1.26	1.19–1.34
6. Medical 8	3.546	4.921	1.39	1.32–1.45
7. Cardiology	2.852	5.300	1.86	1.73–1.98
8. Diabetes	6.604	8.822	1.34	1.20–1.47
9. Chest medicine	7.098	5.557	0.78	0.72–0.85
10. Rheumatology and hematology	6.677	8.692	1.30	1.16–1.44
11. Oncology	4.786	4.932	1.03	0.93–1.13
Medical services	3.896	4.964	1.27	1.25–1.30
Surgical				
12. Orthopedics	2.684	3.679	1.37	1.32–1.42
13. Otolaryngology	3.494	7.281	2.08	1.87–2.29
14. Urology	4.259	4.890	1.15	1.03–1.26
15. Neurosurgery	2.357	4.031	1.71	1.59–1.83
16. Neuromedicine	1.811	4.494	2.48	2.31–2.66
17. General surgery	6.153	10.702	1.74	1.65–1.83
18. Ophthalmology	4.147	6.640	1.60	1.31–1.90
19. Thoracic surgery	7.402	5.610	0.76	0.66–0.85
20. Burn	13.368	25.970	1.94	1.49–2.40
21. Renal Transplant	6.552	19.747	3.01	1.92–4.11
Surgical services	3.596	5.714	1.59	1.55–1.63
General hospital	3.753	5.283	1.41	1.39–1.43
Women's hospital				
22. Gynecology	6.955	8.774	1.26	1.14–1.38
23. Gynecology–oncology	4.142	4.474	1.08	0.95–1.21
24. Sick newborn	57.124	69.815	1.22	0.97–1.47
Women's hospital	11.814	12.540	1.06	0.99–1.13
Pediatric pavilion				
25. Pediatrics	17.040	35.496	2.08	1.89–2.27
26. Communicable diseases	4.705	6.119	1.30	1.18–1.42
Pediatric pavilion	11.531	19.089	1.66	1.55–1.76
LAC+USC Medical Center	4.524	6.153	1.36	1.34–1.38

Distribution of reward

Table 3 presents the weights and the distribution of the shares of the total incentive funds for the 20 clinical services that showed significant decrease in the percentage of denied days or increase of the ratio of approved days to denied days in fiscal year 1986–87. The last column in the table presents the number of rewardable units per 1,000 units of total reward. For this pilot incentive program, $750,000 in incentive funds were approved by administration. The actual dollar amount each department received was derived by multiplying the number of rewardable units in the last column of Table 3 by 750.

THE QUALITY-OF-CARE ISSUE

The quality of patient care at the medical center is monitored by the Quality Assurance Committee, which works specifically to fulfill the quality assurance requirements of the Joint Commission of Accreditation of Healthcare Organizations, the California Medical Association, and the state of California as mandated by Title 22 (Medicaid requirements). The Division of Quality Assurance works with each department, conducts problem-oriented reviews, and provides quarterly progress reports to the Quality Assurance Committee. For fiscal years 1985–86 and 1986–87, the Division of Quality Assurance reported that none of the premature discharges were direct consequences of utilization review activities and no differences in the quality of care were detected.

THE ETHICAL CONSIDERATIONS

The Utilization Review Committee of the LAC+USC Medical Center is responsible for overseeing the ethical and professional integrity of all utilization review activities. It operates under the following ground rules:
1. It will not compromise quality of care for utilization review.
2. It refuses to intervene or insert itself in the judgment of the individual physician.

TABLE 3

DERIVATION OF REWARDABLE UNITS

Rank	Service	Odds ratio (O)	Approved days (A)	Ratio × approved days (O × A)	Rewardable units per 1,000
1	Renal transplant	3.039	819	2,489	8
2	Neuromedicine	2.481	7,392	18,340	57
3	Renal	2.422	2,699	6,537	20
4	Otolaryngology	2.084	5,599	11,668	36
5	Pediatrics	2.083	24,386	50,796	158
6	Medical admitting	2.023	9,974	20,177	63
7	Burns	1.946	3,077	5,988	19
8	Cardiology	1.853	9,832	18,219	57
9	General surgery	1.739	32,288	54,149	168
10	Neurosurgery	1.710	7,051	12,057	38
11	Ophthalmology	1.601	1,733	2,775	9
12	Evaluation ward	1.403	1,437	2,016	6
13	Medical 8	1.388	18,149	25,191	78
14	Orthopedics	1.371	24,511	33,605	105
15	Diabetes	1.336	7,640	10,207	32
16	Rheumatology and hematology	1.302	6,180	8,046	25
17	Communicable diseases	1.300	5,317	6,912	21
18	Medical 7	1.264	11,145	14,087	44
19	Gynecology	1.262	9,520	12,014	37
20	Urology	1.148	5,320	6,107	19
	Total		194,069	321,380	1,000

3. It will not tell anyone to discharge a patient.
4. It is concerned about efficient use of the medical center's resources and maximal possible reimbursement and revenue generation.

The committee adhered to these policies before and during the implementation of the financial incentive program.

The legal aspect of the financial incentive program was attended to by imposing categorized limitations on the expenditure of the rewarded funds. The ethical issues that accompany the program have been given careful consideration. The medical center operates on the premises that (1) operating in a manner that wastes scarce resources and thus potentially limits or deprives access of other patients to proper care is unethical, and (2) positively rewarding efforts that instigate constructive and creative changes to ensure proper delivery of health care—as long as the end-point is guaranteed to be improvement of patient care and not personal gains—is ethical. Both the process and the end-point reached are considered to be valuable, good, and ethical.

FURTHER ISSUES OF THE INCENTIVE PROGRAM

The departmental financial incentive program described here is one of several components envisioned as a comprehensive incentive plan. The incentive program is the component that is measurable by direct inpatient care services. The second component is clinical support services such as laboratory, radiology, and anesthesiology. These services may have contributed to the improved reimbursement of inpatient days from third party payers, and the medical center must take further actions to determine these departments' contributions. One of these actions should be to improve the data-collection system to identify the reasons for denials and the changes that have contributed to improved approved days, with designation and recognition by direct and indirect patient care services.

Another effort should concentrate on designing measurable indicators for each of the supportive services for monitoring revenue-generation activities. A mechanism must be devised to allocate a certain proportion of the surplus of all the revenue generated to each supportive service. Of course, this allocation is fair only if revenue generated by the supportive services by other means can be reversely distributed to the clinical departments.

Another group of clinical departments that provide direct patient care in different settings or receive reimbursement from different third party payers may require different monitoring mechanisms. Medical center departments that provide primary outpatient care include gastroenterology, endocrinology, physical medicine, and emergency medicine. Services that receive reimbursement from third party payers other than Medi-Cal include the center's jail ward, psychiatry, and clinical research. For these departments, indicators such as ratios of approved days to denied days can be developed for revenue generation.

Another effort should concentrate on designing measurable indicators for each of the supportive services for monitoring revenue-generation activities.

Services that fall outside the traditional medical administrative departmental designation represent another potential dimension in a comprehensive incentive program. Nursing, financial personnel, and physical therapists, for example, certainly contribute to the cost savings and revenue generation that result from reduction of the number of denied days or increase of the number of approved days. Specific adaptations may have to be developed for these individual services in order for the comprehensive incentive program to be used institutionwide.

An issue that is sure to come up in the future is how to reward a clinical department that reaches a maximum achievable level in its effort to either decrease denied days or increase approved days. This issue should be taken into consideration in the incentive formulas. A maximum level must be established for each clinical department for prospective monitoring. The departments that have reached and maintained the maximum level may be rewarded for their approved days at a flat rate per approved day. The funds for these departments can be a part of or separate from the allotted incentive funds for the other departments. An upper and lower range around the maximum level may be used as guidelines.

• • •

Public hospitals are under continuous pressure to compete with other governmental services for their survival. Revenue-generation and cost-containment programs have been the major prescriptions for the survival of public hospitals: The LAC+USC Medical Center is no exception. Several years of aggressive utilization review efforts with major emphasis on 100% concurrent review of all inpatients; implementation of a computerized information system; and involvement of the medical staff in reviewing cases, establishing guidelines, and developing solutions has led to the development and introduction of this financial incentive program to maximize reimbursement from Medi-Cal in fiscal year 1986–87. One year's implementation resulted in a significant reduction of denied days (4.1%, from 18.1% in fiscal year 1985–86 to 14.0% in fiscal year 1986–87), amounting to a decrease of 4,027 days denied for reimbursement at the acute care rate by Medi-Cal. More significant was an increase of 48,923 days approved for reimbursement during this one-year period.

The improvement was observed throughout the medical center. Among the 26 clinical departments monitored, 20 significantly improved the ratio of approved days to denied days from fiscal year 1985–86 to fiscal year 1986–87. This result agrees with the conclusion drawn by many reviews[2,6,7,11,12] that, of the many examples of coordinated and concerted efforts, a financial incentive plan is potentially one of the most important features in a successful utilization review program.

The satisfaction level of the clinical departments was high when successful participation resulted in the receipt of a financial reward. Funds from the departmental rewards have been expended to support items related to direct patient care and medical education that otherwise would not have been available under the tight budget of public hospitals. The success of this incentive program is evident in continued support of the incentive plan by Los Angeles County's Department of Health Services, which is rarely seen in a nonprofit hospital setting.[14]

It is important to note that the financial incentive plan was not and should not be implemented in isolation. In order for the plan to be successful, it must be well developed[26] and well supported by aggressive utilization review and monitoring efforts.

The plan described in this article is the foundation for more cost-containment and revenue-generation efforts at the LAC+USC Medical Center and is the first of several components of a more comprehensive incentive program. As pointed out earlier, not all hospital departments are included in the consideration of this incentive plan because the hospital data-collection system does not yet have the ability to define in detail the contributions of other departments to the overall rates of approval and denial of reimbursement for acute hospital care.

As the incentive program continues to evolve over subsequent years, it will be desirable and necessary to include support services that have been excluded from this plan. Furthermore, as individual departments reach their maximum capability to reduce cost or generate revenue, a residual proportion of denials will persist, attributable to "system" problems beyond the control of the individual departments. These system problems will require collaborative involvement and policy decisions from hospital administration and from the Los Angeles County Department of Health Services.

During the planning stage of this project, planners were dubious of the success of a financial incentive program at this medical center because of the center's size and because of limited experience in public hospital settings. To everyone's delight, this project provided tangible rewards to cooperative departments and accomplished major improvement in utilization control. Similar programs providing positive motivation, in public as well as in private hospitals, will enable institutions to accomplish what yet needs to be done.

REFERENCES

1. Vraciu, R. "Hospital Strategies for the Eighties: A Mid-Decade Look." *Health Care Management Review* 10, no. 4 (1985): 9–19.
2. Shyavitz, L. Rosenbloom, D., and Conover, L. "Financial Incentives for Middle Managers: Pilot Program in an Inner City, Municipal Teaching Hospital." *Health Care Management Review* 10, no. 3 (1985): 37–44.
3. Sims, P., et al. "The Incentive Plan: An Approach for Modification of Physician Behavior." *American Journal of Public Health* 74, no. 2 (1984): 150–52.
4. Fraser, C., and Woodford, F. "Strategies to Modify the Test-Requesting Patterns of Clinicians." *Annals of Clinical Biochemistry* 24 (1987): 223–31.
5. Egdahl, R., and Taft, C. "Financial Incentives to Physicians." *The New England Journal of Medicine* 315, no. 1 (1986): 59–61.
6. Jones, D. "An Incentive Management Plan for Hospital-Based Physicians." *Journal of Medical Sciences* 10, no. 1 (1986): 57–63.
7. Wolmering, C. "Incentive Programs: A Way to Cost Containment." *Nursing Management* 18, no. 9 (1987): 49–51.
8. Burchman, S., Dewey, B., and Schneider, C. "Unlocking Employee Performance: Pay for Performance." *Manage-*

ment Solutions 33, no. 3 (1988): 23–29.
9. Watkins, G. "Incentive Plans: Has the Time Come?" *Radiology Management* 9, no. 4 (1987): 50–55.
10. Rowland, D. "Incentive Pay: Productivity's Own Reward." *Personnel Journal* 66, no. 3 (1987): 48–57.
11. Neuhauser, D. "Stimulating Cost Effective Behavior in Hospitals." *Health Policy* 7 (1987): 205–13.
12. Magerlein, D., et al. "New Systems Can Mean Real Savings, Part 2: Preadmission Testing, Concurrent Review and Outpatient Surgery." *Healthcare Financial Management* 32, no. 5 (1978): 18–26.
13. MacStravic, R. "Hospitals' Marketing Challenge: Influencing Physician Behavior." *Health Progress* 66, no. 5 (1985): 54–57.
14. Williams, F., and Anderson, D. "Cost Control Incentive Programs: Appropriate for Non-Profits?" *Healthcare Financial Management* 32, no. 5 (1978): 14–17.
15. Eisenberg, J., and Williams, S. "Cost Containment and Changing Physicians' Practice Behavior." *Journal of the American Medical Association* 246, no. 19 (1981): 2195–201.
16. Nagurney, J., Braham, R., and Reader, G. "Physician Awareness of Economic Factors in Clinical Decision-Making." *Medical Care* 17, no. 7 (1979): 727–36.
17. Maynard, A. "Incentives for Cost-Effective Physician Behavior." *Health Policy* 7 (1987): 189–204.
18. Jay, J. "Furthering Cost-Effective Medical Practice." *Hospital and Health Services Administration* 30, no. 4 (1985): 65–76.
19. Eisenberg, J., and Rosoff, A. "Physician Responsibility for the Cost of Unnecessary Medical Services." *The New England Journal of Medicine* 299, no. 2 (1976): 76–80.
20. Hillman, A. "Financial Incentives for Physicians in HMOs." *The New England Journal of Medicine* 317, no. 27 (1987): 1743–48.
21. Galaif, M., et al. "Creative Revenue Generation and Cost Containment in a Public Hospital." *Quarterly Review Bulletin* 12 (1986): 218–22.
22. SAS Institute. *SAS User's Guide: Statistics, Version 5 Edition.* Cary, N.C.: SAS, 1985.
23. Weissburg, C., and Stern, K. "Can Hospitals Reward Physicians for Reducing Unnecessary Utilization?" *FAH Review* 18, no. 5 (1985): 45–46.
24. Murray, L. "Physician Incentive Compensation Programs." *Healthcare Executive* 1, no. 2 (1986): 33–35.
25. Fleiss, J. *Statistical Methods for Rates and Proportions.* New York: Wiley, 1973.
26. Ewell, C. "Incentives May Become a Popular Way to Spur Aggressive Management." *Modern Healthcare* 13, no. 2 (1983): 139.

APPENDIX

FORMULAS FOR THE ODDS RATIO AND THE DISTRIBUTION OF THE INCENTIVE FUND

ODDS RATIO

- Denote the original data as follows:

Year	No. of approved days	No. of denied days
1986	D1	D2
1985	D3	D4

- The odds ratio (O) is

  $$O = \frac{D1/D2}{D3/D4}$$

- O is estimated by

  $$O' = \frac{(D1 + .5)(D4 + .5)}{(D2 + .5)(D3 + .5)}$$

- The standard error of O' [s.e.(O')] is

  $$O' \sqrt{\frac{1}{D1 + .5} + \frac{1}{D2 + .5} + \frac{1}{D3 + .5} + \frac{1}{D4 + .5}}$$

DISTRIBUTION OF THE INCENTIVE FUND

- Notations:

  k = The number of clinical departments with significant improvement
  i = The ith clinical department, where $i = 1,...k$
  $O(i)$ = The odds ratio of Approved:Denied Days of the new to the old fiscal year
  $N(i)$ = The number of approved days for the ith clinical department
  A = The total amount of incentive funds to be distributed to the k clinical departments
  $A(i)$ = The amount of reward to the ith department

- $A(i)$ is calculated as

  $$A(i) = \frac{O(i) \times N(i)}{\text{SUM}[O(i) \times N(i)]} \times A$$

  where SUM is the summation over the k departments.

  Or, $A(i) = p(i) \times A$

  where $p(i)$ is the proportion of rewardable units for the ith department.

Hospital cost savings: resembling business

Nancy L. Davis
and
Thomas Choi

Hospitals, like other employers, offer health care benefits to their employees. Cost savings can be realized when hospitals offer packages that resemble those provided by business employers. Business packages explain variance in cost savings more than employee characteristics do.

Hospitals, like industrial employers, need to find ways to contain health care costs for their employees. This article addresses two research questions: (1) How do varying hospital, employee, and health care benefit characteristics affect health care costs? and (2) Specifically, does cost decrease when health care benefits and incentives offered by the hospital resemble those offered by business? Results of a recent study the authors conducted are presented.

One of the biggest issues in health care provision in the 1980s has been the need to contain costs. The hospital industry is most often blamed for high costs and has been pressured to make the provision of care more efficient. The problem of health care costs should first be viewed from the financing or payer side rather than the hospital delivery side as nearly 95 percent of all Americans do not personally or directly pay hospital bills.[1] Most health care is financed through insurance purchased at the place of employment. This is due to the favorable tax treatment of health insurance premiums paid by employers and the advantages of group-purchased policies. For the 80 percent of eligible employees in the United States, an estimated 24 percent of health insurance premiums ($60 billion) and 40 percent of overall health care costs are paid for by the employer.[2] This system creates consumer behavior that is less than responsive to normal marketplace forces. Therefore, little incentive exists for consumers to search out efficient providers of care. Consequently, increased costs face the employer who provides health care benefits to employees.

Past studies have shown that employers were not particularly concerned about the cost of health care benefits.[3] Since benefits were a small proportion of total operating costs, were an attractive recruitment tool for employers, and were seen as a way of inhibiting union activity, employers were hesitant to change the funding structure or type of benefit offered. Increasingly higher costs, however, have forced greater awareness on the part of employers about health care

Nancy L. Davis, *M.H.A., is an Assistant Administrator for Presbyterian Healthcare Services, Albuquerque, New Mexico.*

Thomas Choi, *Ph.D., is Associate Professor at the University of Minnesota, Minneapolis, Minnesota, where he holds appointments in the Center for Health Services Research and Policy, Graduate Program in Hospital and Health Care Administration, Department of Graduate Studies in Social and Administrative Pharmacy, and Department of Sociology.*

issues. Recent studies indicate a focus on maintaining benefits at reduced costs through a variety of methods.

Hospitals, while providing a large portion of health care and generating a large amount of the costs, are also employers experiencing the same increasing financial burden of the health care benefit for employees. Hospitals on average spend approximately 60 percent of their total operating budget on human resources costs for employees.[4] Total health care expenses as a percentage of total operating expenses for hospitals have on average increased from 1.7 percent in 1981 to 2.27 percent in 1984.[5] Therefore, it seems logical that the hospital industry will increase its efforts toward the cost-efficient delivery of health-related benefits for employees.

Knowledge of different benefit plans and funding alternatives and their effects on total expenditures will enable hospitals to develop a strategy for cost-effective employee benefit packages.

Knowledge of different benefit plans and funding alternatives and their effects on total expenditures will enable hospitals to develop a strategy for cost-effective employee benefit packages. For example, do an increased number and type of plans, more employee cost sharing, or wellness programs, education, or second surgical opinions have favorable effects on reducing expenditure? Does the composition of the employee work force or size and structure of the organization predispose an employer to higher expenditures?

To examine how hospitals decreased the cost of employee health care benefit expenditure, descriptive information was compiled on the current state of health care benefits offered to hospital employees. In addition, estimates were made of how particular characteristics of hospitals, employees, and health care benefit plans affect the overall expenditure within the hospital sector.

LITERATURE REVIEW

Employee and hospital characteristics

Various employee characteristics have been shown to affect aggregate expenditure through insurance company and other delivery system premiums and through employee response to health care programs offered at the work place. For example, age has been shown to correlate with a desire for extra coverage and fee-for-service plans instead of health maintenance organizations (HMOs).[6] Premium costs for employers have been shown to increase as average age of employees increases.[7] Other findings suggest that current small group rates for insurance are largely based on age of qualified employees.[8-10]

Another employee characteristic affecting expenditure is gender. Approximately 88 percent of hospital employees are females, who generally have a longer life expectancy and also file more health claims than males on the average.[11,12]

Hospital characteristics that affect employee health care expenditures include size, type of ownership, and unionization.[13,14] Group size determines rates. This has two implications. First, size is important because as size of the group decreases, the expenditure per qualified employee increases.[15] Second, hospitals offering multiple plans would have fewer employees in each of the plans offered and therefore would be more likely to experience a greater cost per enrolled employee. The percentage of unionized hospital personnel also can have a significant effect on increasing health care benefit expenditures.[16]

Health care benefits

The available information on health care benefit activity by employers is primarily from the business sector, which seems to spearhead an increasing awareness on the part of employers that competition is a mechanism to control costs in the health care marketplace.[17] Corporate encouragement of alternative delivery systems such as HMOs and preferred provider organizations (PPOs) has grown in recent years, but little conclusive evidence exists indicating specific effects on health care expenses for employers.[18,19] The PPO is a recent option for employers that offers financial savings from prior authorizations, quality assurance systems, concurrent review and discharge planning.[20] The PPO can directly affect the cost of health care benefits through both provider discounts and decreased cost sharing for employees who utilize PPO providers.[21] The reservation with PPOs is that cost savings from discounts can be offset by other related costs unless efficient providers are selected and effective utilization controls are exercised.[22]

Employers offer additional benefits to employees such as dental plans, second surgical opinion programs, utilization review, disability insurance, and wellness programs.[23-29]

To reduce expenditures on health care benefits, hospitals and other employers are looking at alternative funding mechanisms. There are three options to assume financial risk for a health care plan. Fully pooled conventional insurance provides low risk to the hospital and premiums set by pooled experience. Partial self-insurance involves a greater assumption of risk for the hospital and is the most common self-funding alternative. Premiums are determined for the employer based on actual experience for a period usually ending six months prior to the new benefit period.[30] The last option is total self-insurance where the hospital assumes all risk for claims during the period.

The cost savings of self-insurance programs are generally favorable. Most of the hospitals in one survey indicated that their self-funded programs were satisfactory in cost savings and had improved cash flow and service.[31] One hospital in South Carolina saved money over a period from self-funding. The first three years of a partially self-funded program yielded an 11.6 percent decrease in expenditures and an avoidance of $97,000 in increased costs. Under a totally self-funded program this hospital incurred a 41.4 percent claims cost increase from fiscal year 1976-1977 through fiscal year 1980-1981 for an average annual increase of 8.2 percent.[32]

Employer contribution to health care plans is used as a method of cost reduction. One survey of large employers in Minnesota between 1981 and 1982 suggested that indemnity plan premiums are significantly lower when employers make a level dollar contribution rather than paying a level percent contribution or a full amount for a chosen plan.[33]

Increased employee responsibility for health care costs is used by some employers to control costs. A survey by the Health Insurance Association of America (HIAA) indicated that the number of single plan deductibles of at least $100 has risen almost 10 percent since 1982. The same survey found that greater than one and one-half times as many employers, nearly 88 percent, offered full health insurance premiums in 1984 as in 1979.[34] One survey showed that employers preferred raising deductibles rather than coinsurance. Sixty-four percent of respondents increased deductibles within the past two years while 46 percent increased coinsurance.[35]

RESEARCH METHODOLOGY

A study was conducted in 1985 to determine the effect of health care benefit characteristics on costs and the impact of hospitals' offering benefits and incentives similar to those offered by businesses.

Sample and data acquisition

The sample for the study presented here encompassed all 184 Minnesota hospitals listed in the 1984 edition of the *American Hospital Association Guide to Health Care Institutions*. Hospitals in Minnesota were selected because of the state's reputation of having health care competition, innovations in health care financing, a strong business community, and close ties between business and health care.

After the sample was selected, a questionnaire was developed based on relevant information from the literature and interviews with personnel and labor representatives. The questionnaire asked directors of employee benefits from the 184 hospitals for information in three areas: hospital characteristics, employee characteristics, and health care benefits characteristics. The information was requested for the last completed fiscal year. Of the 184 questionnaires, 126 were returned completed, for a 68 percent response rate.

Data analysis

Data from the questionnaire were subjected to two separate sets of analyses. The first explored whether certain hospital and employee characteristics affect health care expenditure. This was done through an evaluation of statistically significant differences using tests and analysis of variance between those independent variables that the literature identified as important and the variable of health care expenditure per employee. An exploratory model made up of all the significant variables was developed based on this set of analyses.

The second part of the analysis used discriminant analysis and multiple linear regressions to assess those characteristics that differentiated hospitals by level of expenditure. The analysis discriminated the lowest from the highest expenditure group.

Operationalization of variables

Table 1 contains all variables used in the study that, based on the literature review, are expected to affect employee health care benefits expenditures. All are self-explanatory except expenditure per enrolled em-

TABLE 1

SUMMARY OF VARIABLES IN STUDY

Code	Description
PCTG46	Employee age
PCTFEM	Percentage female with benefits
PCTINUN	Percentage employees in union
BEDSIZE	Number of hospital beds
AFFIL1	For-profit multihospital system affiliation
AFFIL2	Multihospital system affiliation
CLASS	Hospital ownership classification
UNIONNO	Number of unions
HMO1	HMO offered
PPO1	PPO offered
BCBS1	Blue Cross/Blue Shield offered
COML1	Commercial insurance offered
SHTM	Short-term disability program offered
LGTM	Long-term disability program offered
WELL1	Wellness program offered
DENTAL	Dental program offered
OTHER	Other health care benefits offered
PCTBUD	Percentage of hospital budget for health care
TOTAL	Total number of plans and programs offered
PARTIAL	Partially funded plan
TOTPLANS	Totally funded plan
COPYMT	Change in copayment structure
DEDAMT	Change in deductible structure
SURGOP	Second surgical opinion program
FULLCONT	Full contribution toward premium
EXPEND	Expenditure per enrolled employee
SCORE	Business proximity index

ployee for health care benefits and the business proximity index.

Expenditure per enrolled employee refers to the total average dollar amount spent on each employee enrolled in health care benefits at the hospital. It was calculated by dividing total expenditures on health care benefits offered by the total number of qualified employees receiving health care benefits for the last completed fiscal year. Total expenditures were defined as the cost of all the health care benefits and programs offered at each hospital. Enrolled employees were defined as those who qualified by the individual hospital standards and received health care benefits for the last completed fiscal year.

The business proximity index was developed to measure how closely the sample hospitals resemble business in dealing with health benefits and containment of health care expenditures. This index measures the relative closeness of each individual hospital's health care benefit activity to that offered by business. The business activity index was developed from a survey of Minnesota employers conducted in 1983 by Dowd and Feldman.[36] The components of the index are:

- HMO plan offered;
- at least one totally self-insured plan offered;
- at least one partially self-insured plan offered;
- second surgical opinion program offered;
- full contribution toward premium not offered;
- change in deductible made;
- change in copayment made; and
- wellness programs offered.

The presence or absence of each component was ordinally ranked for each respondent hospital to determine the hospital's proximity to business in terms of employee health care benefits.

FINDINGS

The health care benefits environment

Tables 2 and 3 show an adequate fit between the study sample and the state's population of hospitals

TABLE 2

POPULATION REPRESENTATION

Size (no. of beds)	Population*	Sample	Percent of population
1–24	29	15	52
24–49	66	32	48
50–99	25	26	104
100–299	31	31	100
300 or more	18	22	122
Total	169†	126	74.5

* Data from Minnesota Department of Health Division of Health Resources *Directory of Licensed and Certified Health Care Facilities*, 1984.

† Does not include psychiatric and other specialized hospitals.

TABLE 3

HOSPITAL OWNERSHIP

Ownership	Population*	Sample	Percent of population
Not for profit	86	64	74
For profit	6	6	100
Public	77	55	71
Total	169†	125‡	74

* Data from Minnesota Department of Health Division of Health *Resource Directory of Licensed and Certified Health Care Facilities*, 1984.

† Does not include psychiatric and other specialized hospitals.

‡ One responding hospital did not complete the question on hospital ownership.

in terms of hospital size and ownership. Hospital size ranged from 8 beds to 1,158 beds with a mean of 169 and a median of 79.

The number of unions present in the hospitals studied ranged from 0 to 41 with most respondents indicating that no unions were present. Only 23 hospitals did not offer options that resembled the business menu. The 103 hospital business proximity scores—based on percentage similarity with the health care menu offered by business—ranged from 8 percent to 65 percent with a mean of 27 percent and a median of 26 percent. Those employee characteristics that have been associated with high health care costs are displayed in Table 4 and health care benefit plans offered and their characteristics are displayed in Tables 5 and 6.

Survey results

From t tests and analysis of variance, 6 of 27 independent variables that the literature listed as important were shown to significantly affect hospital expenditure per employee on health care benefits. A summary of the significant variables is presented in Tables 7 and 8. However, these findings represent isolated associations between each independent variable and the dependent variable.

Multiple linear regression and discriminant analysis were then used to test the 6 independent variables simultaneously against the dependent variable. Results are shown in Tables 9 and 10. The final regression equation explained 26 percent of the variance in expenditure per employee. The final discriminant analysis equation explained 73.53 percent of the variance between hospitals in the low-expenditure group and hospitals in the high-expenditure group. The analysis correctly classified 67 percent of the cases falling into the low expenditure or high expenditure group.

The business proximity index, when tested in the competing presence of all the significant variables, was the most powerful in accounting for reduction of expenditure per employee. This suggests that when the elements included in the index are offered together in what is labeled as a "business approach," hospitals can reduce costs.

The summary tables also show the variables total number of plans offered and second surgical opinion to be significantly associated with increased expenditures. Expenditures were significantly lower for hos-

TABLE 4

DESCRIPTION OF EMPLOYEE CHARACTERISTICS

Variable	Mean	Median	Mode	Minimum	Maximum
Percentage older than 46	35.26	34.72	25	1	70
Employees with benefits	333	100	62	0	3,159
Percentage female	83.33	87.50	80	12	100
Percentage in unions	42.04	40.83	60	0	100

TABLE 5

NUMBER OF HOSPITALS OFFERING HEALTH CARE BENEFIT PLANS

Plan	Number offered			
	0	1	2	3 or more
Blue Cross/Blue Shield	43	65	9	9
Commercial insurance	99	23	3	1
HMO	96	12	9	9
PPO	119	6	1	0
Short-term disability	94	24	3	5
Long-term disability	57	51	7	11
Wellness program	103	13	1	10
Dental plan	68	46	5	7
Other	105	16	0	5

pitals that offered one or two plans than for hospitals that offered three or more plans. Significant differences also occurred between hospitals offering one plan and hospitals offering three plans or hospitals offering one plan and hospitals offering four plans.

DISCUSSION

Of the employee attributes tested, the percentage of females with benefits was the only variable that had a significant impact on lowering expenditure per employee. The unexpected inverse relationship with the dependent variable (shown in Table 8) can best be

TABLE 6

NUMBER OF HOSPITALS OFFERING HEALTH CARE BENEFITS PLAN OPTIONS

Option	Offered	Not offered
Totally funded plan	11	84
Partially funded plan	15	74
Change in copayment	18	108
Change in deductible	15	111
Second surgical opinion	46	80
Full contribution toward premium	94	32

TABLE 7

T TEST

Variable	Effect on expenditure	p value
Long-term disability	direct	.001*
Dental plan	direct	.001*
Second surgical opinion	direct	.016†
Hospital classification	inverse	.030†

* Significant at 0.01.

† Significant at 0.05.

explained by the smaller overall contribution for employee premiums for females than for males. This occurs because hospitals generally contribute more money to the family plan than the single plan. The only time females would use the family plan would be if they were unmarried with dependents or if the spouse's plan was less extensive than the one offered by the hospital. Since no other variable significantly affected expenditure per employee it may be concluded that composition of the work force had little effect on premium costs and overall expenditure for the sample hospitals. Thus, the results do not suggest that changing the gender mix of the hospital would help administration provide more cost-effective benefits to its employees.

When tested individually, hospital characteristics had little effect on expenditure per employee. Hospital ownership was the only significant variable and those results suggested that public institutions had a lower expenditure per employee than private institu-

TABLE 8

ANALYSIS OF VARIANCE FINDINGS

Variable	Effect on expenditure	p value
Total number of plans	direct	.041*
Percentage female	inverse	.074

* Significant at 0.05.

TABLE 9

RESULTS OF FINAL MULTIPLE LINEAR REGRESSION ANALYSIS

Overall equation statistics			
F statistic	F statistic p value	R^2	Change in R^2
4.041	0.002	0.2628	0.2628

Variables in the equation					
Variable	Nonstandardized beta	Standardized beta	Beta	t	p value t
PCTBUD	9.786	5.618	.18631	1.742	.086
SCORE	−1032.439	375.078	−.33912	−2.753	.007
BCBS1	−107.243	55.921	−.28972	−1.918	.059
SURGOP	272.980	95.173	.34449	2.868	.005
HMO1	−69.633	54.845	−.18873	−1.270	.208
TOTAL	51.548	15.434	.58443	3.340	.001
(Constant)	1192.180	105.190		11.333	.000

tions. Increasing the percentage of unionized employees did not make a difference in expenditure per employee. This suggests that the presence of unions in the hospital setting does not affect cost of health benefits to the employer, perhaps because the type of health care benefits desired by union members were either already in place or were demanded by nonunionized employees as well. Another explanation could be that more attention was given to managing or reducing the cost of benefits when unionization was present.

The number of hospital beds was not shown to affect expenditure per employee. This could be explained by the fact that even though larger institutions have more employees receiving benefits and fewer employees enrolled per plan, which increases premium expense overall, the strategic advantage of large organizations in negotiating discounts from insurers may have helped to control costs.

In terms of design of health care benefit plans, the individual variable testing showed that the presence of a long-term disability program, dental plan, and second surgical opinion program were associated with increased costs. It can be concluded that either the returns from investment in such programs are not yet realized or are less than the program costs, or these programs are associated with those hospitals with higher than average health care benefits expenditures.

As the total number of plans offered increased, expenditure per employee also increased. This challenges the perspective that offering more plans forces more competitive prices that crowd out inefficient plans and programs. One reason the number of plans

TABLE 10

FINAL RESULTS OF STEPWISE DISCRIMINANT ANALYSIS OF LOW-EXPENDITURE AND HIGH-EXPENDITURE GROUPS

Group centroids	
Low expenditure	High expenditure
−0.32996	0.38496

Results	
Entry order	1
Independent variable	SURGOP
Standardized weight	1.000
Wilks' lambda	0.883
p value Wilks' lambda	0.01
Change in Wilks' lambda	0.116

> *As the total number of plans offered increased, expenditure per employee also increased. This challenges the perspective that offering more plans forces more competitive prices that crowd out inefficient plans and programs.*

may affect expenditure levels is that hospitals with one or two plans may focus on benefits packages while hospitals offering many plans are assuming the competitive theory will work and are therefore less concerned with monitoring the plans. In addition, the findings suggest that hospitals would likely save money by limiting the number of plans to no more than two.

Results of multivariate analyses suggest perhaps the most important finding of the study: In combination, only variables pertaining to health care benefits packages and the business proximity index explain variability in expenditure per employee. Results of multiple regression analysis show that the business proximity index, use of HMOs, and use of Blue Cross/Blue Shield decreased expenditure while the following increased expenditure: use of more than two plans, second surgical opinion programs, and size of hospital budget allocated for health care benefits. Since these factors are all under the control of the hospital's human resources department and administration they could be altered to maintain, decrease, or increase expenditures if desired. The results of the discriminant analysis substantiated the finding that second surgical opinion was a particularly significant discriminating variable. Due to the consistent finding that this particular health care benefit increased expenditure, its overall importance to the hospital organization should be evaluated.

RECOMMENDATIONS

Hospitals

The findings reported here suggest that human resources departments and hospital administrations might wish to reevaluate the health care benefits policy of their hospitals. The additional expense of such programs as long-term disability, dental, and second surgical opinion should be weighed against the desired level of comprehensiveness and quality of employee benefits.

If reduced costs are desired, hospitals might consider adopting a health care benefits policy that resembles the components of the business proximity index. In addition, hospitals might want to limit the number of health care benefits plans and programs to fewer than three to reduce costs.

Policy makers

The study presented here suggests that offering an increased number of plans does not reduce expenditures for the employer. Current legislation encouraging more health care plans as the solution to the high costs of health insurance and health care programs should be reevaluated in light of the information produced by this study.

Further research

Further research should evaluate the effects of the variables on expenditure over time to account for factors outside the control of a cross-sectional survey. Special attention should be paid to insurance company cost efficiencies, changes in hospital work force, labor market effects, and inflationary pressures.

Future studies should also attempt to delineate expenditure by family and single coverage in order to determine the effects of a spouse's benefit plan and dependents on the hospital's expenditure per enrolled employee. Finally, future studies should analyze hospital management philosophy and values toward health care benefits as this is a likely determinant of hospital expenditure.

REFERENCES

1. Miller, J.A. "The Health Care Cost Epidemic." *Best's Review Life/Health Insurance Edition* 85 (July 1984): 26, 30, 32.
2. Gordon, G.L. "Industry's Role in Health Care." *Connecticut Medicine* 46 (1982): 492–95.
3. Sapolsky, H.M., et al. "Corporate Attitudes toward Health Care Costs." *Milbank Memorial Fund Quarterly* 59 (1981): 561–85.
4. Schanie, C.F. "Unionization in Hospitals: Causes, Effects, and Prevention Strategies." *Hospitals and Health Services Administration* 29 (November–December 1984): 68–78.

5. Roesler, B.E. "Factors Associated with Health Care Plan Choice among Hospital Employees." Master's thesis, University of Minnesota, Minneapolis, June 1985.
6. Bentkover, J.D. "Choice, Costs and Competition." *Best's Review Life/Health Insurance Edition* 84 (January 1984): 12, 14, 15, 20, 95.
7. Jensen, G., Feldman, R., and Dowd, B. "Corporate Benefit Policies and Health Insurance Costs." *Journal of Health Economics* 3 (1984): 275–96.
8. MacDougald, J.E. "The Missing Link in Cost Containment." *Best's Review Life/Health Insurance Edition* 84 (October 1983): 36, 112.
9. Williams, C.A., and Heins, R.M. *Risk Management and Insurance.* New York: McGraw-Hill, 1981.
10. MacDougald, "The Missing Link in Cost Containment."
11. Dodge, D.D., and Head, J. "Self-funding for Employee Health Plans." *Southern Hospitals* 49 (July–August): 16–18.
12. Klett, S.V. "Women at Work: The Benefit Implications." *Business and Health* 2 (March 1985): 25–29.
13. Williams and Heins, *Risk Management and Insurance.*
14. Dowd, B., and Feldman, R. "Twin Cities Firms Reveal Recent Changes in Health Benefits Design." Paper sponsored by The Minnesota Coalition on Health Care Costs, with support from the CIGNA Corporation. Minneapolis, Minn.: University of Minnesota, February 1984.
15. Williams and Heins, *Risk Management and Insurance.*
16. Achenbach, S.H., and Swenson, K.S. "Employee Benefits: How Hospitals Can Provide High Quality Packages at Less Cost." *Texas Hospitals* 38 (October 1982): 42–43.
17. Bentkover, "Choice, Costs and Competition."
18. Van Bell, R.J., "The Selling of an HMO." *Business and Health* 1 (January–February 1984): 18–20.
19. Bieber, O.F. "Bargaining for Equitable Cost-Effective Health Care." *Business and Health* 2 (April 1985): 20–24.
20. Kodner, K. "Getting a Fix on PPOs." *Hospitals* 56 (November 1982): 59–66.
21. Fox, P.O., and Spies, J.J. "Alternative Delivery Systems: What Are the Risks?" *Business and Health* 1 (January–February 1984): 5–10.
22. Ibid.
23. Halonen, R.J., and Check, R.C. "Dental Coverage Helps Recruit, Retain Employees." *Hospital Progress* 63 (August 1982): 44–45, 63.
24. Hanley, R.J., and Ayers, J.T. "Second Opinion: A Tool to Save Money, Improve Care." *Business and Health* 2 (March 1985): 22–23.
25. McIntyre, K.J. "Hospital Chain Designs Health Cost-Control Plan." *Business Insurance* 18 (May 28, 1984): 18.
26. Williams and Heins, *Risk Management and Insurance.*
27. Clement, J., and Gibbs, D.A. "Employer Considerations of Health Promotion Programs: Financial Variables." *Journal of Public Health Policy* 4 (1983): 45–55.
28. Sajewski, C.J., "Small Business Develops Its Own Brand of Health Promotion Programs." *Business and Health* 2 (April 1985): 54–55.
29. Dreher, W.W., and Wallace, P.A. "Can Employees Cut the Cost of Health Benefits?" *Management Focus* 30 (March–April 1983): 15–19.
30. Sal, D.E., and Gregory, D.D. "Can Hospitals Succeed at Self-Insuring Employees?" *Hospital Financial Management* 35 (April 1981): 14–18.
31. Ibid.
32. Anderson, J.P., and Hershinger, M.G. "Self Funded Health Insurance." *Southern Hospitals* 51 (November/December 1983): 22, 24.
33. Jensen, Feldman, and Dowd, "Corporate Benefit Policies and Health Insurance Costs."
34. "Rise Seen in Levels of Benefit Coverage, Use of Deductibles." *Hospitals* 59 (April 16, 1985): 39.
35. Gardner, S.F., Kyzr-Sheeley, B.J., and Sabatiav, F. "Big Business Embraces Alternate Delivery." *Hospitals* 59 (March 16, 1985): 81–84.
36. Dowd and Feldman, "Twin Cities Reveal Recent Changes in Health Benefits Design."

Employee suggestion programs in nonprofit hospitals

Susan G. Richer
and
David M. Weiss

In an effort to improve productivity and foster cost consciousness, many not-for-profit hospitals have implemented employee incentive programs. Employee suggestion programs were reviewed, addressing the factors that influence hospital employees' attitudes toward participation in the program and the design and organization of effective employee suggestion programs.

Hospitals today are heavily in debt to the private sector, not financially, but administratively. Over the years, hospitals have relied heavily on the successful managerial theories and techniques of private industry. The language of business has become common parlance in health care institutions. During the 1980s, it is expected that hospitals will continue to adapt and adopt managerial techniques and organizational structures of non-health care businesses to maximize economic survival.

Hospitals' viability will become more critical as reimbursement policies grow more restrictive, as corporations begin to exert more pressure on hospital financial and operational procedures, and as interhospital competition increases. To compete with private industry and with each other, hospitals are starting to implement traditional private sector incentive practices.[1]

One such practice involves employee incentive programs. Hubler and O'Neill[2] suggest that hospitals have been reluctant to experiment with such programs for several reasons: possible deterioration in the quality of care; lack of departmental control over expenditures; difficulties in employee orientation and control of programs; unfair pressure on program participants to improve on previous years' standards; and ineligible employees' resentment toward the hospital and program. More significantly is the question of the incentive plan's effect on the hospital's nonprofit status.

Most nonprofit hospitals are exempt from federal income taxes under section 501(c)(3) of the Internal Revenue Code. To maintain its exemption a hospital must operate to serve public rather than private interests and ensure that no portion of its net income is to inure to the personal benefit of any private individuals.[3] Although the Internal Revenue Service (IRS) formerly held a view that incentive plans constituted "profit sharing" by nonprofit institutions, thus violating tax exemption statutes, the IRS has modified its position.[4] More recently, it has issued private rulings that incentive plans are compatible with section 501(c)(3) under the following conditions: The resultant cost savings must not interfere with the quality of patient care, and

Susan G. Richer, M.P.A., is an Assistant Administrator at Maimonides Medical Center, Brooklyn, New York.

David M. Weiss, Ph.D., is Assistant Vice President for Academic Affairs at Long Island University, Greenvale, New York.

the measures of productivity and/or cost savings must be reasonably related to the productivity and efficiency of the program participants.[5]

Primarily, employee incentive systems are expected to reduce unnecessary or excess costs by offering compensation in exchange for more efficient results. In other words, such plans encourage the achievement of organizational goals at a price commensurate with the benefit to the institution.[6] Institutions offering incentive programs often are more inclined to encourage staff to develop new departmental strategies to function with increased profitability.

The successful incentive program can reduce overtime, decrease personnel turnover, improve attendance, and enhance organizational communication and employee loyalty. Most important, incentive plans get people to think about themselves, their jobs, their work relationships, and their organization as a whole. Incentive plans may focus on productivity,[7] attendance,[8] employee referrals,[9] extracurricular activities,[10] performance,[11] and employee suggestion programs (ESPs). This article highlights ESPs and examines the role these programs play in nonprofit hospital management.

LITERATURE REVIEW

Employee suggestion programs date to the middle 1800s when the Yale and Towne Manufacturing Company used an "idea box," with hopes of increasing communication between management and workers.[12] Eastman Kodak established one of the earliest suggestion systems in the United States in 1898. Its first reward was two dollars to a worker who suggested that production could be increased if the factory windows were washed. That employee's forward-thinking supervisor later became president of Kodak.[13]

The modern suggestion system was born after World War II, when American industry had to rapidly convert to peacetime production. Management recognized that employees could contribute ideas that resulted in increased and efficient productivity, but that more sophisticated methods were needed to administer these new suggestion programs. Thus, in 1947, the National Association of Suggestion Systems (NASS) was established. This Chicago-based organization helps launch suggestion systems, develops model plans, and monitors legal developments.[14] In 1970, the 1,400 member companies of NASS, representing about 7.5 million eligible employees, received over 3.5 million ideas and paid out over $42 million in awards. Participating companies reported an average return of $4.46 for each dollar invested in a suggestion program.[15] By 1973, the savings-cost ratio had risen to $5.70,[16] and by 1974, to $6.52.[17]

Group award systems allow more people to share in the benefits of effort and also bring group dynamics, such as peer pressure, into play.

A variety of options are available for the format of a suggestion system. Curiously, literature is sparse concerning the reasons certain options are preferred over others. For example, some institutions initiate intradepartmental quality circles in which only tangible ideas, measurable in dollars saved and contained within the scope of an employee's job description, are acceptable.[18] Others establish departmental quotas, requiring a specified minimum number of ideas submitted or dollars saved per month. Still others opt for a broader-based program, soliciting cost containment ideas hospitalwide with no restrictions.[19] Burdette Tomlin Memorial Hospital in Cape May, New Jersey, takes its effort one step further. In addition to employees, physicians and volunteers may participate, creating a hospitalwide, comprehensive program.[20] Some firms even extend eligibility to retirees and employees on leave of absence and disability.[21]

Depending on the individual institution's policy and guidelines, supervisory and administrative personnel may be eligible for incentive compensation[22] or specifically excluded from participation.[23,24] NASS reports that managers typically cannot take part in idea plans as they are "paid to have good ideas to begin with."[25] In most cases where supervisory or technical personnel are involved, awards are paid only for suggestions that have impact beyond the domain of one's job.[26]

Methods of award distribution also vary significantly. The most common approach is to award the individual either a single designated payment or a percentage of the first year's net savings resulting from an accepted suggestion.[27] At Johns Hopkins University Hospital in Baltimore, Maryland, suggestions accepted and approved by its IdeaBank Committee return 10% of net savings to the submitter.[28] NASS reports average figures of 15% to 25% of net savings, although some industries pay as high as 40% to 50%.[29]

In contrast, Holzer Medical Center in Gallipolis, Ohio, uses a cash disbursement format that distributes savings among all hospital staff. The Holzer program encourages intradepartmental cost containment, stressing elimination of waste and breakage and improvement of procedures and systems. Productivity and cost standards are set by comparative statistical data reported by the American Hospital Association.[30] In December of each year, Holzer calculates the margin of difference between its cost per patient day and that of neighboring hospitals. One-half of the savings is returned to the patients by way of contained costs for hospital services; the other half is divided among staff according to the number of hours worked. Employees receive their share in the form of a cash bonus: In 1977 the ESP resulted in a bonus of $107.80 per employee.[31]

Lewis suggests that group award systems allow more people to share in the benefits of effort and also bring group dynamics, such as peer pressure, into play.[32] However, one major drawback to a disbursement-based system is that while it rewards cost-conscious workers, it also rewards those who ignore or defy cost-reduction efforts.

A third method of award distribution is to offer gifts or other nonmonetary awards. Nonmonetary incentive awards can include a designated parking space for one month[33]; extra paid vacation days[34]; movie tickets, meal vouchers, or ten gallons of gasoline[35]; and complimentary travel passes.[36]

METHOD OF STUDY

A project was conducted to study employee attitudes toward ESPs. Two questions were addressed:
1. What factors influence hospital employees' attitudes toward participation in ESPs? and
2. How can these findings be applied to the specific design of an ESP?

A major metropolitan New York teaching hospital was chosen as the site of the survey. Permission to conduct the study was obtained from both hospital administration and the Department of Human Resources. A questionnaire was developed to assess whether hospital employees viewed suggestion programs as worthwhile and to what extent they would be willing to participate.

This value/participation index (VPART) was graded into four levels of response:
1. ESPs are a worthwhile idea, the employee would definitely participate;
2. Worthwhile, the employee would probably participate;
3. Worthwhile, although the employee would probably not participate; and
4. Not worthwhile, the employee would definitely not participate.

Additional questions covered experiences in previous suggestion programs, job satisfaction levels and supervisor ratings, and general demographic data. An accompanying cover letter briefly described the project and its purpose, and assured participants of anonymity and confidentiality.

A modified cluster sample was employed for this study. Selection was based on the following process. Departments were classified into five areas: medical, nonmedical, laboratory, support services, and nursing. One to two departments from each category were then selected randomly and their day shift employees were used as respondents. Departments selected included Dentistry (medical), Financial Aid and Cashier's Offices (nonmedical), Chemistry (laboratory), Security (support services), one medical nursing unit, and one surgical nursing unit.

Supervisors in each of these departments were contacted and appointments scheduled to administer the questionnaire. Following a brief explanation of the project, employees were informed that participation was strictly voluntary, and any questions deemed inappropriate or offensive by the respondent need not be answered.

Out of 122 questionnaires, 96 were completed and returned, resulting in a 79% response rate. Responses were subsequently coded and the data analyzed using the updated Statistical Package for the Social Sciences (SPSS-X) software package.

RESULTS

Demographic characteristics of the sample population are shown in Table 1, along with job satisfaction indicators and evaluation of ESP components. Of the population surveyed, 76% were women and 24% were men, which is consistent with the sex distribution generally found in the health care profession. However, 94% of the sample listed "white" as their ethnic background. This is not representative of the field as a whole, since minorities make up a large percentage of the work force. Baccalaureate degrees and high school diplomas were listed most frequently as educational attainment (27% and 22%, respectively), which corre-

TABLE 1

DEMOGRAPHICS, SATISFACTION INDICATORS, AND ESP EVALUATION

Variable	Percentage of respondents	Variable	Percentage of respondents
Sex		**Program rating**	
Female	76	Very worthwhile	36.8
Male	24	Somewhat worthwhile	47.4
		Not worthwhile	15.8
Ethnicity		**Award receipt**	
White	93.8	Yes	26.3
Black	5.2	No	73.7
Other	1.0		
Education		**Type of award**	
High school diploma	21.9	Certificate	40
Associate degree	17.7	Cash	20
Bachelor's degree	27.1	Merchandise	20
Master's degree	8.3	Other	20
Doctoral degree	17.7		
Other	7.3	**Was award enjoyed?**	
		Yes	60
Job satisfaction		No	40
Extremely satisfied	28.4		
Somewhat satisfied	48.4	**Would you reparticipate?**	
Neither satisfied nor dissatisfied	9.5	Yes	78.9
Somewhat dissatisfied	12.6	No	15.8
Extremely dissatisfied	1.1	Do not know	5.3
		Award preference	
Rating of immediate supervisor		Certificate	7.3
Excellent	34.1	Merchandise	2.1
Good	43.9	Paid vacation days	40.6
Fair	14.3	Cash	50.0
Poor	7.7		
		Value/participation index (VPART)	
Prior ESP participation		Definitely participate	48.9
Yes	21.5	Probably participate	43.8
No	78.5	Probably not participate	5.2
		Definitely not participate	2.1

sponds well with job classification data: Nursing staff (25%) and clerical workers (25%) predominate.

The majority of the employees surveyed are more satisfied than not in their present jobs: 28% consider themselves extremely satisfied and 48% somewhat satisfied. Supervisor ratings tended to be skewed toward the higher end of the scale: 78% rated their supervisor good or excellent; however, over one-fifth of the respondents (22%) were displeased with their superiors.

Twenty percent of the sample population had prior ESP participation. Of these individuals, 84% rated their previous program as worthwhile, 37% as very worthwhile, and 47% as somewhat worthwhile. Twenty-six percent received an award in that program: merchandise (40%), cash (20%), certificates of appreciation (20%). Sixty percent were pleased with the award they had received. Seventy-nine percent indicated that they were willing to participate again in another ESP.

Although the majority (78%) of the employees responding to the survey had never participated in an ESP, 93% felt ESPs were worthwhile and would be willing to participate in one (47% would definitely participate, and 42% would probably participate). One-half of the sample selected cash as the most desirable award in such a program, with extra paid vacation days a close second (48%).

A strong Spearman rank-order correlation ($r > 0.5$) is exhibited between the variables of prior ESP rating and the VPART ($r = .619$, $p = .002$). Moderate positive relationships ($0 < r < 0.5$) are shown for job satisfaction levels with both employees' rating of their immediate supervisor ($r = .454$, $p = .000$) and job classification ($r = .193$, $p = .031$). A moderate negative relationship ($-0.5 < r < 0$) exists between age and job classification ($r = -.335$, $p = .000$).

DISCUSSION

Of greatest importance to this study were the findings associated with the VPART. Results show that the likelihood of an individual participating in an ESP is affected by the past experiences he or she had in such a program. This includes their perception and subsequent rating of that program, and whether an award was received. While award receipt is correlated with VPART, VPART itself is statistically independent of the type of award received—there was no significant difference between cash or merchandise awards—and independent of whether the employee liked the award. Merely that an award was received was sufficient to positively influence participation. This stresses the importance of widespread distribution of small gifts to employees for any and all ideas submitted, regardless of its chances of implementation. This becomes especially important at the outset of an ESP.

Spearman correlations further indicate that the higher employees rated their previous program, the more they value an ESP and the more likely they are to participate in future programs. However, it is significant—both in the statistical and nonstatistical senses—that program ratings are independent of award received, type of award, and even whether the employee liked what was received. Employees consistently rated their previous program well, even if they did not get an award or did not like the award presented to them. In addition, reparticipation in an ESP by those with earlier participation was also independent of these variables.

These findings strongly suggest that participation is the key: that employees feel their suggestions have been reviewed, taken seriously, and dealt with fairly; that management is willing to listen, and that employees and their ideas matter. Awards are secondary. Rather, it is the perception of being recognized that prevails. That is, employees appreciate when their superiors publicly and regularly recognize their worth and achievements.

• • •

The increase in incentive plans and ESPs within the health care field reflects the growing influence of private sector management techniques. Although incentive programs afford great potential to an organization, public or private, they are by no means a panacea for its ills. ESPS cannot compensate for poor supervision, inadequate long- or short-range planning, or insufficient budget preparation.

Rather, an ESP is a managerial tool, one of many in the repertoire of a competent administrator. Its success ultimately depends on two factors: (1) the quality of the program design, and (2) the level of sustained effort and long-term commitment that the institution is willing to invest.

The initial step in the introduction of a suggestion system is to specify program design parameters and most important, the rules governing the program. A primary reason for suggestion system failure is the lack of clear policies or ground rules.[37]

Therefore, all rules and policies must be spelled out to employees, detailed in brochures, and included on any suggestion submission forms. Regardless of how the program is designed, its rules need to be stated in clear terms, understandable to all.

In addition, eligibility requirements must be carefully considered. Several basic questions must be addressed:

1. Who is eligible? Will supervisory or managerial personnel be included? Retirees? Should medical staff be involved in the program? What about volunteers, students, and visitors?

2. What is eligible? Most suggestion systems specify certain areas that are not open for suggestions. Programs solely concerned with cost reduction may prohibit suggestions on safety, public relations, or routine maintenance. Other exclusions may cover changes in company policy, employee benefits, and ideas related to the suggester's job description. Ideas already under consideration or adopted by the organization are generally labeled ineligible.
3. What is the award policy? How are the amounts and kinds of awards to be determined? Nearly all suggestion plans base monetary awards for suggestions that result in tangible savings on a portion of the first year's net savings. Intangible ideas may be rewarded with either designated lump-sum amounts or gift items. Small token gifts might be distributed for each suggestion received, to help spur interest in the program. Turnaround time should be rapid; employees should not be discouraged by lengthy delays in check processing or merchandise delivery.
4. What will be the plan for publicity? In order for an employee suggestion plan to succeed, it must be highly promoted by the institution. An in-house campaign should be launched well before the program goes into effect, to explain how it works and its benefits. Promotional activities should be continued after the plan becomes operational, so that a high level of interest is maintained.[38] Regular reminders of the program—posters, pay envelope stuffers, and flyers distributed in the cafeteria—are essential to its viability.

Clearly, there exists no single format for an ESP. Although all ESPs share a general framework, specific design parameters vary according to the needs and goals of the sponsoring institution. Employee attitude questionnaires are useful in defining problem areas and can assist in design-related decisions, especially when used to resolve specific points of the program, such as award preference, and inclusion or exclusion of management personnel. It may be advantageous to introduce simultaneously administered intradepartmental quality circles, to generate an institutional atmosphere of cost consciousness and employee recognition.

One caveat is in order. The possibility of a viable ESP should not be abandoned should the results of the questionnaire indicate an uninterested work force. A well-designed, well-publicized program that is staffed by knowledgeable and enthusiastic promoters can still be successful. Key factors include supervisory support; rapid turnaround time on submitted suggestions; or interdepartmental competition for greatest number of ideas accepted and/or dollars saved. It must be stressed that in a successful ESP, everyone who is involved benefits. The sponsoring hospital is able to control expenditures without reducing patient services, and its employees benefit twice—once from the monetary value of the award received, and once again from the recognition of their participation.

There exists no single format for an employee suggestion program.

Effective incentive programs can permeate every facet of organizational life and can reach individuals at every level of the organization. Employees' need for recognition, as well as organizational needs for cost containment, can be mutually satisfied.

The scope of an incentive plan, however, goes well beyond immediate gratification. A series of motivational and/or participative programs, such as the program discussed here, when designed, implemented, and administered properly, can powerfully shape employee attitudes and behavior, which subsequently affects efficiency and effectiveness. As many have demonstrated, ESPs can significantly enhance an employee's perception of an organization, while maximizing efforts to achieve organizational goals and objectives.

REFERENCES

1. Whitted, G.S. "A Plan to Attract and Retain Qualified Managerial Talent." *Health Care Financial Management* 37, no. 7 (1983): 20-24.
2. Hubler, M.J., and O'Neill, T.P. "A Monetary Incentive Program Could Cut Costs." *Hospital Financial Management* 35, no. 11 (1981): 40-44.
3. Meredith, S.L., and Padget, D.L. "Direct Financial Incentives to Motivate Hospital Employees." *Health Care Financial Management* 37, no. 7 (1983): 26-33.
4. "Employee Incentive Plans: The Newest Idea: Give Employees a Share of What They Save." *Cost Containment* 1, no. 6 (1979): 3-6.
5. Mancino, D.M. "Avoiding the Pitfalls: New Revenue

Sources and Tax-exempt Status." *Hospital Financial Management* 35, no. 5 (1981): 46-50.
6. Showalter, R.H. "Adopting Big Business Attitudes in Health Care?" *Hospital Financial Management* 36, no. 5 (1982): 50-52.
7. Zemke, R. "The 3Rs of Productivity Improvement: Responsibility, Recognition, Reward." *Training* 16, no. 7 (1979): A18-20.
8. Wood, W. "Incentive Programs: Do They Really Work?" *Executive Housekeeping Today* 3, no. 2 (1982): 8.
9. "Employee Incentive Plans."
10. "Motivation Through Recognition." *Hospital Supervisor's Bulletin* 30, (March 1980): 1-4.
11. Cleverly, W.O., and Mullen, M.P. "Management Incentive Systems and Economic Performance in Health Care Organizations." *Health Care Management Review* 7, no. 1 (1982): 7-14.
12. Gregg, G. "Suggestion Boxes." *Venture* 6 (April 1984): 33-34.
13. Reuter, V.G. "A New Look at Suggestion Systems." *Journal of Systems Management* 27, no. 1 (1976): 6-15.
14. Gregg, G. "The Power of Suggestion." *Across the Board* 20 (December 1983): 27-31.
15. Hein, J.E. "Employee Suggestion Systems Pay." *Personnel Journal* 52, no. 3 (1973): 218-21.
16. Reuter, "A New Look at Suggestion Systems."
17. Reuter, V.G. "Suggestion Systems: Utilization, Evaluation and Implementation." *California Management Review* 19, no. 3 (1977): 78-89.
18. St. Joseph's Hospital and Medical Center, Memorandum. Phoenix: St. Joseph's Hospital, December 9, 1980.
19. Traverse City Osteopathic Hospital. *Employee Cost Containment Suggestion Program* (memorandum). Traverse City, Mich.: Traverse City Osteopathic Hospital, February 12, 1982.
20. Buonanni, B.F. "Employees Help Hospitals Save Money." *Hospital Purchasing Management* 7, no. 9 (September 1983): 6-7.
21. Grumman Aerospace Corporation. *Questions and Answers About Project Sterling: An Employee Suggestion Handbook.* Bethpage, N.Y.: Grumman Aerospace Corporation.
22. Munson Medical Center. *Purpose and Objectives of the Employees' Suggestion Plan* (memorandum). Traverse City, Mich.: Munson Medical Center, July 1980.
23. International Business Machines Corporation. *Ideas for Improvement.* Armonk, N.Y.: IBM, September 1976.
24. Memorial Hospital. *The Big M Program* (memorandum). Sarasota, Fla.: Memorial Hospital, December 13, 1977.
25. Gregg, "Suggestion Boxes," 33-34.
26. National Association of Suggestion Systems. *Starting a Suggestion System in Your Organization.* Chicago: NASS.
27. Blue Cross and Blue Shield of Michigan. *Employee Suggestion Program Handbook.* Detroit: BC/BS, October 1981.
28. Nelson, L. "Cost Containment: Why Not Try an Idea Bank?" *Medical Laboratory Observer* 15, no. 3 (1983): 105-8.
29. "Employee Suggestion Systems: How Hospitals Can Capitalize on Employee Creativity." *Cost Containment* 4, no. 17 (1982): 3-6.
30. "Holzer Medical Center's Incentive Plan in Action." *Cost Containment* 1, no. 6 (1979): 2.
31. "Hospital Savings Mean Employee Earnings." *Health Services Manager* 11, no. 12 (1978): 8.
32. Zemke, "The 3Rs of Productivity Improvement."
33. "Employee Suggestion Programs Net Hospital/Patient Savings." *Hospital Topics* 58, no. 1 (1980): 1.
34. Edsel, W.M. "Employee Recognition Award—It Pays." *Medical Group Management* 28, no. 4 (1981): 46-50.
35. Neider, L.L. "Cafeteria Incentive Plans: A New Way to Motivate." *Supervisory Management* 28, no. 2 (1983): 31-35.
36. Zemke, R. "Incentives Provide the Winning Edge." *Training* 17, no. 7 (1980): A1-4.
37. Graf, L.A. "Suggestion Program Failure: Causes and Remedies." *Personnel Journal* 61 (1982): 450-54.
38. "Employee Incentive Plans."

PART IV

DOWNSIZING, ROLE STRESS AND OTHER ISSUES

The hospital merger: Its effect on employees

Marcia K. Petchers,
Sandra Swanker,
and
Mark I. Singer

Hospital closings are occurring with increasing frequency, yet little is known about their impact on employees. A study was conducted to assess the impact of a psychiatric hospital's closing on employees' perceptions of and preparation for the closing process, as well as the impact on employment outcomes.

The occurrence of hospital mergers and acquisitions has dramatically increased over the past five years, a trend that is expected to continue. The resulting closings have far-reaching impacts on the overall health care system, on the organizations directly involved, on employees, and on consumers. In labor-intensive health care organizations, employees are often the most dramatically affected. Careful management of the transition is essential during a merger process to minimize the negative effects on performance and productivity. Williams and Feldman[1] point out that merger deliberations may take place over an extended period of time and may be in the public eye. The maintenance of quality care, acceptable performance, and productivity are particularly difficult given the political, financial, and legal issues that inevitably contribute to the complexity of merger management.

An emerging body of literature has addressed the financial, legal, and corporate implications of mergers and acquisitions. Few studies, however, have directly focused on the impact of mergers on employees. In their review of the literature, Hernandez and Kaluzny[2] conclude that future research should address the entire closure phenomenon. The study on which this article is based was designed to contribute to this goal by assessing the direct impact of hospital closings on employees. Employees' perceptions of the closing process, their preparation for it, and their ultimate employment outcomes were examined. The purpose of surveying employee perceptions is to identify patterns of response that will yield information that can assist administrative staff in orchestrating merger processes in a productive, yet sensitive, manner. It is hoped that merger management strategies in the future can be refined as a result of this and other studies.

Marcia K. Petchers, *Ph.D., is Associate Dean and Associate Professor, Mandel School of Applied Social Sciences, Case Western Reserve University, Cleveland, Ohio.*

Sandra Swanker, *R.N., M.S.N., is Administrative Director, Department of Psychiatry, St. Vincent Charity Hospital & Health Center, Cleveland, Ohio.*

Mark I. Singer, *Ph.D., is an Associate Professor, Mandel School of Applied Social Sciences, Case Western Reserve University, Cleveland, Ohio.*

METHOD AND SAMPLE

The hospital under study, Woodruff Hospital, located in Cleveland, Ohio, was a 98-bed free-standing psychiatric facility acquired by a larger (492-bed) general hospital. As a result of the merger, the free-standing facility was closed, patients in the hospital were transferred, and staff were released on the closing date. The acquiring hospital did give employment preference to the released employees for open positions. Within the first three months postclosing, 43% of the displaced employees were offered positions. Of those offered positions, 87% accepted employment at the acquiring facility.

An unusual complication of the closing process should be noted. The initial merger plans that were under negotiation for over a year with another general hospital fell through. This first negotiation was then followed by a second and quite rapid successful negotiation process with another facility. Staff had been given positive, supportive, and accurate information about the initial intended merger. This merger proposal would have kept Woodruff open and operational. However, management control would have been assumed by the acquiring hospital. The second and accepted proposal closed Woodruff in addition to having the acquiring hospital assume management control. Due to the expedience of the negotiations, staff were not given updated information in a timely manner. Unfortunately, the announcement of the final merger and closing was released by the media prior to the hospital's administration directly communicating adequate information to employees.

Description of sample and research methods

The study's sample included all personnel employed at Woodruff Hospital during the merger period for whom current home addresses were available. A three-page questionnaire was mailed to each individual to elicit his or her perceptions of the merger process. Questionnaires were mailed four months after the physical merger to the two hospitals. Of the 169 employees receiving the instrument, 99 satisfactorily completed and returned their questionnaires. Thus, an overall return rate of just under 60% was achieved.

Description of the sample

Approximately two-thirds of the respondents were female and just over one half of the sample (53.0%) were age 40 or below. Two-thirds (67.3%) of the sample were white and a little more than one-half (53.5%) were married. About seven out of ten respondents considered themselves to be heads of household. One-half reported having children living at home. Slightly more than one-half of the sample was college graduates, 40% of whom held graduate or professional degrees. The largest group of respondents (43.6%) reported household income above $35,000, with 36.1% reporting income between $18,000 and $35,000, and 20.2%, $18,000 or below. Approximately four out of ten (42.4%) of this former employee group were considered professional care providers, 12.1% were non-professional care providers, 14.1% were classified as clerical, 18.2% were unskilled laborers, and 13.1% were managers. Years of service at Woodruff Hospital varied widely from less than 1 to 17, with a median length of employment of 6 years. Finally, at the time of the survey, about three-quarters (76.8%) of the respondents reported being employed; of those, 42% said they had gained employment by the hospital mounting the takeover (see Table 1).

FINDINGS

Initial status of employees' sample

To provide baseline data, the survey included questions about employees' satisfaction levels with their original employer (Woodruff Hospital) as well as their perceptions of the management style of the hospital before it closed. Results indicated a very high degree of satisfaction with work responsibilities (95.9% indicated they were very or moderately satisfied) and with co-workers (89.9% were very or moderately satisfied). While still predominantly positive, employee satisfaction with management staff was not uniformly high (65.7% rated themselves as very or moderately satisfied). One-half of the respondents considered the hospital to have had no consistent management style, while the rest were split between perceptions of an autocratic (10.1%) and a participative (37.4%) management style.

Perceptions of the hospital closing process

The second area of inquiry was aimed at establishing the employees' views of why the hospital closed and their evaluations of how well the closing process had been handled by management. When asked about the reason(s) for the hospital closing, the largest percentage of respondents, almost one-half (48.5%), blamed the

TABLE 1

EMPLOYEE DEMOGRAPHIC DATA

Category	%
Gender	
Female	66.3
Male	33.7
Race	
White	67.3
Nonwhite	32.7
Age	
21–30	17.3
31–40	35.7
41–50	20.4
51–60	18.4
61+	8.2
Income	
$18,000 or below	20.2
$18,001–$35,000	36.1
$35,001 and above	43.6
Job classification	
Managerial	13.1
Professional	42.4
Nonprofessional	12.1
Clerical	14.1
Unskilled	18.2
Marital status	
Married	53.5
Not married	46.5

closing on poor management; one-quarter cited competition as an important reason, while 22.2% felt there was insufficient demand for services. Smaller percentages agreed that problems with third party reimbursement (11.1%) or an unwillingness to cut costs (10.1%) led to the closing. Opinion was quite mixed about whether the closing could have been avoided, with most respondents falling somewhere between the two extremes of definitely yes (22.2%) and definitely not (2.0%). Just over one-half of the respondents said the closing probably could have been avoided, and 20.2% thought probably not.

When asked whether they were kept informed about the closing process, most (71.4%) of the former employees indicated that they had been only slightly or not at all informed. Management was perceived as only slightly helpful or not helpful at all by almost two-thirds (65.7%) of the respondents. A similar percentage of employees, two-thirds (62.6%), did not feel they had been well-prepared for the closing.

Impact of the closing

Over three-quarters of the respondents (76.8%) reported the closing had been a stressful experience. Interestingly, there were widely different areas of loss signified as characterizing this experience. For four out of ten employees, the very tangible loss of wages was the greatest loss resulting from the closing, while 27.3% cited the loss of a satisfying job itself as the salient issue.

The next largest group of respondents signified less tangible areas of loss; 18.2% cited loss of relationships, while 17.2% noted loss of a sense of belonging. Much smaller percentages of respondents chose as significant loss of self-esteem (5.1%) or loss of daily routine (4.0%).

While the short-term impact of the closing appears to have been stressful and engendered a sense of loss, a look at the longer-term results shows the consequences of the closing were not as negative as employees may have anticipated. The study shows the majority of respondents were reemployed and resumed their previous work status and satisfaction levels. Four months following the closing, about three-quarters of the respondents reported they were reemployed. Of those, about one-half had made lateral job moves, while 29.2% experienced a demotion and 18.1% a promotion. New positions were judged to be very to moderately similar to old ones by over one-half of the respondents, while one-quarter said they were slightly similar. However, when it came to working conditions, only two out of ten respondents found them to be similar to their original

---

While the short-term impact of the closing appears to have been stressful and engendered a sense of loss, a look at the longer-term results shows the consequences of the closing were not as negative as employees may have anticipated.

place of employment; for 80.3%, the degree of change was judged to be great or moderate.

Last in this section is the analysis of work satisfaction levels for the reemployed respondents. It was heartening to find that despite the stress and changes associated with job transition, satisfaction with new positions was indeed high. The majority (68.9%) were very to moderately satisfied with their new work responsibilities; an even higher percentage (87.8%) were very to moderately satisfied with relationships with new coworkers; and two-thirds (67.5%) were very or moderately satisfied with their new salaries. The area with the lowest satisfaction level related to management staff, where one-half was very or moderately pleased with their management staff.

At the time of the study, just under one-quarter of those surveyed indicated that they were currently unemployed, which is likely to be attributed to the hospital layoff. Almost one-half of this group reported current household incomes of $18,000 or under. Two-thirds of this group of former employees were women, which is comparable to the proportion of women in the total sample; one-half of the unemployed group were nonwhite, which is a disproportionately high representation of minorities in comparison to the sample.

Unemployed workers did not perceive the probability of averting the merger any differently than those who were reemployed and unemployed respondents reported similar perceptions. The same was true for the amount of information employees perceived they had been given about the merger; the reemployed and unemployed respondents reported similar perceptions.

Differences were observed in the degree to which employees felt they were prepared for the merger. Of those who were unemployed at the time of the study, 27.2% reported having been very or moderately prepared as opposed to 39.4% of those who ended up reemployed by the time of the study.

A significant correlation between employment status and self-reported stress levels was noted in the study. About 60% of the unemployed workers rated their resultant stress as very high, whereas 35.5% of the reemployed workers placed themselves in this category. However, the modal response by reemployed workers to the stress item did fall into the moderately stressful category, thus suggesting that considerable stress was associated with the merger experience regardless of postmerger employment status. Finally, 40.9% of the unemployed respondents indicated that the loss of the job itself was their most critical loss, but this reason was cited by only 23.7% of reemployment respondents. Figure 1 shows a comparison of characteristics for those who achieved reemployment over the four-month period following the closing and those who did not.

Problem identification/areas for improvement

The major problem area in the closing process identified by respondents was the lack of communication of believable information as the closing transpired. Apparently, most employees did not feel they had been engaged in the process even though they were key stakeholders. They believed they were not given access to the information necessary to make informed decisions. A feeling of bitterness seemed to accompany this perception. In particular, resentment surrounded the fact that official news of the final sale of the hospital came through the media, not firsthand from the hospital administration.

Employees also believed their interests were not well represented in the negotiation of terms and conditions of the merger. For example, the severance pay plan was thought to be unfair as was the nontransference of seniority and the procedures for applying to transfer. There was a perception that since management had handled the merger, employee rights had not been protected and management staff had brokered only in their own best interest. Suggestions for improvement included the hospital providing information sessions about the merger, employee rights, benefits, job options, job-hunting skills, and support groups to aid with the emotional consequences of the transition.

SUMMARY AND DISCUSSION

A major dilemma highlighted by this study is the need to strike a balance between employee concerns about receiving timely, accurate information and the organization's need to maintain confidentiality in conducting negotiations. While a hospital undergoing a takeover will not be able to provide full disclosure as negotiations are underway, every effort must be made to furnish employees with complete and timely information, in order to mitigate the potential negative effects of the negotiation process.

The authors would recommend that the hospital hold regular information meetings to apprise employees of the status of the situation, minus details that would compromise negotiations. Even when new information is not available, this type of meeting provides a process through which employees can air their feelings and ask questions, and in a structured manner, release anxiety and receive support.

Concern for the impact on employees is especially great in a hospital setting where patient care must be preserved at all times, while this stressful experience is unfolding. The attendant risks of sabotage make efforts to ease the tensions all the more critical.

FIGURE 1

4-MONTH REEMPLOYMENT RATE

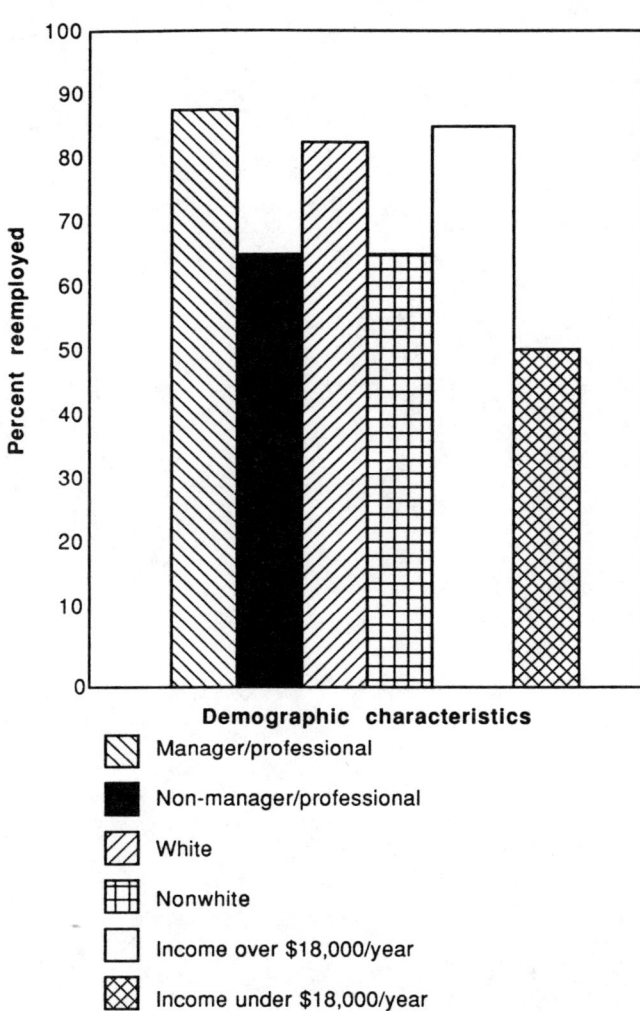

Demographic characteristics
- Manager/professional
- Non-manager/professional
- White
- Nonwhite
- Income over $18,000/year
- Income under $18,000/year

These results also suggest the need for management to make supervisory staff available on all shifts both to serve in an information-giving role and to intervene when emotions run high. A 24-hour management presence may be essential to contain the situation and to avert problems before they manifest themselves. Regularly scheduled management meetings should accompany this presence in order to provide information and help management fulfill supervision responsibilities, which will be considerably more stressful during the merger process. Sensitivity will be needed in preparing management staff to serve as a support resource with nonmanagement employees since both groups are undergoing stress and uncertainty. Personalized memorandums updating employees on the status of the merger would also serve as a mechanism to fill the information void. Coping by all employees would be enhanced by the employer maximizing the amount and frequency of information that is provided.

The related issue of employee representation in the merger process also requires striking a delicate balance. Although employees are key stakeholders in the outcome, technically they are guaranteed no representation, continuation of benefits, or jobs when the process is completed. Clearly, the status of these major issues will depend on the particular situation at hand, financial resources, and the bargaining power of the negotiating institutions. While complete employee representation may not be feasible, it would be appropriate for the hospital to involve the personnel director as negotiations are proceeding and operational details are being worked out. Employees should be advised that their needs and options are being considered as an important element in the process. Where unions are concerned, attorneys should be involved regarding benefits and other legal agreements.

Another major issue to emerge from this study is the need for employers to provide support resources and services during the closing process. These ranged from addressing tangible needs such as job-hunting skills, social services, and benefits availability, to the less tangible need to air feelings and discuss problems.

A final area of attention arising from this study is the need for early identification of the people most likely to become unemployed as a result of the merger. Clearly the situation will differ depending on the local economy and employment outlook; however, special services and concern should generally be directed toward unskilled, nonprofessional employees. This group of employees will often contain a disproportionately high percentage of low-income employees and/or minorities. Social and employment support services should be brought to bear on behalf of such employees as early as possible in order to increase their likelihood of resuming employment.

• • •

In conclusion, this study has highlighted some of the difficulties encountered during the closing of a psychiatric hospital. Through employee self-report, the study documented the need for timely dissemination of information to employees, adequate employee representation in selected aspects of the closing process, and the provision of supportive resources to employees who are likely to be most adversely affected by the closing. In the labor-intensive field of health care, hospital closings are becoming increasingly common; new ways must be sought to preserve the integrity of employees and management within the emotionally ladened context of the closing process.

REFERENCES

1. Williams, J.B., and Feldman, M.L. "Life After Merger: The Human Resource Factor." *Healthcare Forum* 29 (September–October 1986): 33–37.
2. Hernandez, S.R., and Kaluzny, A.D. "Hospital Closure: A Review of Current and Proposed Research." *Health Services Research* 18 (1983): 421–75.

Substance abuse and mandatory drug testing in health care institutions

John Tanner,
Jerry Kinard,
Sam Cappel,
and
Peter Wright

Substance abuse in the workplace is a problem of enormous proportions. Hospitals, like other organizations, must recognize the potential risks posed by substance abuse and effectively deal with the problem. A nationwide study was conducted of hospital personnel managers' attitudes toward drug abuse and mandatory drug testing. The study focuses on the policies hospitals have formulated and reveals hospital administrators' concerns over the legality of mandatory testing.

One of the most controversial issues facing human resource managers today is mandatory drug testing. Advocates of mandatory testing claim that substance abusers have more accidents, turn out inferior products or services, make poor decisions, and are more likely to steal than other workers. Critics, on the other hand, contend that mandatory drug testing is an unwarranted invasion of privacy. They say that tests are not 100% accurate and argue that employers often cannot show a relationship between drug use and work performance.

Statistics show that drug abuse on the job may be a much bigger problem than people think. In one survey conducted by the New Jersey-run Cocaine National Help Line, 75% of the 227 drug users admitted to using illegal drugs on the job. Sixty-four percent said that drugs hindered their performance; 44% said they sold drugs to fellow workers; 18% admitted to having a drug-related accident; and 18% said they had stolen from their employer to support their habit.[1] Since 1975, the number of firms reporting substance abuse problems has increased dramatically. Federal experts now estimate that 10% to 23% of all American workers use dangerous drugs on the job.[2] One report gives the following characteristics of drug users at a dozen major corporations: (1) they are late three times as much as their fellow workers; (2) they ask for early dismissal or other time off 2.2 times as frequently as their fellow workers; (3) they have 2.5 times as many absences totalling eight days or more per year; (4) they are five times more likely to file a worker's compensation claim; and (5) they are involved in accidents 3.5 times more than their fellow workers.[3]

John Tanner, Ph.D., is a Professor of Finance at Western Kentucky University, Bowling Green, Kentucky, and teaches in the production management and management science area.

Jerry Kinard, D.B.A., is a Professor of Management and Chairman of the Management Department at Western Kentucky University and teaches in the personnel and organizational behavior area.

Sam Cappel, M.B.A., is a doctoral candidate in management at Memphis State University, Memphis, Tennessee, and has extensive experience as personnel manager at several hospitals in Louisiana.

Peter Wright, Ph.D., is a Professor of Management at Southeastern Louisiana University, Hammond, Louisiana, and teaches in the business policy and strategy area.

COPING WITH SUBSTANCE ABUSE IN THE WORKPLACE

No industry or socioeconomic stratum is immune to the problem of substance abuse. Many tend to think the problem is relevant only to those in the lower organizational levels or the lower-income classes, but this is certainly not the case. In fact, there has been a 100% increase in the number of top executives coming to treatment facilities over the past 5.7 years.[4] Also, it has been noted that with the big salaries and high pressures associated with Wall Street executives, drug addiction has been a subject of much concern for this influential group.[5] In the nursing profession alone, some studies say that as many as 40,000 nurses are alcoholics, and one out of every three disciplinary cases among nurses is drug related.[6]

Job applicants are typically required to provide a history of drug use on the employment application form, and they may be subjected to urinalysis or some other type of testing as a routine part of the pre-employment screening process. Thus far, the testing of applicants is much more sound legally than the testing of present employees who were hired under an agreement (written or verbal) that did not address drug testing specifically.

Detection of drug use by current employees is usually achieved through searches of employees' lockers and personal belongings and through unannounced testing. Although searches can lead to lawsuits for invasion of privacy, employers can successfully defend their actions by showing that their "reasonable need to know" outweighs an employee's "reasonable right to privacy." Probably the most commonly asked question about drug testing is, "Is it legal?"[7] The Fourth Amendment to the United States Constitution protects persons from search and seizure without probable cause, but this protection applies only to government employees, not to private employees. Currently, there is no federal or state legislation that prohibits drug testing directly, but case law may rely, in some instances, on indirect forms of legislation.

Employers also may require current employees to undergo drug testing if a supervisor suspects they are under the influence of alcohol or drugs, or when employees are involved in work-related accidents or altercations. In addition, the employer may establish a policy of unannounced, mandatory drug testing in an effort to detect drug users.

The detection of illegal substance use by job applicants normally precludes their employment. Dealing with illegal substance use by current employees is not as simple, but it can be done if company policy is designed so that it will stand up in court. Most companies find it advantageous to develop a system of progressive penalties that are applied to habitual drug users, as well as employee assistance programs that provide counseling and guidance to employees who have substance abuse problems. In addition to the counseling function provided by employee assistance programs (EAPs), they can serve as a useful tool in improving employee morale and demonstrating an organization's commitment to its employees.

SUBSTANCE ABUSE IN HEALTH CARE ORGANIZATIONS

Not as much information is available as to the extent of substance abuse in health care facilities as in other organizations, but this situation is apparently changing, as seen from the emerging statistics on nurses, and it will continue to change in the future. Yet, because of the critical nature of health care, and because some health care employees have access to controlled drugs, it is important for hospital administrators to be aware of potential problems in their organizations.

The following information is needed by hospital administrators: extent of drug abuse in health care organizations; drug screening practices initiated by hospitals; attitudes of human resource managers toward substance abuse and its detection; and specific ways hospitals cope with substance abuse problems. The results of a nationwide survey of human resource managers in health care provide this information.

THE STUDY

The purpose of the study discussed here was twofold: (1) to assess the problem of substance abuse in health care facilities throughout the United States, as perceived by hospital human resource managers; and (2) to examine the attitudes of these same hospital human resource managers toward mandatory drug testing of hospital job applicants and employees.

Since human resource/personnel managers are instrumental in promulgating personnel policies, their perceptions and attitudes are reflected in the policies and practices of their employer(s). Policies governing substance abuse are no exception. This study examined the ways health care institutions prefer to deal with substance abuse, in the hopes of providing hospital administrators with more information on which to base decisions in this crucial area.

STUDY METHODOLOGY

A questionnaire comprised of 32 Likert-type statements using a five-point scale (1 = strongly agree, 2 = agree, 3 = neutral, 4 = disagree, 5 = strongly disagree) was developed and the pilot tested on seven hospital human resource managers and three university professors whose primary teaching and research responsibilities are in the area of human resource management. The results of the pilot test were used to modify the instrument to secure clarity.

Thereafter, the instrument was mailed to a random sample of 600 hospital personnel directors throughout the United States. The sample was taken from the roster of the American Society of Hospital Personnel Administration of the American Hospital Association. Certainly there are more hospitals than those listed in this roster, but this listing includes hospitals of all sizes (one bed to more than 1,650 beds, and from one to more than 10,000 full-time employees), and would thus appear to be a relatively representative body from which to sample. One hundred eighty-three usable questionnaires, representing hospitals in 41 states, were returned, yielding a response rate of 30.5%.

While this response rate might appear to be somewhat low on first consideration, the actual number of 183 usable instruments seems to be about average with respect to surveys of this type, and the fact that a good cross-section of hospitals across the country are represented would seem to make the return rate more acceptable. The breakdown of states by region (West Coast, Northeast, Southeast, and Midwest) and the number of respondents from each region are as follows: West Coast, 8 states, 20 hospitals; Northeast, 9 states, 44 hospitals; Southeast, 9 states, 48 hospitals; Midwest, 11 states, 53 hospitals; and Southwest 4 states, 18 hospitals.

The results and conclusions represent the findings of those responding. The researchers considered addressing the problem of nonresponse bias, but because of the delicate nature of the subject matter (controlled substance/alcohol abuse and mandatory drug testing of hospital employees), they decided to keep the respondents absolutely anonymous. To remain true to this anonymity, no procedures were established to check which hospitals returned the questionnaire and which did not, thereby making a second-mail follow-up difficult, if not impossible. Respondents were traceable only by state, as shown above.

FINDINGS

Before examining the responses of human resource managers to the specific statements contained in the questionnaire, a general observation should be made. Only 9% of the employers represented in the survey have mandatory drug screening policies governing job applicants, and only 7% have policies requiring the screening of current employees. Because hospitals are generally regarded as being responsible for the lives and health of others, these figures appear to be alarmingly low. Compared to American industry in general, they are low. According to the United States Drug Enforcement Administration, almost 30% of Fortune 500 companies now screen job applicants or current employees for signs that they use illegal drugs.[8] One positive aspect of these figures, however, is that they may underline the need for hospital administrators to address the problem of substance abuse in their organizations.

Table 1 presents the responses of the survey participants to 11 of the Likert-type statements contained in the questionnaire. These statements examined human resource managers' general philosophy regarding substance abuse throughout the country and pinpointed their attitudes toward drug testing. The statements in all subsequent tables are ranked with respect to their average level of agreement (1 = most agreement and 5 = least agreement) and are presented in that order. Each table includes the mean, standard deviation, and percentage of people responding to each statement at each level of agreement.

Only 9% of the employers represented in the survey have mandatory drug screening policies governing job applicants, and only 7% have policies requiring the screening of current employees.

Respondents agreed most with the statement that employers should not have to wait for an accident to require drug testing and disagreed with statements suggesting limits on drug testing, indicating that they believe a program of preventive maintenance toward drug testing should be followed, that is, testing should be done before, not after, an accident occurs.

The second highest level of agreement was shown toward the statement that substance abuse has reached

TABLE 1

HOSPITAL PERSONNEL ADMINISTRATORS' GENERAL PHILOSOPHY ON DRUG TESTING

			Percentage of responses at each level					
Statement	Mean	Standard deviation	Strongly agree	Agree	Neutral	Disagree	Strongly disagree	No response
1. Employers should not have to wait until an accident happens before requiring drug testing.	2.339	.885	8.7	62.4	18.0	8.2	2.7	
2. Substance abuse has reached epidemic proportions in the general population of this country.	2.457	.975	16.4	41.6	24.6	16.9	0.5	
3. In my opinion, drug testing is not an invasion of privacy.	2.705	1.148	7.6	51.4	14.2	16.4	10.4	
4. Mandatory drug testing of all job applications should be an integral part of the job application process.	3.137	1.321	12.6	25.1	16.4	27.9	18.0	
5. Mandatory drug testing suggests a lack of trust on the part of the employer.	3.208	1.144	7.7	21.3	26.2	32.2	12.6	
6. Drug testing is unfair because it subjects the innocent as well as the guilty to intrusive body searches.	3.227	1.168	9.8	19.1	18.6	41.6	9.8	1.1
7. I believe that legal opinions support mandatory drug testing for both job applicants and employees.	3.297	1.056	3.3	23.5	24.6	36.6	11.5	0.5
8. Mandatory drug testing should be required of job applicants but not employees.	3.568	1.045	4.9	12.6	18.0	49.7	14.8	
9. Drug testing should be strictly voluntary.	3.612	1.042	4.9	10.9	18.6	49.2	16.4	
10. Mandatory drug testing will help dry up the supply of illegal drugs.	3.732	.691	1.1	15.3	66.7	12.0	3.8	1.1
11. Employers should require drug testing only after a work-related accident.	3.858	.903	1.6	9.3	10.9	57.9	20.3	

epidemic proportions in this country. Certainly other experts would concur with this idea. According to one article, "Public and private efforts to eliminate, or even reduce, the problem seem to be impotent."[9] However, when the respondents of this study were asked their perception as to whether mandatory drug testing would help dry up the supply of illegal drugs, most of them (66.7%) remained neutral. This neutrality should not necessarily be mistaken for a defeatist attitude toward drug testing in each of the respondents' hospitals, however. The fact that these hospital personnel administrators perceive the seriousness and pervasiveness of this problem is a positive sign and may be one of the first steps in working toward a resolution within their organizations.

The issue of privacy must be faced by each organization when it contemplates mandatory drug testing. While not completely silent, case law has been relatively silent on the subject, because it has not been adequately tested in the courts yet.[10] But some people think the courts and arbitrators are beginning to say that an employee's civil rights do not extend to endangering the safety and welfare of others.[11] Apparently, the respondents of this study feel somewhat the same way, as indicated by the fact that 67% believe that mandatory drug testing is not an invasion of privacy, even though they are somewhat uncertain as to the legality of testing for both applicants and current employees, as indicated by the level of disagreement with statement 7.

Responses to statement 4 indicate somewhat of a split among these personnel administrators as to whether all job applicants should be tested. They also do not agree that only applicants (and not current employees) should be tested (see statement 8). However, some experts have said that employees have more protection against testing than applicants[12]; thus, these respondents must exercise much care when developing drug testing policies for current employees.

Further evidence that the respondents perceive a need for drug testing policies is indicated by their responses to statements 5 and 6. Almost half did not agree that the implementation of a drug testing policy suggests an employer's lack of trust, and more than 50% disagreed (or strongly disagreed) with the statement that drug testing is not fair because the innocent as well as the guilty are subjected to the tests. This perception may have some basis in fact, as one study showed that 72% of all full-time workers surveyed would volunteer for drug testing.[13]

The five statements contained in Table 2 focus on the respondents' attitudes about the organizational impact of mandatory drug testing programs. The responses to the first two statements provide further evidence that these hospital personnel administrators have what could be interpreted as a positive view of mandatory drug testing for organizational members. A large majority (62.3% and 60.1%, respectively) agreed that testing will enhance safe working conditions and will have a significant impact on drug use in the workplace. This

TABLE 2

HOSPITAL PERSONNEL ADMINISTRATORS' ATTITUDES TOWARD THE ORGANIZATIONAL IMPACT OF DRUG TESTING

	Statement	Mean	Standard deviation	Percentage of responses at each level					
				Strongly agree	Agree	Neutral	Disagree	Strongly disagree	No response
1.	Mandatory drug testing of employees will have a significant impact on drug use in the workplace.	2.415	1.007	15.8	46.5	20.8	14.2	2.7	
2.	Drug testing is primarily a means to ensure a safe work environment.	2.536	1.083	12.0	48.1	22.4	9.3	8.2	
3.	The implementation of a policy requiring mandatory drug testing causes employee morale to decline.	2.929	1.048	8.7	27.9	29.5	29.5	4.4	
4.	As a result of drug testing in American industry, the number of job-related accidents is beginning to decrease.	3.022	.691	1.1	15.3	66.7	12.0	3.8	1.1
5.	A policy of mandatory drug testing will result in fewer job applicants.	3.264	.944	2.7	21.9	25.7	44.8	4.4	

opinion must be held by other executives, because research shows that some of the companies involved in drug testing today range from IBM and Exxon to the U.S. Postal Service and the Department of Defense.[14] It is interesting to note, however, that even though these human resource managers seem to see a need for drug screening, they are not prepared to make a judgment as to whether drug testing has decreased job-related accidents, as evidenced by the large neutral response to statement 4.

Even though the responses to statement 3 of Table 2 revealed a mean that was slightly in the agreement area,

TABLE 3

HOSPITAL PERSONNEL ADMINISTRATORS' ATTITUDES TOWARD THE IMPLEMENTATION OF DRUG TESTING

	Statement	Mean	Standard deviation	Percentage of responses at each level					
				Strongly agree	Agree	Neutral	Disagree	Strongly disagree	No response
1.	Employees should be given a written notice of rules governing drug use on the job.	1.544	.572	48.2	49.7	0.5	1.1	0.5	
2.	Job applicants should be required to sign a consent form prior to a urinalysis or blood test.	1.989	.858	26.8	56.3	9.8	5.5	1.6	
3.	Employees should be screened for substance abuse only if there is reasonable cause to suspect drug use.	2.276	1.096	24.6	45.4	7.7	19.7	1.6	1.0
4.	A system of progressive penalties should apply to substance abuse cases, just as it applies to other violations of workplace rules.	2.434	1.088	14.8	53.0	8.2	20.2	3.8	
5.	Employers should have the right to conduct workplace searches for illegal drugs, but this right should not extend to personal searches of employees.	2.575	1.065	7.7	56.8	10.4	18.0	6.0	1.1
6.	EAPs for employees who have substance abuse problems should be voluntary.	2.776	1.222	13.0	42.1	5.5	32.8	6.6	
7.	Penalties for drug use on the job should range from moderate to severe, depending on the substance.	2.809	1.219	11.4	43.2	16.0	31.7	7.7	
8.	Employers should develop standards for acceptable alcohol and drug levels.	3.077	1.281	8.7	32.8	18.6	20.8	18.6	0.5
9.	If mandatory drug testing is required for employees, the actual testing should be unannounced.	3.087	1.388	12.6	32.8	9.3	24.0	21.3	
10.	Penalties should be imposed against employees who use drugs only if the employer can show a relationship between drug use and adverse job behavior.	3.220	1.206	6.6	30.8	9.8	39.3	13.0	0.5
11.	Termination is the best solution to problems involving substance abuse.	4.082	.937	3.3	3.8	8.7	49.8	34.4	

the respondents were relatively evenly dispersed in their opinions as to whether drug testing policies would cause employee morale to decline. Ultimately, their opinions would probably depend on the formulation, wording, and structuring of such policies. Perhaps the responses to this statement and statement 10 of Table 1 indicate that these administrators realize there is no simple solution to the problem.

Still more evidence that the respondents support mandatory drug testing is indicated in statement 5 of Table 2. Almost half of the respondents disagreed that mandatory drug testing would have an adverse impact on the number of job applicants, while fewer than one-fourth showed any level of agreement.

Table 3 contains 11 statements related to the respondents' attitudes toward the implementation of drug testing and drug testing programs. The highest level of agreement (97.9% either agreeing or strongly agreeing) was exhibited with the statement that a written notice of rules pertaining to drug use on the job is advisable; this plan of putting company policies on drug use and abuse in writing, in the context of concern for job safety, is strongly recommended and is used by those Fortune 500 companies with drug abuse policies.[15]

Along these same lines, 83.1% of these hospital personnel administrators agreed, at some level, that consent forms should be signed prior to any kind of drug test. Even though the most common and least expensive type is the urinalysis, it cannot always distinguish between cough syrups, decongestants, pain killers, and illegal drugs.[16] Also, since the Centers for Disease Control has stated that these tests can be inaccurate as much as 66% of the time,[17] the signing of a consent form is advisable. Since human resource managers are often responsible for some degree of representation in litigation related to such matters, it is understandable that they would strongly support heavy documentation before problems arise.

Seventy percent of the respondents showed some level of agreement with the statement that reasonable cause should precede screening for drugs (statement 3), and almost 65% agreed that searches should not include personal searches of employees (statement 5). These perceptions appear to be in line with those relating to documentation. Reasonable cause is important, as evidenced in one case where the judge said a company must have a "reasonable suspicion based on specific, objective facts" before demanding that a worker be tested.[18] Interestingly, however, more than half of the respondents did not agree that there is a need to show a relationship between drug use and work performance before imposing penalties (statement 10), a perception that may not be shared by some experts.[19]

Respondents were almost equally divided on the issue of unannounced testing, although they disagreed that the testing should be strictly voluntary. Thus, it appears that these human resource managers favor mandatory testing but would not favor unannounced testing, again supporting the ideas of advanced notification and documentation. The respondents were also about equally divided in their levels of agreement and disagreement with respect to whether employers should be responsible for developing standards for acceptable alcohol and drug levels. Perhaps those who disagreed thought their responsibilities were more in the area of sound policy development after the acceptable levels have been exceeded.

Most of the respondents (67.8%) favored a system of progressive penalties for violations of workplace rules governing substance abuse (statement 4), and more than half (54.6%) agreed these penalties should range from moderate to severe, depending on the substance. Almost 85% disagreed that termination is the best solution to drug abuse problems. This opinion is shared by

A majority of the respondents agreed that a higher standard of accountability should be in effect for those having responsibility for the lives of others.

many experts in the field; in fact, one says that employees should not be terminated on the basis of one positive test, and companies with good EAPs should follow up with a more sophisticated test.[20]

With respect to EAPs, more than 55% of these hospital personnel administrators thought such programs should be voluntary. This opinion appears to be in line with current practice, according to one expert who said that about half of all such programs were initiated by employees, and the other half resulted from supervisory action.[21] EAPs are usually worthwhile if an organization can afford them, as it has been said that the recovery rate is between 35% and 60% of treated employees, and a good EAP returns five dollars for each dollar invested.[22]

Five statements included in the questionnaire dealt specifically with hospital personnel administrators'

TABLE 4

HOSPITAL PERSONNEL ADMINISTRATORS' ATTITUDES TOWARD SPECIFIC ISSUES PERTINENT TO DRUG TESTING IN HOSPITALS

				Percentage of responses at each level				
Statement	Mean	Standard deviation	Strongly agree	Agree	Neutral	Disagree	Strongly disagree	No response
1. People who hold the lives of others in their hands should be held to a stricter standard than other workers.	2.230	.979	19.7	55.2	9.3	14.2	1.6	
2. Mandatory drug screening in hospitals is more critical than testing in other organizations because of the basic nature of some hospital jobs.	2.769	1.167	12.0	38.8	15.3	26.8	6.6	0.5
3. Substance abuse among employees at my hospital is limited to only a few individuals.	2.834	1.052	9.8	29.0	33.3	21.3	5.5	1.1
4. Mandatory drug testing will yield improved health care to hospital patients.	3.093	.967	4.4	22.4	38.8	27.3	6.6	0.5
5. Only those hospital employees (or applicants) directly involved in patient care should be subjected to mandatory drug testing.	3.127	1.300	14.8	20.2	14.8	36.1	13.1	1.0

attitudes toward mandatory drug testing of hospital employees and job applicants. As shown in Table 4, a majority of the respondents agreed that a higher standard of accountability should be in effect for those having responsibility for the lives of others; this statement had the highest level of agreement. Likewise, most respondents (more than 50% agreed that mandatory drug screening in hospitals is more critical than in other types of organizations.

As shown in statement 3, the respondents were about evenly divided as to their opinions about the number of abusers in their own organizations, with the largest percentage occurring in the "neutral" response. Apparently, this uncertainty or denial that a problem may exist is not unusual, as one study stated that many hospital administrators seem to think their hospital has no problem.[23] Perhaps this attitude explains the 30.5% response rate for this survey. Some of the hospital personnel administrators might not have responded because they were unaware that problems existed, or they were unwilling to admit the existence of problems and thus felt no need to respond. In fact, should one wish to carry this line of thought one step further, some might not have responded because they did have drugs or alcohol abuse problems.

More of the respondents disagreed than agreed (33.9%, as opposed to 26.8%, respectively) that mandatory drug testing programs would lead to improved patient health care, but most were neutral with respect to this statement (38.8%). Almost half of these human resource managers (49.2%) showed some level of disagreement with the statement that only those employees directly involved in patient care should be tested. This opinion again suggests that they believe drug abuse is a problem that permeates the entire organization, but they have mixed feelings about the solution to this problem.

● ● ●

Drug and alcohol abuse in the workplace is a problem of enormous proportions and can be blamed, in many organizations, for accidents, poor performance, employee theft, and other work-related problems. Unfortunately, this problem appears to be getting worse instead of better. The number of major companies with some type of drug screening program is somewhat low, and the results of this study show that for hospitals, the figures are even lower.

However, the hospital personnel administrators responding to this survey do show some positive attitudes. They realize the seriousness and the magnitude of substance abuse in this country, and they show some positive attitudes with respect to the initiation of drug detection policies and procedures for their organizations. It is possible that they have taken somewhat of a "wait and see" attitude toward this implementation, due to the hazy legal decisions reached thus far. But they favor a nonvoluntary drug testing program, and they believe that such a program can be effected without a decline in employee morale or a loss of job applicants. Apparently they realize that as employees, they do not have to stand by while time, money, and productivity are lost.

These respondents also favor workplace searches for illegal drugs, but not personal searches. They are also strongly in favor of good written documentation for all employees. This interest is some evidence of concern about legality, but it also could be evidence of an underlying desire for fairness, especially when combined with their attitudes toward penalties for abusers. They favor a progressive penalty system ranging from moderate to severe and do not see termination as the only solution. Also, they appear to have a favorable attitude toward voluntary EAPs.

Overall, these human resource managers believe that because of the nature of health care institutions, these organizations operate under a higher standard of accountability than other organizations. Therefore, mandatory drug testing policies are more critical, even though the managers are uncertain as to whether they should develop acceptable drug and alcohol standards or whether mandatory drug testing will yield better patient care. Still, if sound policies are determined as to how to detect, penalize, and assist drug users and abusers, new and current employees will have a clear picture of rules. Certainly these respondents seem to understand that if an organization condones (even secretly) its employees' drug problems, these employees will be less willing to get treatment, and the company will remain in an impaired state.

Since the respondents' attitudes appear to be right for the implementation of drug testing policies, they should make sure their fellow executives and supervisors at all organizational levels are trained to identify drug use symptoms. Upon such identification, they should confront the abuser instead of covering up for him or her. And, upon detection, the executives and supervisors should be aware that there are other solutions besides humiliation and professional ostracism and termination. Employers' attitudes should be positive and helpful toward errant employees; based on the findings in this survey, such attitudes currently appear to exist. Therefore, this may be the time for the implementation of drug testing programs in these employers' health care institutions.

REFERENCES

1. O'Boyle, T.F. "More Firms Require Employee Drug Tests." *Wall Street Journal*, 8 August 1985, p. 6.
2. Castro, J. "Battling the Enemy Within." *Time*, 17 March 1986, p. 53.
3. Hoffer, W. "Business' War on Drugs." *Nation's Business* 74 (October 1986): 19.
4. Flax, S. "The Executive Addict." *Fortune*, 24 June 1985, pp. 24–31.
5. "War on Drugs." *Fortune*, 29 September 1986, p. 8.
6. Lachman, V.D. "Why We Must Take Care of Our Own." *Nursing* 16 (April 1986): 41.
7. Anagarola, R.T. "Drug Testing in the Workplace: Is It Legal?" *Personnel Administrator* 30 (September 1985): 79.
8. O'Boyle, "More Firms Require Employee Drug Tests."
9. Muczyk, J.P., and Heshizer, B.P. "Managing in an Era of Substance Abuse." *Personnel Administrator* 39 (August 1986): 91.
10. Ibid., 91–96.
11. Hoffer, "Business' War on Drugs," 24.
12. Schein, D.D. "The Work Environment: How to Prepare a Company on Substance Abuse Control." *Personnel Journal* 65 (July 1986): 30–38.
13. "War on Drugs."
14. "Difficulties Cited in Combating Employee Drug Abuse." *Journal of Accountancy* 162 (October 1986): 52.
15. "Fortune 500 Firms Use Urinalysis Tests to Stem Employee Drug Abuse." *Administrative Management* 47 (January 1986): 12.
16. Hoerr, J. "The Drug War Will Be Won with Treatment, Not Tests." *Business Week*, 13 October 1986, p. 52.

17. "Worker Drug Tests Spread Despite Deep Controversy." *Money*, October 1985, p. 13.
18. Muczyk and Heshizer, "Managing in an Era of Substance Abuse."
19. Anagarola, "Drug Testing in the Workplace," 88.
20. Hoerr, "The Drug War Will Be Won with Treatment."
21. Hoffer, "Business' War on Drugs," 18–26.
22. Hoerr, "The Drug War Will be Won with Treatment."
23. Cherskov, M. "Substance Abuse in the Workplace." *Hospitals* 1 (June 20, 1987): 72.

Downsizing: How one hospital responded to decreasing demand

Anne D. Mullaney

However difficult, organizational downsizing will be a reality for many hospitals throughout the next decade. The experience of one hospital system is described in this case study for the purpose of sharing the lessons learned during the process.

It is an understatement to characterize the environment in which today's health care manager functions as "challenging." For a myriad of reasons that are far beyond the scope of this article, overall utilization of a large percentage of U.S. hospitals has declined steadily over the last decade. This phenomenon has caused tremendous pressure for them to become more efficient and more competitive. Management must pay strict attention to each cost incurred by the institution. Because hospitals are so labor-intensive, with a large percentage of the overall budget going to pay salaries, it has become absolutely critical to pay close attention to staffing patterns, employee productivity, and overall operational efficiency in order to ensure that staffing is commensurate with the utilization and financial profile of the operation.

Traditionally, health care institutions have been very stable employers, albeit not the most remunerative. Those who have chosen to work in the health care industry have often seen stability as an acceptable tradeoff for higher salaries. That employment stability is being called into question. Because of declining utilization, hospitals no longer require the intensity of staffing in many areas of their operation that they once did. Formal staff reductions have already been necessary in many hospitals and hospital systems across the country, and many more institutions are grappling with the issue at present.[1]

One such system is the Ohio Valley Health Services and Education Corporation (OVHS&E), a three-hospital system with two hospitals in northern West Virginia and one in eastern Ohio. It has a total of approximately 1,000 licensed beds and operates in a region with a large elderly population, high unemployment, and significant outmigration.

Over the past three years, OVHS&E has been forced to examine its operation, its place in the market, and its need for personnel relative to the current and projected future demand for its services. This analysis, coupled with the severe economic pressures existing in the geographic region, led to the conclusion that a serious reduction in force was mandatory if the institutions were to survive and continue to operate in the market. This article discusses the process by which OVHS&E reduced its work force from approximately 2,600

Anne D. Mullaney, *M.H.A., J.D., is currently an Associate Attorney with the law firm of Nash & Company, P.C., Pittsburgh, Pennsylvania. She formerly served as Administrator of Peterson Hospital, one of the member hospitals of OVHS&E.*

employees to approximately 2,000. The description of one experience may be of use to other hospitals contemplating a similar reduction.

Before any details of the OVHS&E experience are presented, it must be unequivocally stated that no matter how carefully a reduction in the work force is implemented and no matter how effective communication between and among affected parties may be, layoffs are a traumatic occurrence for a hospital and its employees. The mission of a hospital is to deliver quality care to its patients. If it is to achieve this mission, it needs some level of positive employee morale and professional satisfaction. Downsizing undermines positive morale and creates fear and dissension.[2] Although a reduction may be indisputably necessary for the continued health of the institution, it is not possible for the process to be painless. The institution must be sure that it has the collective resolve to accomplish what needs to be accomplished with the least amount of damage. A good planning process, effective communication, and a comprehensive plan that avoids a piecemeal response will go a long way toward preventing irreparable wounds.

ASSESSING THE PROBLEM

Snapshot of the organization

Before making any attempt to reduce the size of the work force, management must have the most accurate understanding possible of the current and projected financial and operational state of the institution. The services of an outside consulting group may be required to provide a reliable and independent assessment of the hospital's current status as well as its short- and mid-term prospects.

This assessment will be an invaluable tool in management's struggle to adjust its human resources to its service utilization over time. Strategic planning decisions, including staffing decisions, can and should be made proactively as the hospital responds to a changing environment. The literature suggests, however, that administrators tend to deny the need to retrench until the evidence of that need is overwhelming.[3]

Such was the case of OVHS&E. Unfortunately, decisions were delayed in the hope that conditions would improve. A valid criticism of management's response to its situation is that it waited too long to take definitive action. The development of productivity standards and a plan to reduce the work force should not be implemented simultaneously if at all possible.[4] Ideally hospitals should assess their staffing needs periodically, so that when they find themselves in a crisis situation, they can respond appropriately. However, given today's environment, this is not always feasible.

With the "operational snapshot" of the institution complete, plans to deal with the financial condition of the hospital can be formulated. If expense reductions include elimination of personnel, management must be careful not to reduce staff so drastically that patient care, and thus the reputation and future of the hospital, are jeopardized. Reaching a balance between the goals of financial stability and quality of care is often difficult.

Prelayoff reductions

Before implementing a full reduction-in-force program, OVHS&E pursued several less drastic forms of employee reductions—namely, a system-wide hiring freeze, a delay in salary increases, and a voluntary early retirement program. Much was gained from such efforts. One cost-saving measure used during the three summer months prior to the layoffs was a mandatory time-off program. All employees' time, including management's, was reduced by 10%; nine days were worked and paid for out of the ten-day work period. In addition, department managers were asked to be more diligent than usual in reducing hours worked as utilization waned. Those departments that were not directly utilization dependent, such as housekeeping and maintenance, were also asked to reduce as much as possible, even though it was more difficult to assign staffing on the basis of patient load.

The array of prelayoff reduction methods can be effective for several reasons. First, the urgency of the hospital's or system's financial situation is made real to the employee work force in a very personal way that may lead to an emotional preparation for the reduction,

---

Ideally hospitals should assess their staffing needs periodically, so that when they find themselves in a crisis situation, they can respond appropriately.

---

should one be concluded to be necessary. Second, although no one relishes a smaller paycheck, management can, and in the case of OVHS&E did, gain some degree of credibility and sympathy among employees by pursuing milder options before considering the most radical, mass layoffs. Third, and most obvious, such efforts can save a significant amount of money.

An interesting tension surfaced at OVHS&E during the prelayoff period. Initially most employees supported the concept of "shared suffering"; that is, everyone tolerated a 10% pay cut in order to prevent the loss of jobs by some. Over time, however, the tolerance of the majority lessened, and the sentiment surfaced that if permanent staff reductions were necessary, they should be made, and the hospital should move on. Although management had already concluded that such a measure was inevitable, it was helpful to have the work force reach the same conclusion.

DESIGNING THE PLAN

Commitment

The critical elements of any downsizing effort are (1) the development of a logical and workable plan of action with appropriate accountability, (2) good communication among the key parties, and (3) a commitment to the goal of fiscal responsibility. It must be stressed again that by the time the organization has reached the point of developing a reduction-in-force plan, all must share the resolve necessary to carry out the difficult tasks ahead. The board of trustees must understand the seriousness of the situation and be willing to take the criticism from the community that will surely ensue. Management, on both upper and middle (i.e., department head) levels, must be willing to work hard to achieve a balance between financial responsibility and quality of care. Physicians must be educated early in the process regarding the need for reductions and be included in the reduction process as appropriate.

Although the employee reduction process will certainly affect the entire hospital community, it is imperative that the approach taken be flexible enough to respond to the staffing challenge on a department-by-department or service-by-service basis. A strategy that calls for a 15% across-the-board reduction in staff is doomed to failure, because efficient departments are penalized while inefficient departments are not forced to correct deficiencies. In addition, such an approach slights departments or services that show increasing utilization, have growth potential, or contribute in a special way to the overall hospital image.[5] An across-the-board reduction is thus to be avoided.

Personnel management committee

Critical to a downsizing effort is the functioning of a highly visible steering committee that is responsible for developing and implementing the overall reduction plan. The committee should be composed of senior management and should represent all major areas of operation within the institution or system. The future direction of the hospital, the possible elimination of services or departments, acceptable quality thresholds, the institution's place in the community, and the role of educational programs are but a few of the fundamental philosophical issues that this group will have to debate as it tackles the extremely difficult task of reducing the work force. Although there will never be total agreement on these and other basic points, the discussion and exchange among the top managers will at least encourage each to see the other's position. Operations people must be made sensitive to the concerns of finance, and financial people to operational and quality issues. This kind of pressurized values analysis is what makes hospital staff reductions so much more difficult than staff reductions in other industries.

Critical to a downsizing effort is the functioning of a highly visible steering committee that is responsible for developing and implementing the overall reduction plan.

OVHS&E established such a steering committee, calling it the Personnel Management Committee (PMC). Its members were the chief executive officer, the chief operating officers of all three hospitals, and the system's medical director, vice presidents for human resources and fiscal affairs, and directors of nursing. The vice president for marketing was also included from time to time so that he would be better able to respond to media inquiries, confidentiality being an important consideration throughout. This group became the repository of all pertinent information, the ultimate decision makers, and the catalyst for implementing changes. Although it may appear that the group was too large to be efficient, it actually functioned quite well. Without such broad representation, important perspectives would have been lost, communications hindered, and regrettable mistakes more easily made.

Perhaps the first major decision that a PMC needs to make is how many positions must be deleted from the overall operation. It must calculate a target staffing goal that, if achieved, will ensure financial stability and an acceptable standard of care. OVHS&E set as its overall target 3.25 full time equivalents (FTEs) per occupied bed.

Measuring productivity

Two of the three OVHS&E hospitals lacked a system to measure employee productivity or to set clearly defined standards for employee performance; thus it was difficult to decide which areas of the operation should in fact be cut.

Department managers had no tools with which to measure the productivity of employees, compare that measurement against a standard, and then staff according to utilization requirements. If the system was to downsize in an effective, well-reasoned manner, and if future staffing was to be made a function of utilization, such tools had to be available to middle management. Unfortunately, developing such standards while planning for staff reductions engendered suspicion among employees.

The PMC commissioned a private consultant to help the hospitals establish a baseline staffing measure and appropriate productivity standards for use by department managers. Clearly this process demanded considerable education of and support for the department managers, since it was they who would be responsible for developing and utilizing this sophisticated management tool, which was to be customized for each department. It was made clear to the department managers that the standard developed would be not only an ongoing tool to assist them in staffing their departments but also the efficiency measure against which each manager would be judged.

Although it is beyond the scope of this article to describe in meticulous detail how productivity standards are developed, a brief discussion of the process employed by OVHS&E is appropriate. The germane factor is that department managers must be involved and, in fact, must spearhead the development of such standards within their departments. Staff-wide support of the productivity concept will be easier to achieve if front-line managers are included from the start.

With the help of the outside consultant and often of department members, managers of each department or service developed productivity/efficiency standards upon which staffing patterns would later be built. The instructions given to the managers were as follows:

1. List all tasks performed by department members, task times, and volumes, as appropriate.
2. List the tasks that could be eliminated and describe how they could be eliminated.
3. List the programs that could be eliminated and the reasons why.
4. Define the amount of improvement (reduction) in hours and FTEs that could be achieved.
5. Define the ways in which such improvements could be realized.
6. Define the productivity index that might be appropriate for the department.
7. List any barriers to change, including personnel policies, physician preferences, employee issues, work methods, and computer support arrangements.

The information and recommendations that were generated through this process allowed management to compare OVHS&E's productivity and staffing patterns with industry norms. These norms were obtained from such sources as Monitrend, standards generated by various professional organizations, and statistics provided by the consulting firm from other hospitals with which it was associated.

Industry norms can be useful; however, productivity within a department is dependent upon many factors that are not easily standardized.[6] Therefore, statistics should be used for comparison only, and hospital managers should scrutinize closely the particulars of their own environment.

A brief example of how a department manager established productivity standards and then used such standards to make staffing decisions may be illustrative. The department head of the physical therapy department kept track of all tasks performed within the department for a two-week period, how often they were performed, and how long it took to perform them. She then analyzed the data, eliminated those tasks that could be eliminated, and scrutinized the work flow for improved efficiency. Knowledge of the volume of work within the department and the time it took to perform it permitted staffing (and consequently reduction)

Industry norms can be useful; however, productivity within a department is dependent upon many factors that are not easily standardized.

decisions to be based upon empirical information and thus better controlled.

One caveat for those who may wish to apply the process set forth in this article to their own situations is this: Having requested the input of department managers, be sure to respect it.[7] It is easier for department managers to accept unilateral instructions from those to whom they report than to be told that their opinions and perspectives are valued, only to have them disregarded. Much confidence must be placed in middle management if this cooperative approach is to be taken; however, if it is taken, and middle management responds in a professional and competent manner, adjustment after the downsizing will be easier due to the mutual respect that is gained during the process.

Recommendations for reductions

The major focus of the department manager in the course of downsizing is to examine present staffing patterns and corresponding work volume in order to determine which jobs can be eliminated and which done more efficiently. Although the natural tendency is to justify every task and every hour worked, enough pressure and professional accountability can be built into the system to overcome this tendency. In the case of OVHS&E, middle management was given approximately six weeks to develop productivity standards and to recommend to senior management the number of positions that could be eliminated. This was the most serious flaw in the process. It would have been far better to have had accurate productivity measures in place long before any downsizing occurred. Nevertheless, time in this particular case was of the essence.

Creativity and flexibility should be encouraged.[8] Productivity standards can help to identify areas of operation that are less efficient or overstaffed. Well-reasoned and defensible reduction decisions can be based on empirical data on the utilization and operation of the particular institution. Areas can be, and in the case of OVHS&E were, examined individually, thus permitting specific reductions to be made in areas where operational adjustments were warranted.

On more than a few occasions, attempts to eliminate positions resulted in recommendations requiring major operational adjustments—for example, requests for other departments to pick up additional work, changes in scheduling that sometimes required the reduction of hours, and the elimination of services altogether. Physician practice patterns and the manner in which physicians ordered services were often found to impede departmental efficiency. Obviously department managers could not unilaterally make the operational adjustments they believed to be necessary with no regard for physician reaction, and so such dilemmas were passed on to the PMC for resolution.

As has been noted, the PMC served as a repository of information. It collected and reviewed the middle management recommendations for reductions. Often the reductions recommended by the department managers were not dramatic enough to satisfy the PMC, and it pressured the managers to find ways to become still more efficient. This process continued until a sufficient number of reductions were identified.

The PMC resolved as many of the dilemmas referred to it as was possible, primarily those pertaining to "hospital-oriented" issues. For obvious reasons it was neither willing nor able to tell physicians how to practice medicine. It did need the assistance of the physicians, however, to achieve the necessary reductions. With the assistance of the medical director, who was a member of the PMC, a physician task force was established to review those areas in which physician use of services caused staffing problems and those in which the elimination of services and scheduling changes would affect a physician's ability to meet the needs of patients. The task force's recommendations were made to the PMC.

Reduction-in-force plan

As middle management worked on the tasks just described, senior management struggled with the issues of seniority, bumping between departments, exclusions to the reduction-in-force plan, layoff pay, recall procedures, continuation of benefits, and full-time vs. part-time employees. When resolving such issues, management must be sensitive to legal concerns as well as to the unique environment in which the hospital functions. This may also be an opportunity to reemphasize to department heads and supervisors the importance of job evaluations. If thorough and dependable job evaluations are in the employee files, reductions can more easily be justified on the basis of performance. Legal review of the completed policy agreed to by senior management should be sought, and the board of trustees should be asked to approve the parameters of the new policy as well.

Each institute must examine the specifics of its own situation as it develops a reduction-in-force policy, for such a policy will in fact be a reflection of its own organizational philosophy. It must keep one eye on quality of care and the other on fairness to employees. The hospital must examine itself, its position on employee-related issues, the legal ramifications of its various options, and, above all, how the decisions will affect the patient care to be rendered by the downsized work force.

Another caveat must be articulated at this point: It is absolutely crucial that as the PMC finalizes the number of employee reductions to be made, cuts be made deeply enough so that this traumatic process must be endured only once. The months and years after downsizing must be devoted to rebuilding. If the specter of additional layoffs hovers over the heads of the work force, such rebuilding cannot take place. It is better to make cuts that are too deep and rehire personnel later than to experience a second layoff.

IMPLEMENTING THE PLAN

With the preparations all complete, the truly hard work begins. The target number of FTEs per department has been identified, and now names and faces must be assigned to each reduction. Faced with the highly emotional task of laying off people with whom they have worked, managers may experience a collective sense of grief. Having to dismantle programs that they themselves helped to build is a most difficult assignment. Thus emotional hardship will be experienced not only by those who are laid off but also by those who remain employees of the institution.[9]

The most important element of the implementation phase of the reduction-in-force plan is to develop a timetable and to communicate it effectively. Once the design of the reduction-in-force policy is complete and the final determination about numbers made, action should be swift.

It is up to the PMC to design the timetable and to set in motion the reduction-in-force policy. If the board of trustees has been kept informed throughout, there should be no unexpected delay at this point. The trustees must be officially notified of all final decisions and the agreed-upon time frame.

Although compliance with the schedule is important at this stage of the process, care must also be taken to ensure effective communication. Department managers should be apprised collectively of the details of the reduction plan and given an opportunity to ask questions and demand justification. As the people who will be on the front line of implementation, they must be comfortable with the final decisions. They must also be able to answer the questions that employees will surely pose and to support their answers to at least some extent. Once again, mutual respect between senior and middle management is important. Withholding information serves no useful purpose.

Once the board of trustees and department managers have been informed, the actual layoff process can commence. Management must work closely with the personnel department to ensure that the correct employees are selected for termination. Dates of hire and job classifications must be confirmed in order to be certain that decisions involving seniority and bumping are properly implemented.

The process of employee notification that OVHS&E used was quite straightforward. The personnel department prepared packets of information for department managers to distribute to employees who were targeted for layoffs. Included was information on benefits, layoff pay, and outplacement services. Department managers were coached on interpersonal techniques and the need to respect the emotions and dignity of the dismissed employees. They were required to notify personally all those affected by the reduction within 48 hours of the department managers' meeting. Management made every effort to inform employees of the reductions personally before they were reported by the media, but the effort was not entirely successful due to the media's interest and some premature leaks.

The issue of severance pay can be a difficult one. Financial constraints may prohibit generous separation pay, yet senior management may have a strong feeling of obligation to employees. This feeling of obligation,

---

Employees whose positions were eliminated were given considerable assistance with locating other employment.

---

coupled with the realization that some of the operational adjustments occasioned by the downsizing might well require a period of transition, led management at OVHS&E to create a "transfer pool" for one month. All terminated employees were given one month's pay, with the stipulation that they could be called in during that month to work in any department with staffing deficiencies. Fortunately, no one from the transfer pool was ever called upon to work; the one month's salary was, in effect, severance pay. Thus, although employees had not been forewarned of their termination, department managers were able to offer a month's salary to cushion the blow. The lack of warning could be considered to be harsh, but management concluded that for employees to know that they were "short termers" could only be disruptive.

One aspect of the OVHS&E reduction-in-force plan that was particularly well received was the outplacement service made available through the personnel department at each hospital to those who lost their jobs. Employees whose positions were eliminated were given considerable assistance with locating other employment. Job recruiters were on site to interview certain classifications of employees, particularly professional and technical people. Assistance was given with résumé writing, and free long-distance telephone lines were made available to those looking for work outside the area. In addition, counseling services were offered to those who expressed an interest.

This effort helped a sizable number of employees find new jobs. An unexpected side effect of the effort was that it also helped the hospitals. It lessened the trauma and guilt experienced by management, improved the community reaction to the layoff, and lessened the hospitals' unemployment compensation expenses. Any hospital proceeding with a large-scale staff reduction should consider such a program.[10]

POSTLAYOFF PERIOD

Once the positions targeted for elimination have in fact been eliminated, the process of rebuilding can commence. The rebuilding will take two forms—an actual restructuring of the workload in order to accomplish the same amount of work with fewer people, and coping with the emotional and morale dimensions of downsizing.

When the size of the work force is reduced by any significant percentage, it is obvious that major operational adjustments will have to be made. Department managers and the senior management staff should have planned for the necessary changes prior to the layoff date and be ready to implement the adjustments immediately. What works on paper, however, is not always practical. Thus aggressive hands-on management is critical at this stage to ensure that departments find valid ways to cope with the new staffing patterns.

At OVHS&E, those departments that found themselves in unanticipated distress (perhaps staffing cuts were too deep or volume unexpectedly increased) could petition the PMC for additional personnel. The decisions of the PMC rested primarily on an analysis of the department's productivity standards vis-à-vis its utilization. If the numbers indicated a need for additional staff, the request was usually granted. However, more often than not the department manager was instructed to find a way to make do without an increase in hours worked. As time went on, department managers were given more and more authority to manipulate their staffing patterns, as long as they operated within the internally generated productivity parameters.

Although it cannot be correctly categorized as a function of rebuilding, a second major concern relative to operational adjustments has to do with those services that require more fundamental changes than the simple redistribution of work. As was noted earlier in this article, some of OVHS&E's overstaffing resulted from physician practice patterns and some from interdepartmental relationships.

Effecting change in such situations required input from physicians, and compromises had to be reached among departments as to the distribution of tasks, authority, and responsibilities. This restructuring took time and required diligent oversight by the PMC. Overall, it had a positive effect. It forced departments to talk with one another and made each realize that it was not the only one suffering from the downsizing. It also drew physicians into the process and made them aware of the various exigencies of the situation. Through their participation they could see first-hand that quality of care issues were being taken seriously and could thus feel comfortable with referring their patients to the institution. The importance of physician involvement cannot be overstated.

The first item of the weekly postlayoff PMC meeting was a report by each member of the committee on actions taken to make the necessary operational adjustments and to achieve the prescribed efficiencies. Timetables were developed to ensure that target dates were established and that accountability was built into the new system. A significant percentage of the staff reductions had been contingent upon these adjustments. Without the necessary postlayoff diligence and follow-up, the financial objectives that were set early in the process would never have been realized.

The need for good communication and for high visibility on the part of management remains even after layoffs have actually occurred. Information as to what has been accomplished and what still needs to be done is crucial. Department managers should be kept informed of the status of the downsizing efforts and should pass the information along to the people they supervise. Management on both senior and departmental levels must be highly visible for two reasons: First, to quell the notion that the institution is falling apart and that no one is in charge; second, to show by their physical presence that management is concerned and is sharing the burden. Such visibility will also give employees an opportunity to vent their frustrations and their concerns.

• • •

As utilization of traditional hospital services continues to decrease, more and more hospitals will face the necessity of reducing their work force. The purpose of this article was to share one hospital system's experience with downsizing in the hope of guiding other institutions that are contemplating a similar effort.

Although downsizing is a difficult task, it is nonetheless critical to the long-range health of the organization. Only if hospitals are diligent in matching staffing to utilization will they be able to survive the pressures of the current health care environment and continue to provide their services.

OVHS&E management learned during the process that downsizing requires the application of sound techniques and that effective management can preclude much long-term organizational trauma. Of utmost importance is that management define clear goals, formulate a comprehensive plan to achieve the goals, communicate effectively both the goals and the process for achieving them, and that management be prepared to deal with both anticipated and unanticipated consequences of its action. A piecemeal approach or an approach that is not well thought out, communicated, and executed will hinder not only a hospital's attempt to downsize but also the rebuilding process that must follow.

REFERENCES

1. Kazemek, E.A., and Channon, B. "Avoiding the Trauma of Organizational Downsizing." *Healthcare Financial Management* 42, no. 5 (1988): 40–48.
2. Fottler, M.D., and Schuler, D.W. "Reducing the Economic and Human Costs of Layoffs." *Business Horizons* 37, no. 4 (1984): 9–15.
3. Fottler, M.D., Smith, H.L., and Muller, H.L. "Retrenchment in Health Care Organizations: Theory and Practice." *Hospital and Health Services Administration* 31, no. 5 (1986); 29–43.
4. Suver, J.D., and Neuman, B.R. "Resource Management by Health Care Providers." *Hospital and Health Services Administration* 31, no. 5 (1986): 44–54.
5. Kazemik and Channon, "Avoiding the Trauma."
6. Ibid.
7. Suver and Neuman, "Resource Management."
8. Fottler and Schuler, "Reducing the Economic and Human Costs."
9. Fottler, Smith, and Muller, "Retrenchment in Health Care Organizations."
10. Ibid.

Role stress in hospital executives and nursing executives

George C. Burke, III
and
Cynthia C. Scalzi

Two recent studies report for the first time on role stress in hospital executives—those in general administration and those in top nursing positions. Analysis reveals some similarities and differences, which provide insight into the difficult jobs of these executives.

The literature predicts role stress among hospital administrators—those in general administrative positions as well as those in top nursing positions. Peter Drucker once claimed that hospitals are among the most complex organizations to manage.[1] White and Wisdom stated, "The management of interpersonal relations, the importance of the services offered, and ever-changing technology in the hospital all contribute to a highly stressful environment."[2] In 1974 Leininger described the problems facing nurse administrators as extremely complex, diverse, and weighty.[3] Hospital executives must make management decisions within an extraordinarily complex and stressful environment. Yet until recently there was a void in the literature concerning role stress in hospital executives. The two studies reported here were conducted to begin to fill that gap.

Role theory evolved in the late 1920s and early 1930s from anthropology, sociology, and psychology, and has been addressed extensively in the literature. Within an organizational context, role is defined as a set of expectations applied to the incumbent of a particular position by others both within and beyond an organization's boundaries.[4-6]

Yet roles can sometimes present problems for members of an organization. Role stress is a general concept encompassing those conditions in which role demands are vague, difficult, or impossible to meet. Within the concept of role stress are two subconstructs: role conflict and role ambiguity. Role conflict was described by Rizzo, House, and Lirtzman (1) as conflict between an individual's internal standards or values and the defined role behavior; (2) conflict between the time, resources, or capabilities of the individual and the defined role behavior; (3) conflict in the form of incompatible organizational policies and requests; and finally (4) role overload, meaning conflict between several roles for the same person that require different or incompatible behaviors.[7] In more familiar terms, role conflict is that feeling people get in the pit of their stomachs when the boss asks for the impossible, and needs it immedi-

George C. Burke, III, *Dr.P.H., F.A.C.H.E., is an Assistant Professor of Health Care Administration at Southwest Texas State University, San Marcos, Texas. He has served in various administrative positions in hospitals.*

Cynthia C. Scalzi, *R.N., Ph.D., is an Associate Professor at the University of Texas at Austin School of Nursing, and Director of the Doctoral Program in Nursing Service Administration.*

ately, or when they are caught in a crossfire between two senior managers, or when three deadlines are staring them in the face and the family vacation starts tomorrow.

Role ambiguity has been defined as the extent to which clarity is lacking regarding job performance expectations, methods for carrying out the job, and consequences of performance.[8] A person who is given insufficient information regarding his or her organization is likely to use coping behavior to solve problems or avoid stress. The person may also use defense mechanisms to avoid reality and its uncomfortable situation.

These behaviors may result in dissatisfaction, anxiety, and ultimately less effective work.[9] Role ambiguity is the feeling that people get when they play a game without knowing the rules—they have to get zapped numerous times before they can play, and they know the rules can change without warning.

LITERATURE REVIEW AND SETTING

Numerous studies have been conducted regarding role conflict and role ambiguity, although very few have been directed to executive health care managers. The best documented outcomes of role conflict are job dissatisfaction and job-related tension, which have been isolated in a variety of occupational groups.[10-16] Other studies have demonstrated correlations between role conflict and other organizationally dysfunctional outcomes, such as unsatisfactory work group relationships,[17] slower and less accurate group performance,[18] lower commitment to the organization,[19] and lower performance evaluations.[20] Within the health care setting, role conflict and role ambiguity among staff nurses have been investigated in a limited number of studies.[21-24] No studies of role conflict and role ambiguity among hospital executives have been located in the literature prior to the present studies.

Results of studies of effects of role ambiguity indicate that lack of clarity about performance expectations is associated with a greater concern with performance, lower actual and perceived group productivity, less concern or involvement with the group, lower job satisfaction, unfavorable attitudes toward role senders, and increased tension, anxiety, depression, and resentment.[25] Johnson and Graen have linked role ambiguity to job turnover.[26]

The present studies were based on the research and instrument designed by Rizzo, House, and Lirtzman to measure role conflict and role ambiguity. Rizzo and

It is an interesting finding that the mean role conflict score of the hospital executives in the present study was lower than that reported by Rizzo et al. in their original study, given the frequently heard complaint that hospital management is a "stressful" field.

colleagues administered their questionnaire to a 35% random sample of the central offices and main plant of a major firm and to a 100% sample of the research and engineering division. The total number of respondents in Sample A, discussed in this article, was 199. All respondents were salaried managerial and technical employees. The authors defined role ambiguity as the predictability of the outcome responses to one's behavior and as the existence or clarity of behavioral requirements to guide behavior. The instrument employed a seven-point scale ranging from 1 to 7, with a higher score indicating higher role conflict or role ambiguity.[27]

In the present studies, Scalzi, interested in effects among top-level nurse executives, used the original seven-point scale. The population for Scalzi's study consisted of all top-level nurse executives (124 potential subjects) in general medical-surgical hospitals in Los Angeles County. The hospitals were taken from the annual survey compiled by the American Hospital Association (1982). The response rate was 60%. Among the respondents, 39 (52%) held the title of Director of Nursing Services; 10 (13%), the title of Vice President for Nursing; 9 (12%), the title of Assistant Administrator/ Nursing; and 17 (23%), various other titles indicative of a top-level nurse executive. As to size of the institution, 12 were employed in institutions with fewer than 100 beds, 29 in institutions of 100 to 250 beds, 25 in institutions of 251 to 600 beds, and 9 in institutions with more than 600 beds. Scalzi's questionnaire, and subsequent personal interviews, also included questions related to depressive symptoms and potential mediating variables such as job satisfaction and sources of job-related stress.

Burke's questionnaire, based on the Rizzo et al. instrument, employed a four-point scale, with a range from 1 to 7 (1, 3, 5, 7) to allow comparison with Scalzi's and other previous studies. Additionally, he included an instrument developed by Porter and Lawler to measure job dissatisfaction.[28] The population for Burke's study included all affiliates of the American College of Healthcare Executives (ACHE) in Houston and San Antonio, Texas, employed in executive positions in hospitals. The total number of questionnaires mailed out was 169; 119 (70%) were returned. Of the respondents, 33 (28%) were chief executive officers of their hospitals; 45 (38%) were one level below the chief executive; 32 (27%) were two levels below the chief executive; and 9 (7%) were three or more levels below the chief executive. As to size of the hospital, 11 respondents (9%) were employed in hospitals with fewer than 100 beds; 50 respondents (42%) were employed in hospitals of 100 to 500 beds; and 58 (49%) were employed in hospitals with more than 500 beds.

RESULTS

With regard to role conflict and role ambiguity scores, the scores reported by Burke for hospital executives, with no nurses included in the study, were lower than the scores reported by Scalzi in her study of top-level nurse executives. The lower scores are indicative of lower role conflict and role ambiguity. The differences, however, are not substantial. Role ambiguity scores for Burke's and Scalzi's studies are significantly lower than the scores reported by Rizzo et al. in their original research. It is an interesting finding that the mean role conflict score of the hospital executives in the present study was also lower than that reported by Rizzo et al. in their original study, given the frequently heard complaint that hospital management is a "stressful" field. Table 1 shows the comparative mean scores for the three studies.

Of particular interest was the single scale item that dealt with the issue of role overload. The item was, "I work with two or more groups who operate quite differently." Burke found the mean score for that item to be the highest of all items, indicating a high level of role overload among the hospital executives. Scalzi's population also scored that item highest among all role conflict items. In the original research by Rizzo et al., role overload did not emerge as a predominant concern. Kahn described this item in the following way:

> A very prevalent form of conflict in industrial organizations is role overload. Overload could be regarded as a kind of inter-sender conflict in which various role senders may hold quite legitimate expectations that a person perform a wide variety of tasks, all of which are mutually compatible in the abstract. But it may be virtually impossible for the focal person to complete all of them within given time limits. He is likely to experience overload as a conflict of priorities.[29]

Also of interest was the relationship between role conflict, role ambiguity, and job satisfaction or dissatisfaction. The aim was to determine the relationship of job dissatisfaction to role conflict and ambiguity. Scalzi reported a $-.4$ ($p < .01$) correlation between role conflict and job satisfaction and a $-.3$ ($p < .01$) correlation between role ambiguity and job satisfaction. Burke found a correlation of .36 ($p < .001$) between role conflict and job dissatisfaction and a correlation of .39 between role ambiguity and job dissatisfaction. Rizzo et al., in their original study, reported correlations somewhat higher than those found by Burke and Scalzi.

DISCUSSION

Principles of classical organizational theory have implications for role conflict and role ambiguity. Ac-

TABLE 1

MEAN ROLE CONFLICT SCORES FROM THREE STUDIES

Variable	Burke's study of hospital executives (N=119)	S.E.M.*	Scalzi's study of nurse executives (N=75)	S.E.M.*	Rizzo et al. study of industrial firm (N=199)	Standard deviation
Role conflict	3.4	0.077	4.0	0.14	4.2	1.21
Role ambiguity	2.4	0.080	2.7	0.11	3.8	1.08

*Standard error of the mean.

cording to the chain of command and unity of command principles, organizations with a clear and single flow of authority from the top to the bottom should be more satisfying to members and result in more effective performance than organizations without such clear authority. Role theory suggests that when the behaviors expected of an individual are inconsistent, he or she will experience stress, become less satisfied, and perform less effectively than if the expectations were consistent.[30]

Classical organizational theory also prescribes that every position in a formal organization should have a specified set of tasks or position responsibilities. Role theory suggests that if an employee does not know what he or she has the authority to decide, what he or she is expected to accomplish, and how he or she will be judged, he or she will hesitate to make decisions and will perform less effectively. This phenomenon is known as role ambiguity.[31]

The authors hypothesized that the amounts of role conflict and role ambiguity experienced by hospital and top-level nursing administrators would be comparable to amounts experienced by managers in industry generally, and that the role conflict and role ambiguity (generally referred to together as role stress) would lead to job dissatisfaction.

The hypothesis, diagrammed in Figure 1, shows organizational factors (including unclear structure, unclear goals, minimal feedback, and boundary position) leading to role stress (including role conflict—a component of which is role overload—and role ambiguity), leading to adverse consequences (including job dissatisfaction, avoidance of risk taking, poor performance, job turnover, increase in depressive symptoms, and psychosomatic complaints).[32-35] Of these adverse consequences, the present studies included only job dissatisfaction.

The findings are consistent with the model shown in Figure 1. Role ambiguity was lower in the studies of Burke and Scalzi when compared with Rizzo's research. This is somewhat surprising and cannot be explained without further study. The authors doubt that role stress is actually lower in health care executives than in managers as a whole. A possible explanation is that Rizzo's instrument may not be sensitive to the type of stress experienced by health care executives, namely role overload. By developing additional questions related to role overload, future researchers may be able to obtain a more valid measure of role stress among health care executives. The scoring of the Rizzo instrument involves averaging, which has a tendency to obscure individual item differences. The single factor that was extremely high was the form of role conflict known as role overload. To practicing hospital and top-level nurse executives, this finding may not be surprising. While the demands of the job, if taken one at a time, may be quite manageable, it is the sheer volume of demands and associated time limits that may make the job seem overwhelming. As predicted, role conflict and role ambiguity were positively related to job dissatisfaction.

The existence of role conflict and role ambiguity among hospital and top-level nurse executives is not surprising. The hospital is sometimes cited as an example of an organization that deviates from the classical model. Multiple authority often exists due to the administrative and medical hierarchies, with nurses as a prime example of groups who are caught between the two lines of authority.[36]

Boundary positions are another potential source of stress common in hospitals and may result in role stress for the incumbents of those positions. Boundary positions are those for which some of the direction comes from another unit in the organization or from another organization entirely. Examples would be managers or others who serve in a liaison position linking two departments. Such liaison or "linking pin" functions[37] are generally carried out in hospitals by the executives who are administratively responsible for those areas—departmental managers, assistant or associate administrators or vice presidents, and nurse executives. Boundary positions are essential for hospitals due to the need for coordination between diverse departments.[38]

FIGURE 1

RELATIONSHIP OF ORGANIZATIONAL FACTORS, ROLE STRESS, AND ADVERSE CONSEQUENCES

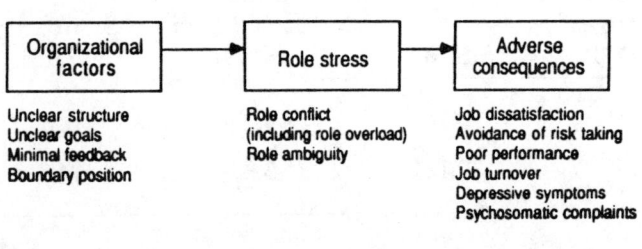

Kahn indicated that boundary positions constitute a major battleground of intergroup conflict. As Adams explains, the occupant of a boundary-spanning position is subject to behavioral expectations that arise from role senders located in separate social systems.[39] The activities performed by the boundary spanner—linking and coordination, information transfer, and receipt of feedback—contribute to role conflict.[40] The number of boundaries and the amount of diversity within the role can increase the potential for role stress.

IMPLICATIONS

The authors were interested in how senior executives can help their subordinates deal with role stress in the organization. A certain amount of role stress is inevitable and goes with the territory. Managers are, by nature, responsible for linking or coordinating diverse functions and are therefore subject to role stress. Yet there are certain practices that senior managers can employ to minimize the adversarial effects of role stress among their subordinate levels of management.

Hospital chief executives must recognize that their subordinates, senior and junior executives whether line or staff, often possess different visions of what the institution is and should be. These different perceptions lead to role ambiguity. Drucker likens the differences to the old story of the blind men meeting up with an elephant. Each level of management and different specialist sees the same "elephant"—the business—from a different angle of vision. The result is that each sees only a portion of the business and therefore sees different priorities. It is the chief executive officer's role to see and articulate the big picture so that his or her subordinates can all pull in the same direction—together instead of apart.[41]

One way to clarify the vision is through an emphasis on goal setting. Systematic, periodic goal setting results in improved performance with an emphasis on results. As Drucker states:

> Constant effort alone can counteract the inherent tendencies toward diffusion and misdirection. The superior needs to understand what he expects of his subordinates. The subordinate, in turn, needs to be able to know what in the way of results he should hold himself accountable for. Without special efforts, superior or subordinate will not know and understand this and their ideas will not be compatible, let alone identical. Each manager, from the "big boss" down to the production foreman or chief clerk, needs clearly spelled out objectives. Otherwise confusion can be guaranteed.[4]

Another way that executives can reduce role stress is by providing frequent opportunities for feedback with their subordinates. More frequent interaction between superior and subordinates results in improvement in goal clarity, as well as a feeling of participation and satisfaction with the superior. Frequent feedback is particularly important to managers who are cautious in decision making and who are in jobs involving frequent change, such as the health care industry.[43] Opportunities for feedback may be through weekly conferences or less formal breakfast discussions. Unless the meetings are held on a regular basis, chances are the communication will be insufficient.

Another valuable yet difficult service that the hospital executive can perform to reduce role stress for subordinates is the role of mediator. Turf battles are inevitable in a hospital. While some interdepartmental conflict or conflict between line and staff managers may be acceptable, if it escalates it becomes destructive and demoralizing, and results in role stress for the incumbents. Some chief executives may choose to let the incumbents fight it out. Yet it is the chief executive's responsibility to settle conflict and restore order before the situation becomes destructive.

Finally, the hospital and top-level nurse executive can affect the role stress of others through their own personal example. The senior manager who makes casual remarks that conflict with established goals, spends lavishly on travel while preaching cost containment, and is discourteous while he or she publicly advocates marketing and good service, is creating an environment of role conflict and role ambiguity for his or her subordinates. They do not know what to believe. Personal example may be the most powerful message managers can convey.

REFERENCES

1. Drucker, P. Management. New York: Harper & Row, 1974.
2. White, D., and Wisdom, B. "Stress and the Hospital Administrator." Hospital & Health Services Administration 30 (September-October 1985): 112-19.
3. Leininger, M. "The Leadership Crisis in Nursing: A Critical Problem and Challenge." Journal of Nursing Administration 4 (1974): 28-34.
4. Neiman, L.H., and Hughes, J.W. "The Problem of the Concept of Role—A Re-survey of the Literature." Social Forces 30 (1951): 141-49.
5. Gross, N., Mason, W.S., and McEachern, A.W. Explora-

tions in Role Analysis: Studies of the School Superintendency Role. New York: Wiley, 1958.
6. Banton, B. *Roles: An Introduction to the Study of Social Relations.* New York: Basic Books, 1965.
7. Rizzo, J.R., House, R.J., and Lirtzman, S.I. "Role Conflict and Ambiguity in Complex Organizations." *Administrative Science Quarterly* 15 (1970): 150-63.
8. Graen, G. "Role-making Processes Within Complex Organizations." In *Handbook of Industrial and Organizational Psychology,* edited by M.D. Dunnette. Chicago: Rand McNally, 1976, pp. 1201-1245.
9. Rizzo, House, and Lirtzman, "Role Conflict and Ambiguity in Complex Organizations."
10. Beehr, T., Walsh, J., and Taber, T. "Relationship of Stress To Individually and Organizationally Valued States: Higher Order Needs as a Moderator." *Journal of Applied Psychology* 61 (1976): 41-47.
11. Brief, A., and Aldag, R. "Correlates of Role Indices." *Journal of Applied Psychology* 61 (1976): 468-72.
12. Brief, A., et al. "Anticipatory Socialization and Role Stress Among Registered Nurses." *Journal of Health and Social Behavior* 20 (1979): 161-66.
13. Gross, Mason, and McEachern, *Explorations in Role Analysis: Studies of the School Superintendency Role.*
14. House, R., and Rizzo, J.R. "Role Conflict and Ambiguity as Critical Variables in a Model of Organizational Behavior." *Organizational Behavior and Human Performance* 7 (1972): 467-505.
15. Miles, R. "A Comparison of the Relative Impacts of Role Perceptions of Ambiguity and Conflict by Role." *Academy of Management Journal* 19 (1976): 25-35.
16. Oliver, R., and Brief, A. "Determinants and Consequences of Role Conflict and Ambiguity Among Retail Sales Managers." *Journal of Retailing* 53, no. 4 (1977-78): 47-58, 90.
17. French, R., and Caplan, R. "Organizational Stress and Individual Strain." In *The Failure of Success,* edited by A.J. Marrow. New York: AMACON, 1972, 30-65.
18. Liddell, W., and Slocum, J. "The Effects of Individual-Role Compatibility upon Group Performance: An Extension of Schutz's FIRO Theory." *Academy of Management Journal* 19 (1976): 413-26.
19. Baird, L. "A Study of the Role Relations of Graduate Students." *Journal of Educational Psychology* 60 (1969): 15-21.
20. Haas, J.E. *Role Conception and Group Consensus.* Columbus, Ohio: Bureau of Business Research, 1964.
21. Brief, et al. "Anticipatory Socialization and Role Stress Among Registered Nurses."
22. Hardy, M.E. "Role Stress and Role Strain." In *Role Theory,* edited by M.E. Hardy and M.E. Conway. New York: Appleton-Century-Crofts, 1978, 73-109.
23. Seybolt, J., Pavett, C., and Walker, D. "Turnover Among Nurses: It Can be Managed." *Journal of Nursing Administration* 8, no. 9 (1978): 4-19.
24. Szilagyi, A. "An Empirical Test of Causal Inference Between Role Perceptions, Satisfaction with Work, Performance and Organizational Level." *Personnel Psychology* 30 (1977): 375-88.
25. Caplan, R., and Jones, K. "Effects of Work Load, Role Ambiguity, and Type A Personality on Anxiety, Depression, and Heart Rate." *Journal of Applied Psychology* 60 (1975): 713-19.
26. Johnson, T.W., and Graen, G. "Organizational Assimilation and Role Rejection." *Organizational Behavior and Human Performance* 10 (1973): 72-87.
27. Rizzo, House, and Lirtzman, "Role Conflict and Ambiguity in Complex Organizations."
28. Porter, L., and Lawler, E. *Managerial Attitudes and Performance.* Homewood, Ill.: Irwin, 1968.
29. Kahn, R.L., et al. *Organizational Stress: Studies in Role Conflict and Ambiguity.* New York: Wiley, 1964, p. 20.
30. Rakich, J., Longest, B., and Darr, K. *Managing Health Services Organizations,* 2d ed. Philadelphia: Saunders, 1985.
31. Rizzo, House, and Lirtzman, "Role Conflict and Ambiguity in Complex Organizations."
32. Brief, A., and Aldag, R. "Correlates of Role Indices."
33. Miles, R. "A Comparison of the Relative Impacts of Role Perceptions of Ambiguity and Conflict by Role."
34. House and Rizzo, "Role Conflict and Ambiguity as Critical Variables in a Model of Organizational Behavior."
35. Kahn, et al., *Organizational Stress: Studies in Role Conflict and Ambiguity.*
36. Perrow, C. "Hospitals: Technology, Structure, and Goals." In *Handbook of Organizations,* edited by F. March. Chicago: Rand McNally, 1965, 148.
37. Rakich, Longest, and Darr, *Managing Health Services Organizations.*
38. Ibid.
39. Adams, J.S. "The Structure and Dynamics of Behavior in Organizational Boundary Roles." *Handbook of Industrial and Organizational Psychology,* edited by M.D. Dunnette. Chicago, Rand McNally, 1976, 1175-99.
40. Miles, R. "Boundary Reference: Dimensionality As A Basis for Discrepant Research Findings." In *Proceedings of the Thirty-Sixth Meeting of the Academy of Management,* edited by R.L. Taylor, et al. Kansas City: Harper & Row, 1976.
41. Drucker, *Management.*
42. Ibid., 433.
43. Kolb, D., and McIntyre, J.M. *Organizational Psychology: A Book of Readings.* Englewood Cliffs, N.J.: Prentice-Hall, 1979.

Employee assistance programs in the hospital industry

John C. Howard
and
David Szczerbacki

The health care literature describes the industry's need for employee assistance programs (EAPs). New research results show the degree to which EAPs are used in the industry and ways to determine their success.

Employee assistance programs (EAPs) are a growth industry in the field of human resource management. While such programs share basic dimensions, it is generally recognized that individual EAPs should reflect the uniqueness of their organizational setting. This article reports the results of a survey of EAP utilization in hospitals in New York. Survey questions focused on extent of EAP use, program characteristics, and management satisfaction. To create a frame of reference, the results of a companion survey of EAP utilization in private firms are also reported.

EAPs DEFINED

It is estimated that there are at least 5,000 and perhaps 10,000 EAPs. Generally, EAPs identify troubled employees, motivate them to resolve their troubles, and provide access to counseling or treatment as needed. EAP referrals access professional care for such problems as alcoholism, emotional difficulties, stress, drug dependence, financial troubles, legal complications, and family discord.[1] EAPs providing wellness and life style counseling are surfacing.[2]

EAPs have evolved from a tradition of industrial welfare rooted in the belief that the personal problems of employees have a direct bearing on workplace productivity.[3] Properly designed, EAPs are integrated with such traditional human resource management systems as performance appraisal, progressive discipline, and employee benefits.

EAPs have a number of basic components, including a statement of policy and procedures, delineation of supportive administrative functions, a supervisory and employee education and training component, working knowledge of relationships with appropriate commu-

---

John C. Howard, *M.B.A., is an Associate Professor of Business and Administration in the College of Business and Administration, at Alfred University, Alfred, New York, and Director of the university's Small Business Institute. He is involved in designing a multicounty substance abuse program in Southwestern New York.*

David Szczerbacki, *Ph.D., is also an Associate Professor of Business and Administration in the College of Business and Administration, Alfred University, Alfred, New York. He is working with John Howard on the multicounty substance abuse program.*

The authors acknowledge the assistance of Charles Caputo, Elaine Horton, and Theresa Anne Feltham in completing the survey and literature review for this article.

nity resources, and a program evaluation component.[4] An understanding of the extent, nature, and effectiveness of such programs in hospitals is important given the dynamics of health care delivery. Not only is the efficiency of such delivery often vital to the preservation of life and health, but also growth in health care expenditures continues to outpace that in the economy at large.

PROGRAM PENETRATION

Most corporate EAPs have been instituted in the post-1940 period. Recently programs have emerged in public sector organizations,[5] but program penetration in the hospital industry has tended to lag behind other sectors of the economy.[6] One observer has noted that this industry "curiously lacks representation in employee assistance programs."[7] In commenting on the fact that fewer than one-third of all hospitals provided EAPs as of 1981, one prominent researcher noted that this "may be due to management's view that such programs are unnecessary for health care workers" but that the "evidence indicates otherwise."[8]

HOSPITAL ENVIRONMENTS

While current reports suggest that hospitals may be increasing their EAP activity,[9] the historic lag in programming is disturbing given that environmental forces in hospitals appear to demand a formal structure for counseling services. Indeed, well-publicized workplace problems such as substance abuse and stress appear to be magnified in hospitals.

It is estimated that between 10 percent and 32 percent of workers in the United States use drugs on the job.[10] The estimated economic costs of drug and alcohol abuse is $140 billion.[11] Direct costs to employers from drug abuse due to absenteeism, sick leave, and drug-related accidents and deaths were nearly $17 billion dollars in 1983.[12] Hospitals share this burden.

Estimates indicate that at least 40,000 nurses in the United States are alcoholics; narcotic addiction in health professionals is thought to be 30 to 100 times greater than it is in the general population.[13] Problems appear to be particularly serious in nursing. "Stress and anxiety related to nursing" has been cited as one of the major causes of nurse shortages.[14] As hospitals move to redesign the roles played by professional and paraprofessional staff to accommodate cost-containment strategies, such stress is likely to intensify. One study indicates, for example, that nurses participating in such reorganization plans "were experiencing a significantly higher level of role tension than the nurses who were not experiencing change."[15]

While nursing may bear a large share of the stress in hospitals, the problem is more pervasive. Research indicates that approximately 15 percent of all physicians have a chemical dependency.[16] Moreover, a study by the National Institute on Occupational Safety and Health has found that "incidence of occupational stress per 100 full-time employees is 58 percent higher for health care employees than for other service industry employees."[17]

The data support the contention that EAPs need to be designed and evaluated in the context of the host organization.[18] Hospital environments, for example, create stressors for employees that are not present in most fields. The continual pressure of life-and-death crises, exposure to death and grief, and demand for a 24-hour commitment to the job contribute to stress. These problems, coupled with ease of access to drugs and a "physician heal thyself" mentality (applying to physicians and nonphysicians alike),[19] make it less than surprising that the substance abuse problem is as pervasive as it seems to be in this industry.

PROGRAM SUCCESS: TEMPORAL DIMENSIONS

Despite growth in the number of EAPs, evidence is mixed regarding their effectiveness. What success is reported tends to be expressed in three forms: case studies,[20] level of satisfaction as reported by program managers or immediate supervisors,[21] and productivity studies.[22]

Organizations successfully using EAPs report decreased turnover rates, lower unemployment and insurance costs, and decreased use of sick time. Supervisors at one major hospital reported that job performance improved 50 percent among those employees counseled through an EAP. Eighty-nine percent of these supervisors reported that the service was a "very important resource for managers" and all supervisors reported they would refer employees to the program again.[23]

Despite evidence of positive results of EAPs, drawbacks have also been reported. Managers and human service professionals are often in conflict due to opposing professional judgments. As one manager has complained, "Once in a long while, I just want to get rid of an employee, and the counseling group says to give him one more chance because their professional evaluation is that they can help him. On the other hand it seems like it has gone on forever."[24]

Such situations suggest that organizations face considerable pressure to demonstrate results from EAPs. Unfortunately a number of forces contribute to a relatively flat and difficult learning curve. Indeed, persistence and commitment over time may be among the most critical predictors of EAP success.

Among these forces is the very nature of EAP treatment strategies. Both constructive confrontation and counseling demand considerable time to develop a successful track record.[25] Constructive confrontation, for example, is based on feedback provided from supervisors over time. Continuous discussion, reinforced via a system of progressive discipline, is designed to motivate behavioral change on the job. While counseling shifts the focus of intervention from supervisory practices to the therapeutic milieu, this strategy is no less time consuming. Case histories need to be developed, and therapies need to be designed, implemented, and evaluated.

In addition to the nature of treatment strategies, organizations must confront the problem of entropy in designing and implementing EAPs. Hospitals typically have competing goals and indeterminate technologies.[26] Under such conditions, hospitals lack integrated authority systems due to multiple and competing group identifications.[27]

Professional fiefdoms, departmental rigidity, suspicion of management's motives, and an entrenched tradition of "taking care of one's own" exist as barriers to program success.[28] Successful implementation, from this view, is a function of management's ability to orchestrate change in the face of an often intractable system.

It is perhaps because of the recalcitrant environments found in hospitals that the history of organizational development in the hospital industry has met with mixed success.[29] This environment poses a challenge to those designing and implementing EAPs.

PROGRAM SUCCESS: OPERATING CHARACTERISTICS

Reports of successful EAPs are also clouded by concern that there is a "need for more rigorous and carefully conceived evaluations of counseling programs."[30] One analyst has argued that EAPs face an "identity crisis": little is known about the optimal mix of pro-

Organizations must confront the problem of entropy in designing and implementing employee assistance programs.

grams, or about which programs work, let alone why some are successful while others are not.[31]

There does appear to be agreement on the standards that should define EAPs. The Association of Labor-Management Administrators and Consultants on Alcoholism, the professional organization of EAP providers, has developed a detailed set of EAP standards. While a useful starting point in designing an EAP, these standards lack specificity given the need to tailor programs to specific environments. With respect to location of the EAP, for example, the standards recommend that "the physical location... should facilitate easy access while ensuring confidentiality."[32] With respect to responsibility for EAPs, the standards recommend that the EAP "...be positioned at an organizational level high enough to insure the involvement of senior management and/or union leadership...."[33]

Given the ambiguity of these standards and the paucity of evaluation studies in the literature, work needs to be done in evaluating characteristics of successful programs

QUESTIONS REGARDING EAPs

This discussion of EAPs in the health care industry leads to four interesting questions whose answers affect not only the health and well-being of the industry's employees but also the profitability and viability of many hospitals. In an age of fierce competition, especially in a service business, no organization can survive unless its employees function at peak efficiency most of the time, which is what EAPs aim to ensure. It is at this juncture that many health care administrators must determine whether an EAP is advisable or how it should be structured. Of specific interest are these questions:

1. To what extent have EAPs penetrated the health care industry? The answer to this question allows administrators to compare their operations to others based on whether they have an EAP. If the hospital is one of few to have an EAP it could be labeled foolish or progressive depending on the success rate of EAPs. If the organization has not

yet instituted an EAP, it may be viewed as judicious or backward. In either case, an industry comparison is necessary.
2. Does the hospital environment present employee problems unique in the industrial United States? Given the stress and the singular composition of a hospital's work force, it is logical to wonder whether people working there will evidence problems different from those of employees in the rest of the economy. The answer to this question will dictate the services an EAP should provide and assist the administrator in choosing the correct EAP configuration for the health care operation.
3. Is there a waiting period before a hospital EAP shows results? This question addresses the way an investment in an EAP should be considered. If successful EAPs usually pay for themselves immediately, then individual EAPs should too. If, however, there is a learning curve phenomenon where the EAP tends to increase productivity over time, then early expenditure on such a program should be viewed as necessary to elicit the high yields of the mature EAP. The benefit will be long term rather than short term. Not only will this knowledge reduce the professional concern of the contracting officer, but also it can be used to temper the expectations of trustees who are feeling increasing pressure to observe immediate financial results from all expenditures.
4. Are there certain EAP characteristics associated with success? Uncertainty is an unfortunate feature of EAPs. In human resources management, there are precious few performance signposts, so it should be helpful to the administrator to know the characteristics of a successful EAP. Once an administrator implements an EAP or decides to adopt one, the next decision is where to seat its responsibility. Hospitals, with their in-house talent, must decide whether to institute their own EAP, saving money but sacrificing the employee's feeling of security that comes with confidentiality, or whether to hire an outside agency that will certainly cost more but will return confidentiality as well as professional experience.

METHODOLOGY

Two survey questionnaires were developed to generate primary data from a cross-section of organizations and health care facilities. The first questionnaire was sent in spring 1986 to 600 New York state organizations stratified by type of industry: manufacturing, service, retailing, and distribution. The survey response rate was 28 percent. The sample population for the second questionnaire was chosen randomly from health care facilities in New York state listed in the 1984 American Hospital Association (AHA) guide. The second survey was sent to 200 institutions in spring 1986. The survey response rate was 55 percent including partially completed questionnaires.

Survey questions were worded to facilitate constructing a computer database useful for statistical analysis using the SPSS-X package. Survey questions were written to identify dependence among characteristics of hospitals surveyed, the existence of some form of EAP, and program characteristics.

HYPOTHESES

The four questions of interest may be restated as hypotheses that can be statistically tested. The first is that a majority of the hospital industry has a functioning EAP. The findings verified this hypothesis. In the health care industry sample, 61.1% had formal EAPs while 39.9% did not. This results in 99.9% statistical confidence that well over 50 percent of hospitals have operating EAPs. Additionally, the study found, conclusively, that larger hospitals are more likely to have an EAP than smaller hospitals (χ^2 = 117.70, df = 3, significance = 0.0). Of the 39.9% of health care managers who do not have formal programs, 35% are considering instituting one. Of the latter group 76% are considering a formal program while the remainder are considering an informal one. In light of past research,[34] it can be

TABLE 1

AVERAGE RANK OF SIGNIFICANT PROBLEMS

Problem	Health care rank (average)	Other industry rank (average)
Marital	1 (4.33)	3 (2.89)
Financial	2 (3.83)	2 (3.18)
Drugs and alcohol	3 (3.61)	1 (3.85)
Job related	4 (3.44)	4 (2.69)

TABLE 2

HOSPITAL EMPLOYEE PROBLEMS BY TYPE AND RANK

Problem	Percentage of time ranked 1 or 2
Alcohol abuse	26.2
Marital	26.2
Need for career counseling	16.4
Job related	13.1
Drug related	11.5
Financial	3.3
Other	3.3

TABLE 3

RANK OF PROBLEMS REGARDLESS OF INDUSTRIES

Manager's tenure	Rank of problems	(weighted average)
< 1 year	Financial	5.0
	Marital	4.0
	Drugs and alcohol	3.0
	Job related	3.0
1–3 years	Job related	4.0
	Financial	2.8
	Drugs and alcohol	2.6
	Marital	2.0
3–5 years	Drugs and alcohol	4.33
	Marital	3.67
	Financial	3.33
	Job related	2.50
Over 5 years	Drugs and alcohol	3.83
	Financial	3.04
	Marital	2.86
	Job related	2.37

concluded that almost half of hospital EAPs have been started in the past five years.

The second hypothesis is that employees in the hospital industry display a set of personal problems that are unique to the Industry. The findings verified this hypothesis. Health care workers experience various problems to a degree that is different from the workers in other industries. The data indicate that the number one problem for health care workers is marital discord, while this is only the third-ranked problem for the rest of U.S. industry. This finding contradicts the literature on drug abuse by hospital workers. The relative rankings and average scores for each of the four major problems reported are presented in Table 1. Table 2 illustrates health care employee problems reported by administrators.

A comparison of Tables 1 and 2 leads to the conclusion that when drugs and alcohol are a problem they are severe ones, ranked one and two in importance. It should be noted that regardless of industry type, the perception of the relative severity of employee problems differs among personnel managers according to the managers' tenure. For example, as shown in Table 3, longer-tenured managers see drug and alcohol abuse as more of a problem than their shorter-tenured colleagues who perceive employee financial problems as relatively more grave.

The third hypothesis is that the success of EAPs in hospitals increases with the length of time the EAPs have been in operation. The findings verified this hypothesis to an extent. While the data indicate that the longer an EAP is in place the more likely that contracting managers will consider it a success, there is only 91.3% confidence in this result ($\chi^2 = 8.1$, df = 3, significance = .087). While this confidence level certainly does not dispute the EAP learning curve theory, it does not verify it to the extent that all doubt is removed. Internal statistics, however, tend to bear out the theory (e.g., of those programs rated successful, 40% had a program for between five and ten years, but this group accounted for only 27.8% of the sample). The sample additionally shows that EAPs are relatively new to hospitals. (See Table 4.)

Of interest is the finding that 100% of the hospitals responding rated their EAP as at least "successful," while 37.5% rated them "highly successful." None of the respondents checked the "somewhat unsuccessful" or "very unsuccessful" alternatives in the forced-choice question.

The last hypothesis is that EAPs that are administered by the hospital staff are more successful than those that

TABLE 4

LENGTH OF TIME EAP SERVICES HAVE BEEN PROVIDED IN HOSPITALS

Length of time	Percentage of hospitals
> 10 years	5.6
5–10 years	27.8
3–5 years	27.8
1–5 years	22.2
< 1 year	16.7

are contracted to outside agencies. The findings verified this hypothesis. Analysis shows with 99.9% confidence that in-house-administered programs are perceived by managers using them as successful more often than community-based programs are seen as successful by their users ($\chi^2 = 12.22$, df = 1, significance < .001).

Of the respondents, 56.3% reported that their programs are administered by in-house staff while 43.7% rely on direct referral to appropriate community agencies. Twenty-five percent of in-house EAPs are administered by a social service office, 16.7% by a personnel office, and 58.3% by a number of other departments such as employee assistance coordinators, employee health offices, labor-management committee offices, or education offices. The data showed no evidence whatsoever that the choice of department for EAP administration had anything to do with the success of the EAP ($\chi^2 = .95$, df = 2, significance = .62).

It should be noted that there are hybrid EAPs that may employ both outside counselors and hospital staff as referral dispatchers. The success of this type of operation should be studied in the future and compared to that of the more pure in-house and outside agency EAPs.

IMPLICATIONS FOR ADMINISTRATORS

The findings lead to some clear managerial conclusions and raise some questions that administrators must ask about EAPs. First, EAPs work. With unanimity, administrators see clear benefit to this type of program. It no doubt lifts the responsibility for counseling from the administrator's or supervisor's shoulders and gives it to professionals trained to do it. Larger hospitals, with relatively rich resources, were the first to begin such programs, but now smaller ones see the necessity and long-term benefit. There may be a wait for that benefit but it clearly accrues to the contracting organization.

Second, EAPs can be tailored to the hospital work force. By tailoring the EAP to the health care facility's needs, the hospital can ensure faster payback of EAP expenditures. Perhaps because of the large representation of women in hospitals, marital problems are of tantamount importance.[35] The EAP should concentrate on prevention and education concerning the problem. Theory states that a troubled home environment detracts from the involvement necessary for concentration at work.[36] Additionally, while there is an industry-wide uniqueness in personal problems, there may be an institutional idiosyncrasy that dictates the EAP services offered. An inner-city facility, for example, may insist that its EAP concentrate on drug and alcohol problems, while a suburban installation may concern itself with employee financial problems. The administration, however, should be certain that personal prejudices do not enter into the choice of EAP area of concentration. The study clearly shows length-of-tenure differences in problem perception. Perhaps there are other internal biases that affect perception, such as the administrator's age, sex, income, family status, or personal values.

Finally, EAPs have optimal organizational characteristics. In-house-administered EAPs clearly have a greater opportunity for success than community-based agencies. This fact may account for the reticence that small hospitals have for starting an EAP. They simply do not have the resources necessary to administer the more successful in-house EAPs, so they do not bother starting one. Managers should remember that it does not matter which department is responsible. This idea interacts with the implication that EAPs can be tailored to a work force in that the in-house program can be more sensitive to the particular needs of the installation's employees. The EAP that serves real needs will meet with real success.

The implementation of EAPs can be compared with the advent of computers years ago. Organizations that bought prepackaged installations found them to be more costly and time consuming than the manual operations they were hoping to replace. Computer buyers who performed careful systems analysis, however, experienced success that grew over time because the computers were purchased and programmed in accordance with the organization's needs. EAPs are similar. A prepackaged EAP may suite the average industry but health care is different. Similarly, an EAP may suit the average health care provider, which no provider is. The cost recovery period of an EAP can be substantially reduced if administrators are able to answer employee needs and problems and configure the EAP to meet them. This requires time and effort during start-up but will yield far greater returns in the long run.

REFERENCES

1. Sonnenstuhl, W.J., and Trice, H.M. *Strategies for Employee Assistance Programs—The Crucial Balance*. Ithaca, N.Y.: ILR Press, 1986.
2. Shain, M., et. al. *Healthier Workers*. Lexington, Mass.: Lexington Books, 1986.
3. Blair, B.R. *Hospital Employee Assistance Programs*. Chicago: American Hospital Publishing, 1985.
4. Ibid.
5. Kemp, D.R. "State Employee Assistance Programs: Organization and Services." *Public Administration Review* 45 (May-June 1985): 378-82.
6. Herzlinger, R.E., and Calkins, D. "How Companies Tackle Health Care Costs: Part III." *Harvard Business Review* 64 (January-February 1986): 70-80.
7. Featherstone, H.S., and Bednarek, R.J. "A Positive Demonstration of Concern for Employees." *Personnel Administrator* 26 (September 1981): 43.
8. Blair, *Hospital Employee Assistance Programs*.
9. "Corporations Whittle Mental Health Benefits." *Hospitals* (September 20, 1986): 50.
10. "Sylvester Stallone Joins Teamsters Crusade for Drug-Free America." *International Teamster* 84 (November 1986): 4.
11. Hoffer, W. "Business War on Drugs." *Nation's Business* 74 (October 1986): 19.
12. Schacter, V., and Geidt, T.E. "Cracking Down on Drugs." *Across the Board* 22 (November 1985): 28-37.
13. Robinson, R.S. "On the Scene: The Troubled Nurse at the University of Cincinnati Hospital." *Nursing Administration Quarterly* 6 (Winter 1985): 35.
14. O'Donovan, T.R., and Bridenstine, T.P. "The Handmaiden Revolt: The Nursing Staff Crisis." *Health Care Management Review* 8 (Winter 1983): 77.
15. Brosnan, J., and Johnston, M. "Stressed but Satisfied: Organizational Change in Ambulatory Care." *Journal of Nursing Administration* 19 (November 1980): 46.
16. Keeve, P.J. "Physicians at Risk—Some Epidemiological Considerations of Alcoholism, Drug Abuse and Suicide." *Journal of Occupational Medicine* 26 (1984): 503.
17. Calhoun, G.L. "Hospitals Are High Stress Employers." *Hospitals* 54 (June 16, 1980): 171.
18. Johnson, A.T. "A Comparison of Employee Assistance Programs in Corporate and Government Organization Contexts." *Review of Public Personnel Administration* 6 (Spring 1986): 28-42.
19. Blair, *Hospital Employee Assistance Programs*.
20. Ibid.
21. Ford, R.C., and McLaughlin, F.C. "Employee Assistance Programs: A Descriptive Survey of ASPA Members." *Personnel Administration* 26 (September 1981): 29-36.
22. Shain, et al., *Healthier Workers*.
23. Akabas, S.H., and Akabas, S.A. "Social Services at the Workplace: New Resource for Management." *Management Review* 71 (May 1982):20.
24. Ibid., 20.
25. Sonnenstuhl and Trice, *Strategies for Employee Assistance Programs*.
26. Perrow, C. "The Analysis of Goals in Complex Organizations." *American Sociological Review* 26 (1961): 854-66.
27. Golomboski, R. *Public Administration as a Developing Discipline: Organization Development as One of a Future Family of Miniparadigms*. New York: Marcel Dekker, 1977.
28. Blair, *Hospital Employees Assistance Programs*.
29. Weisbord, M. "Why Organization Development Hasn't Worked (So Far) in Medical Centers." *Health Care Management Review* 1 (Spring 1976): 17-28.
30. Cairo, P.C. "Counseling in Industry: A Selected Review of the Literature." *Personnel Psychology* 36 (Spring 1983): 1-18.
31. Roman, P. "Pitfalls of Program Concepts in the Development and Maintenance of Employee Assistance Programs." *Urban and Social Change Review* 16 (Spring 1983): 9-12.
32. Blair, *Hospital Employee Assistance Programs*.
33. Ibid.
34. Ibid.
35. Ibid.
36. Ibid.

Commitment and discipline in hospitals: Leadership protocols and legal precedents

Sandra L. Gill,
Eric W. Springer,
and
André L. Delbecq

Leadership protocols may be used to develop commitment to organizational goals and norms, and to maintain disciplinary policies and procedures in health care organizations. Recent legal decisions and research from management and social psychology support these protocols.

It is becoming clear that hospital management is liable for all professional activity that takes place under its jurisdiction. Yet, few hospitals can afford the time and resources needed to gain commitment and compliance from the disruptive professional or egregious coalition. In this era of rapid change and competitive forces hospital leaders need to make timely decisions establishing protocols based on recent legal decisions[1] as well as sound management principles.

The focus of this article is on the development of organizational commitment to norms and the application of disciplinary policies and procedures in hospitals and medical centers, consistent with recent legal decisions and social psychological research. In this article the term hospital management will be used to denote not only the chief executive officer and the administrative staff but also the medical staff leadership consisting of officers, department or service chiefs, and chairpersons of committees. The term, of course, also includes hospital boards of trustees and directors.

The emphasis of this article is on physicians and dentists. However, the term health care professional is more accurate because nurse practitioners, psychologists, podiatrists, and similar health care professionals other than physicians are subject to the same stresses and pressures and act out in the hospital environment; moreover, medical staff appointments and clinical privileges are being opened to a variety of health care professionals.

Sandra L. Gill, M.A., is President, Performance Management Resources, Inc., a health care leadership development firm. She is a doctoral candidate at the Fielding Institute, Santa Barbara, California.

Eric W. Springer, Esq., is Principal in Horty, Springer and Mattern, P.C., a law firm in Pittsburgh, Pennsylvania, specializing in hospital and health care law.

André L. Delbecq, D.B.A., is Dean, Leavey School of Business and Administration, Santa Clara University, Santa Clara, California. Each of the authors is also a faculty member at the Estes Park Institute for Medical Staff Leadership.

The authors thank Henry Casale, J.D., for his extensive and insightful legal research and editorial assistance on this article, and David Caldwell, Ph.D. and Kay Currey, M.A., for the helpful comments on an earlier version of this manuscript.

LEGAL AND MANAGERIAL PERSPECTIVES

No longer is professional conduct considered separate from the impact of one's behavior on others in the hospital environment. As Skillicorn has stated, "Behavioral incompetence can be as detrimental to patient care as professional incompetence."[2] Legal decisions have clearly established

> ... that the hospital's legal responsibility for the quality of patient care embraces the quality of medical care as well, and that the hospital, therefore, has not only the right but the duty to establish and enforce standards of competence in physicians who practice in the hospital. The quality of care cannot be divorced from the hospital's environment in which it is provided and therefore the hospital must require that physicians, no less than hospital employees and visitors, meet reasonable standards of personal behavior in the hospital.[3]

Managerial and social psychological studies have also established the link between individual behavior and organizational circumstances and events.[4-6] Organizations are known to be powerful influences upon the behavior of new members, just as individual behavior of one member may impact on others in the same setting.[7] This article examines a series of managerial interventions that could precede if not preclude formal legal actions regarding the behavioral competence of health care practitioners. After summarizing some of the relevant research on such interventions, disciplinary guidelines in four hypothetical circumstances will be presented. While the guidelines are consistent with recent court decisions, it is recommended that hospital management develop and review with legal counsel its bylaws, policies, and procedures regarding the expected and approved behavior of health care practitioners. Indeed, the legal climate toward hospitals has shifted from total immunity from liability to one of almost strict liability for all tortious acts committed in hospitals.

DEVELOPING COMMITMENT

From a management perspective, health care organizations, like universities and law firms, have been identified as a special kind of entity, i.e., a "loosely-coupled" system.[8] The hospital is more often treated like a workshop rather than one's primary organization,[9] a view that weakens the influence of the hospital on health care practitioners.

Under these circumstances, the contemporary hospital leader needs to establish and employ the three key factors that build commitment to organizational goals and norms: volition, visibility, and irrevocability.[10]

First, opportunities for voluntary choice regarding organizational entry must be provided, i.e., once credentials and character have been determined to meet standards, the individual must perceive a sense of free choice to join the organization and behave in a compatible way.[11,12] Second, visible leadership and mentoring are important to provide ongoing reinforcement and illustration of voluntary, constructive organizational behavior.[13] Symbolic rewarding of individuals who meet the hospital's standards of excellence is one visible and dynamic way to increase others' sense of commitment to the organization.[14] Third, a system of conformance must be established to provide restraint against disruptive, noncompliant acts. The irrevocable consequences of failing to observe the established behavioral norms should be delineated in a clear code of conduct.[15] That this is a useful, even powerful, mechanism for hospital leadership is strongly supported by a growing body of legal decisions. Springer and Casale conclude that "the clear trend of the legal cases supports the view that a hospital may deny initial appointment, terminate an appointed practitioner or deny reappointment, solely on the basis that the practitioner's behavior is disruptive."[16] In this regard, a record of disruptive conduct is not different from a record of substandard clinical performance.[17]

Considering that commitment to the organization may be gained by allowing its members voluntary choice, by offering visible leadership, and by having a clear code of expected behavior, what can hospital management do if faced with instances of noncompliant behavior, evidencing lack of commitment to the organization's protocols and practices?

THE PSYCHOLOGY OF REWARDS AND PUNISHMENT

One of the consistent findings in social psychology over the past 20 years is that punishment, i.e., negative sanctions, compared to rewards and incentives as a primary mode of changing individual behavior, is rarely effective in the long run.[18] Individuals who perceive their leaders to be rewarding and capable of recognizing constructive behavior and commitment show a willingness to be influenced by the leader, express higher rates of attraction and commitment to the organization, engage in more self-motivated problem solving and constructive behavior, become more self-disciplined, and have less need for formal control because informal norms maintain the high rates of performance.[19] Contrast the findings regarding positive leadership with those that result when leaders take a problem-oriented focus, often the prevailing modus operandi in the clinical domain. At best, negative sanctions tend to cause a temporary lull in the punished behavior and may simply result in more covert acts, or in more egregious behavior that challenges the authority of the leader. The leader who uses recognition and incentives for the desired behavior will less often need to resort to negative sanctions, though such secondary intervention may sometimes be necessary.[20]

DIAGNOSTICS FOR DISCIPLINARY INTERVENTIONS

The leader who uses a positive incentive style of leadership may still need to consider disciplinary intervention and will need to be able to diagnose the appropriate response to the undesirable behavior.

Delbecq proposed a matrix for the analysis of appropriate managerial interventions in health care circumstances, along with specific intervention guidelines.[21] His model is augmented here to incorporate recent court decisions regarding disruptive health care practitioners (see Figure 1).

In the work place, individual behavior is a function of perceived role(s) and individual style or preference. Managerial intervention must consider not only the behavior but also the rights of the individual who may exhibit stylistic or idiosyncratic behaviors that do not affect his or her professional competence.[22] Thus, two diagnostic questions may be used to help select an appropriate intervention: (1) Is the (alleged) behavior role related, or is it a matter of personal style that does not inhibit the individual's professional competence; and (2) is the behavior observable, that is, can it be corroborated? In the Delbecq model, a corroborated behavior is deemed public and an allegation of behavior is called private, e.g., suggested through rumor or hearsay.

A second leadership act to promote the development of commitment and "fit" between individual values and organizational philosophy is the demonstration of visible, symbolic communication.[23] As symbolic communicators, health care leaders need to provide at every opportunity an articulation of the philosophy, vision, long- and short-term goals, and examples of desirable conduct and standards of excellence. This combination of explicit statements and symbolic action provides a benchmark for emerging leaders and other members to reinforce desired behavior patterns. Some examples of such hospital efforts include the clarification of organizational values; identification and reward of individuals who manifest these in their daily actions; and role clarification to develop a consistent management behavior pattern with the stated hospital philosophy.[24]

Finally, the leadership acts of solving problems, establishing priorities, adjusting to new circumstances around key projects, and linking priorities to organizational goals and norms remain effective strategies for developing organizational commitment to philosophy, inasmuch as they are used as arenas for discussion and clarification.

IRREVOCABILITY: THE LEGAL IMPERATIVE FOR ORGANIZATIONAL BEHAVIORAL COMPETENCE

While social psychological studies have demonstrated the importance of volition to organizational commitment, legal decisions have sounded a much greater note in the orchestration of behavioral competence among health care practitioners. Most of these cases have dealt with the disruptive person, i.e., the individual whose behavior interferes with what may

FIGURE 1

DELBECQ MATRIX FOR DISCIPLINARY INTERVENTIONS

Observability	Task Outcomes	
	Instrumental	Stylistic
Public	1. Sanction Discipline Steps	2. Give Information
Private	3. Cite Norms Obtain Data	4. Avoid

> *While social psychological studies have demonstrated the importance of volition to organizational commitment, legal decisions have sounded a much greater note in the orchestration of behavioral competence among health care practitioners.*

otherwise be acceptable—if not superior—clinical competence. As noted by Springer and Casale, action may be taken against a practitioner solely on the basis that the practitioner's behavior is disruptive.[25] In this regard, a record of disruptive conduct is not different from a record of substandard clinical performance.[26] Thus, legal precedents support the managerial need for a clear code of professional conduct, for example, bylaws that include the specification of consequences for noncompliant behavior. Protocols for intervening in disciplinary circumstances are detailed in the second section of this article.

Three elements have been discussed as they relate to developing a leadership strategy for organizational commitment. First, opportunities for voluntary choice regarding organizational entry must be provided, that is, once credentials and character have been determined to meet standards, the individual must perceive a sense of free choice to join the organization and behave in a compatible way. Second, visible leadership and mentoring are important to provide ongoing reinforcement and illustration of voluntary, constructive, organizational behavior. Third, a system of compliance must be established to provide restraint against disruptive, noncompliant acts. Commitment to the organization depends on the presence of all three.

In the following section four specific protocols are presented for intervening in hypothetical yet conceptually distinct behavioral events. These guidelines are consistent with both social psychology and legal findings.

CASE STUDIES

Case number 1: "public" role-related behavioral infractions

When infractions of clearly role-related expectations are corroborated, the imposition of negative sanctions is a managerial imperative. Social psychological studies have demonstrated that a laissez faire style of leadership, in which normative and overt deviance is tolerated, contributes to an unproductive, neurotic organizational climate. When there is clear evidence that misconduct has occurred, sanctions are recommended as soon as possible after the event to optimize the corrective impact. Such interventions should be decisive, specific, and clear in terms of the consequences for the present misconduct and future consequences for any additional misconduct.

One approach is offered through the managerial method of "progressive discipline,"[27] which suggests that a verbal reprimand be followed by progressively more severe disciplinary actions (see boxed material).

Traditional Progressive Discipline Sequence

Assumptions

Review of organizational norms (formal, informal), policies, and procedures:
- upon interview
- upon employment
- after employment interim period

Actions taken upon "violation" of job-related behavior

1. Verbal reprimand, followed by
2. Written reprimand, followed by
3. Temporary discharge/pay penalty, followed by
4. Permanent discharge

This is not an easy decision to make, because the hospital is at risk whatever action it takes. If the hospital does nothing in the face of disruptive behavior, it contributes to a laissez faire atmosphere. If managerial action is taken, legal counteraction may be threatened or pursued. However, the courts have dealt with and upheld hospital intervention regarding a variety of disruptive behaviors in hospitals. Some are selected here to help illustrate this cell of the matrix in Figure 1, including: impertinent and inappropriate comments written (or "cute" illustrations drawn) in patient medical records or other official documents that impugn the quality of care in the hospital or attack particular practitioners, nurses, or hospital policy; sexual harassment of nurses, other hospital employees, or patients; refusal to accept medical staff assignment, or to participate in committee or departmental affairs on anything but one's own terms, or to do so in a disruptive manner; rude or abusive conduct to nurses or other hospital employees; threats and/or physical assaults on physicians, employees, or others on hospital property.[28]

It may be noted that the courts have upheld hospital decisions to act in three points in the cases of disruptive practitioners. First, a series of cases have established the right of the hospital to deny initial appointment to the medical staff on evidence of or reasonable doubts about an applicant's prior conduct. However, bylaw language should be clear on the issue of behavioral competence to defend against the inevitable suits that will emerge from denied applicants. Second, courts have upheld denial of reappointment of disruptive health care practitioners: "Every court which has reviewed a hospital's denial of reappointment due to the physician's disruptive behavior has found such misconduct to be a sufficient basis for the decision not to reappoint."[29]

When disruptive behavior occurs during the term of appointment, both wisdom and decisive action are required. A single episode of disruptive behavior, which taken alone might not appear to be sufficiently serious to warrant revocation of privileges, may actually be one in a series of similar episodes that render the practitioner's continued presence in the hospital intolerable. In fact, practitioners may well claim that the absence of previous intervention precludes the consideration of their history in current decisions. However, "no court has yet held, that by failing to take action at some particular point in a continuing course of disruptive conduct the hospital has, in effect, waived earlier incidents and is therefore foreclosed from considering them in revoking an appointment."[30]

What about the single episode of disruptive behavior—can summary suspension be invoked? The courts have held in support of suspension in single episodes of clinical misbehavior, but decisions are split regarding disruptive behavior. Presumably, disruptive behavior would have to be shown to impose a severe threat to patient safety or directly impair the orderly operation and management of the institution.

In summary, individual disruptive behavior that impairs the competence of the health care practitioner or burdens the effectiveness of others requires decisive disciplinary action. Such protocols should be clearly stated in the bylaws of the hospital and medical staff. Hospital management and medical staff leaders should be thoroughly coached and provided legal counsel in such events. While protocols may not, in fact, curb the disruptive behavior of the individual, they do protect the rights and responsibilities of the hospital to provide quality medical care free from either clinical or behavioral incompetence.

Case number 2: "private" role-related behavioral infractions

As noted, the hospital is at risk for all professional activity that takes place under its jurisdiction. While this risk may support hospital management's efforts to rid itself of disruptive practitioners, the nature of legal proceedings requires a full, often tedious and redundant development of evidence to support facts in the adversarial format of law. Thus, the managerial imperative to intervene on rumors of misconduct may be outweighed by the potential of a court battle for injunctive relief and money damages for wrongful charges. One result is a kind of institutional paralysis in all but the most exceptional cases. Maintaining fundamental due process considerations and a full and fair opportunity for allegedly disruptive practitioners to avoid onerous discrimination is as important in the management of discipline and sanctions as the appropriate imposition of disciplinary action.[31] Perhaps nowhere does the dilemma become as challenging as in this second circumstance, where a role-related allegation of misconduct is made, but not immediately corroborated. What protocol can hospital management follow in this risky circumstance?

Where an allegation of role-related behavior is made, but defies immediate corroboration, the leader is advised to take the earliest possible opportunity to restate institutional bylaws and policies that have allegedly been violated, along with the penalties for such behavior. In this fashion, information is communicated and the opportunity for volitional self-correction is provided. Of equal importance, restatement of policies reinforces the grounds for the work group to intervene with corrective action, in the event such an infraction is inadvertent.

If the severity of the allegation is great, such as a threat to patient safety or to other hospital practitioners and employees, an immediate investigation may be appropriate. If a reasonable doubt exists regarding an individual's competence, further actions may be taken, such as suspension. Of particular importance here is that no individual be subjected to unreasonable treatment or castigation, lest it create a witch-hunt atmosphere and consequent threat of wrongful charge litigation.

Case number 3: individual style versus conformance—respecting rights

Respecting the practitioner's due process rights is perhaps most clearly illustrated in this third circumstance, where the behavior at question is a matter of individual style that does not interfere with clinical or behavioral competence. However, what happens if it creates an irritation or wake of public interest with which hospital management would rather not have to deal?

Hospital management must respect the rights of practitioners who are "different" to the same extent that they invoke action in the midst of disruptive individuals.

Hospital management must respect the rights of practitioners who are "different" to the same extent that they invoke action in the midst of disruptive individuals. Expressing unorthodox opinions, displaying unusual tastes, or manifesting alternative life styles are often seductive areas of inappropriate—and risky—management intervention. While the board, medical staff, or hospital management may be in a majority whose preferences run counter to the life style in question, the hospital has a positive obligation to treat such practitioners with the same degree of tolerance and acceptance it affords others. So long as personal idiosyncrasies as expressed in word or deed do not affect the ability of others to do their jobs, nor impinge on others' rights to complete their business free of burdensome harassment, and the practitioner continues to perform well professionally, negative sanctions should not be involved.[32] What course of action is appropriate?

Such a circumstance requires a combination of mentoring and constructive feedback for the purpose of providing information and support to solicit the voluntary cooperation of the individual at issue. Delbecq and Ladbrooke have developed the theoretical rationale for this protocol elsewhere, so only the action steps will be summarized here.

First, in meeting with an individual whose preferences are different, a clear statement of the issue is required. This includes what is not at issue, that is to say, the individual's clinical or professional competence. In fact, a statement of his or her value to the organization is recommended to provide a testament to the true context of the meeting. Second, a specific behavioral description of the event should be made, for example, seemingly impatient or curt telephone manner, a pattern of interrupting coworkers or subordinates in meetings, a manner of making comments that seem cynical or particularly harsh. Along with this specific description of behaviors should follow a description of the impact of such conduct, for example, apparent anger, embarrassment, or pain, on the immediate and indirect participants in the situation. The unintended consequences of such conduct should be discussed, such as alienation and lack of cooperation from other members or groups in the organization. Finally, an opportunity for problem solving and follow-up support or coaching should be provided, should the individual express the desire for such assistance.

The purpose of such feedback is not to change behavior through coercive action; rather, it is to provide accurate, unambiguous information about the consequences such behavior has on the organization, and to provide an opportunity for the health care practitioners to engage in more compatible behaviors if the individual so chooses. Of utmost importance is the element of volition in this event, since the person at issue may well decide that the price of conformity is too great. At least he or she has an accurate account of the apparent impact the behavior in question has on others, and vice versa.

Case number 4: "private" behaviors—avoid interventions

Organizational leaders may well be tempted to intervene when coworkers and subordinates appear to be experiencing personal problems or when rumors imply that someone ought to be concerned. Common examples include marital difficulties, family distress, and age-related life transitions. Organizational leaders can ill afford the role confusion or ambiguity that may accompany such concerns.[34] As leadership interventions, such acts often contribute to the recipient's confusion about management's intent, credibility, or capacity to provide appropriate help. Furthermore, management assistance in such personal matters may confuse the role relationship should subsequent events require disciplinary action due to work-related problems resulting from these matters. While management may intend to be helpful and friendly, the cost to subsequent credibility and capacity to maintain appropriate role authority is often quite high. Instead, many hospitals provide professional staff, special programs, and counseling services that are usually better equipped to manage such potentially complex situations.

• • •

Court decisions have established that a hospital is at risk for all its professional acts, whether clinical or behavioral. Inappropriate action and failure to take appropriate action are both part of the complex context of risk affecting contemporary hospitals and medical centers.

This article has provided a brief review of social psychological research and legal precedents for an integrated managerial strategy that enhances organizational commitment and maintains appropriate standards for behavioral competence. Commitment derives from free choice and volition; visible examples of constructive role-related behaviors; and irrevocable parameters to competence, such as a compliance system of bylaws, policies, and procedures for the maintenance of quality care and an orderly institutional environment.

Four circumstances for managerial interventions were described, along with protocols for leadership actions that are consistent with recent court decisions. Beyond these guidelines, the development and review of corporate bylaws with adequate legal counsel are advocated to provide a bulwark against continuing legal involvement in health care organizational management.

Finally, it should be noted that a variety of leadership development strategies have already been established to augment these recommendations. Training in role clarification and negotiation; appraisal interviews; feedback and effective communications; and problem-solving techniques are already available for leaders.[35-39] Because turnover of departmental and medical staff officers, board members, and chief executive officers in hospitals and medical centers attenuates leadership and disciplinary continuity, it is suggested that these guidelines be supplemented with annual leadership development programs, which include a review of current and emerging legal and disciplinary issues affecting the hospital.

REFERENCES

1. Springer, E., and Casale, H. "Hospitals and the Disruptive Health Care Practitioner—Is Inability to Work With Others Enough to Warrant Exclusion?" *Duquesne Law Review* 24, no. 2 (1985): 377–423.
2. Skillicorn, S. "Peer Group Committee Tackles Physician Behavior Problems." *The Hospital Medical Staff* 10, no. 7 (1981): 2.
3. Horty, J. "The Disruptive Physician." *Hospital Law* 2 (December 1984): 1.
4. Reis, M. "Surprise and Sense Making: What Newcomers Experience in Entering Unfamiliar Organizational Settings." *Administrative Science Quarterly* 25, no. 2 (1980): 226–51.
5. Van Maanen, J., and Schein, E. "Toward a Theory of Organizational Socialization." In *Research in Organizational Behavior*, edited by B. Staw. Vol. 1. Greenwich, Conn.: Jai Press, 1979, pp. 209–64.
6. Van Maanen, J. "Breaking In: Socialization to Work." In *Handbook of Work, Organization and Society*, edited by R. Dubin. Chicago: Rand McNally, 1976, pp. 67–138.
7. Springer and Casale, "Hospitals and the Disruptive Health Care Practitioner."
8. Weick, K. "Management of Organizational Change among Loosely Coupled Elements." In *Change in Organizations*, edited by P. Goodman. San Francisco: Jossey-Bass, 1982, pp. 375–408.
9. Gallagher, T. "The Hospital: Is It a Place or a Thing?" *Health Care Management Review* 9, no. 1 (1984): 71–76.
10. Angle, J., and Perry, J. "An Empirical Assessment of Organizational Commitment and Organizational Effectiveness." *Administrative Science Quarterly* 26 (March 1981): 1–14.
11. Calder, B., and Staw, B. "Self-Perception of Intrinsic and Extrinsic Motivation." *Journal of Personality and Social Psychology* 31 (1975): 599–605.
12. Ross, M. "Salience of Reward and Intrinsic Motivation." *Journal of Personality and Social Psychology* 32 (1975): 245–54.
13. O'Reilly, C. "The Creation of Corporate Culture: Lessons from Cults and High Technology Firms." Paper presented at Santa Clara University Executive Seminar in Corporate Excellence, Santa Clara, California, February 1985.
14. Brozovich, J., and Shortell, S. "How to Create More

Humane and Productive Health Care Environments." *Health Care Management Review* 9, no. 4 (1984): 43–44.
15. Delbecq, A. "Influence of Professional Behavior: The Management of Norms and Behavior." In *The Physician in Management*, edited by R. Schenke. Tampa, Fla.: American Academy of Medical Directors, 1980, pp. 179–92.
16. Springer and Casale, "Hospitals and the Disruptive Health Care Practitioner."
17. Horty, "The Disruptive Physician."
18. Luthans, F., and Kreitner, R. *Organizational Behavior Modification*. Glenview, Ill.: Scott, Foresman, 1975.
19. Delbecq, "Influence of Professional Behavior."
20. Ibid.
21. Ibid.
22. Springer and Casale, "Hospitals and the Disruptive Health Care Practitioner."
23. O'Reilly, C., "The Creation of Corporate Culture."
24. Delbecq, "Influence of Professional Behavior."
25. Springer and Casale, "Hospitals and the Disruptive Health Care Practitioner."
26. Delbecq, "Influence of Professional Behavior."
27. Sartain, A., and Baker, A. *The Supervisor and the Job*. New York: McGraw-Hill, 1978.
28. Springer and Casale, "Hospitals and the Disruptive Health Care Practitioner."
29. Ibid.
30. Ibid.
31. Ibid.
32. Ibid.
33. Delbecq, A., and Ladbrooke, D. "Administrative Feedback on Behavior of Subordinates." *Administration in Social Work* 3 (Summer 1979): 153–66.
34. Katz, D., and Kahn, R. *The Social Psychology of Organizations*. 2nd ed. New York: Wiley, 1966.
35. Harrison, R. "Role Negotiation: A Tough-Minded Approach to Team Development." In *The Social Technology of Organization Development*, edited by W. Burke and A. Hornstein. La Jolla, Calif.: University Associates, 1972, pp. 84–96.
36. Koontz, H. "Making Managerial Appraisal Effective." *California Management Review*, 15 (Winter 1972): 46–55.
37. Gordon, T. *Leader Effectiveness Training*. New York: Wyden Books, 1980.
38. Delbecq, A., Van de Ven, A., and Gustafson, D. *Group Techniques for Program Planning: A Guide to Nominal Group and Delphi Processes*. Middleton, Wis.: Green Briar Press, 1986.
39. Ulschak, R., Nathanson, L., and Gillan, P. *Small Group Problem Solving*. Reading, Mass.: Addison-Wesley, 1981.

Negligent hiring and retention: Some evidence of hospital vulnerability

James W. Fenton, Jr.,
Jerry L. Kinard,
and
Fred R. David

Hospital human resource managers were surveyed to determine their understanding of negligent hiring employment law and the tools used in employment screening. This article describes the results, which indicate that hospital human resource managers understand the law but that there are gaps in the use of some employment screening tools. The authors make recommendations for future research.

One of the significant developments in employment law during the last decade has been the expanding recognition of the tort of negligence in hiring decisions.[1,2] Negligent hiring is a legal doctrine under which an employer is charged with breaching a duty to use reasonable care in the hiring and/or retaining of persons who later cause harm to third parties.[3] Due to the sensitive nature of their services, both public and private hospitals have higher levels of exposure to negligent hiring and retention liability.

In negligent hiring cases, the relationship between a third party plaintiff and an employer defendant often drives the outcome. In some cases, there appears to be a higher degree of duty or care required between the employer and the third party due to the nature of the business product or service provided. Hospitals and religious institutions, for example, have an elevated level of duty or care to the public because they invite the public onto their premises. In fact, courts have found that hospitals enter into an implied contract with their patients whereby the hospital assures the patient that it will employ competent and reliable employees.[4]

In cases alleging negligent hiring, a plaintiff must first establish negligence on the part of the employer. Once this is established by the plaintiff, courts look critically at the employer's pre-employment process—specifically the tools of the process. Pre-employment tools include a job application form, a job interview, the reference check procedure, a check for criminal convictions, and a physical examination. In short, for an employer to defend itself effectively against a negligent hiring/retention charge, an array of pre-employment tools designed to inquire into the background of job applicants should have been part of the selection process.

While courts of law are quite willing to entertain negligence suits involving hiring and retention actions, they are also hearing cases from plaintiffs complaining

---

James W. Fenton, Jr., *Ph.D., is an Associate Professor of Business Administration at Francis Marion College in Florence, South Carolina. He also is a labor arbitrator.*

Jerry L. Kinard, *D.B.A., is Dean and Palmetto Professor of Business at Francis Marion College in Florence, South Carolina.*

Fred R. David, *Ph.D., holds the TranSouth Endowed Chair in Strategic Management in the School of Business Administration at Francis Marion College in Florence, South Carolina.*

The authors wish to acknowledge the helpful comments of Professor Gary Miller on earlier drafts of this article.

about employer actions in matters involving alleged employment discrimination,[5] invasion of privacy,[6] and defamation.[7(p.774)] Each of these areas and the expectations of the courts appear to run counter to the demands involving the doctrine and common law precedents of negligent hiring and retention. As such, it is not surprising to hear hospital human resource professionals conclude about this apparent incongruity: "I'm damned if I do and damned if I don't!"

For employers to avoid significant financial risk, they should have a working understanding of the legal doctrine known as negligent hiring/retention. In addition, employers should have pre-employment policies and practices in place to help reduce the risks associated with the negligent hiring/retention doctrine while simultaneously being cognizant of constitutional, common law, and statutory mandates that govern individual rights. Due to the recency in the popular use of the legal theory of negligence in employment, only scant prior research exists on this topic. The purpose of this research, therefore, is to empirically identify and study the association between employers' knowledge of the doctrine of negligent hiring/retention and the pre-employment tools used by these same employers in considering job applicants for employment. All employers surveyed in this study are hospitals.

HISTORICAL DEVELOPMENT OF THE LEGAL DOCTRINE

The doctrine of negligent hiring/retention grew out of the common law fellow servant rule which dealt with absolving an employer of any liability where employee negligence resulted in injury(ies).[8] As the law evolved, exceptions were created by the judiciary. Courts began to recognize an employer's duty to provide a safe place to work. In time this notion was extended to providing *safe employees*.[9]

Negligent hiring liability is positively determined when, prior to the time the employee is actually hired, the employer knew or should have known of the employee's unfitness. The court's focus in such a case is upon the adequacy of the employer's *pre-employment investigation* into the employee's background. Negligent retention, on the other hand, occurs when during the course of employment, an employer becomes aware or should have become aware of problems with an employee that would indicate unfitness and the employer fails to take action to effectively correct or eliminate the problem.[10] It is expected that if employers understand the negligent hiring/retention doctrine,

> *It is expected that if employers understand the negligent hiring/retention doctrine, they will use pre-employment tools in scrutinizing job applicants. Not using such tools could open "windows of opportunity" for potential law suits.*

they will use pre-employment tools in scrutinizing job applicants. Not using such tools could open "windows of opportunity" for potential law suits.[11] Two cases will provide a better understanding of the concepts.

In *Joiner v. Mitchell County Hospital Authority*,[12] a negligent hiring claim was the issue before the court. It was alleged that the hospital was negligent because its employment procedure did not require proof of a physician's qualifications. In this case, the plaintiff's spouse was brought to the hospital with chest pains. A staff hospital physician examined the patient and released him to return home, announcing that the patient's condition was not a problem. Upon arrival home, the patient's chest pains intensified. The family decided to return the patient to the hospital. On the return trip, the patient died.

In her suit, the wife claimed that the hospital failed to require proof of the physician's professional qualifications and that simply relying on the fact that the medical doctor was licensed in the state was not enough. The Georgia Appeals Court agreed, saying that hospitals have an affirmative duty to conduct independent investigations into a physician's professional competence.

In contrast, the case of *Pruitt v. Pavelin*[13] addressed negligent retention. In this case, a realtor hired an agent to sell listed properties. The company became aware of certain improprieties on the part of the agent after hiring the person. Examples of some past indiscretions included forging documents for a former employer, passing bad checks, and lying about obtaining a realty license. Nevertheless, the realtor vouched for the newly hired agent's character. Subsequently, the agent committed an illegal act. After hearing the case, the court concluded that the firm was liable for the consequence of the employee's misconduct. In arriving at the award, the court emphasized the firm's acquiring knowledge of the employee's numerous past indiscretions *after* hiring the person and that such knowledge *should have been* acted on by the employer to protect potential third party victims.

METHOD

Respondents

Using the 1988 membership directory of the American Hospital Society of Healthcare Human Resources Administration of the American Hospital Association, 394 human resource managers (HRMs) were randomly selected. All of these persons either were actively involved in personnel selection and development of selection criteria or were human resource executives who acted in an "overseer" capacity of the employment function for the hospital.

Instrument

A questionnaire was developed to assess the degree and type of screening that health care institutions use to detect "high-risk" job applicants. A high-risk job applicant was defined as having a condition or history of past activities, including criminal convictions, the nature of which could represent potential harm to a third party if the job applicant was employed by the hospital. The questionnaire also was designed to elicit information that would help determine the respondents' understanding of the doctrine of negligent hiring/retention.

The instrument was divided into two parts. The first half contained a list of 14 job categories and a list of pre-employment screening tools. Respondents were instructed to identify the tool(s) used in pre-employment screening by position. The second half of the questionnaire had a series of five-point Likert statements designed to assess respondent knowledge and sensitivity to the doctrine of negligent hiring/retention.

The research instrument was pretested. Modifications were made to several questions as a result of the pretest. The instrument was subsequently mailed first class with a letter on university letterhead signed by one of the authors explaining the purpose of the study.

RESULTS

Ninety-three responses to the survey were received, representing a response rate of 23.5 percent. Fifty-eight of the respondents represented private for-profit hospitals, 32 represented public facilities, and 3 provided no indication.

Hospitals of various sizes were represented in this study. The smallest hospital employed 130 persons and housed 27 beds, while the largest employed 15,000 persons and housed 2,789 beds.

Respondents' use of selection of tools

Respondents were asked to indicate the types of selection tools or procedures used in evaluating job applicants for 14 selected hospital job categories. A summary of these findings appears in Table 1.

Hospital HRMs responding to this survey relied on four screening tools in considering applicants for employment: job application forms, pre-employment interview, letter of reference, and telephone calls to prior employers. For some job categories (e.g. nurse, pharmacist, therapist), scrutiny of licenses, diplomas, and certificates is an important factor in screening job applicants. Interestingly, in the anesthetist job category, only two-thirds of the respondents evaluate the diploma, license, or certificate. Considering the emphasis some courts have placed on this type of screening device for this type of position, this result seems to run counter to prudent expected hospital employment policy.

Table 1 reveals that approximately 20 percent of the respondents routinely check criminal conviction records of job applicants, and relatively few require drug testing as part of the pre-employment screening process. Security job applicants' criminal records were investigated the most often of all hospital jobs. However, only 48 percent of those responding to this survey who had to fill security positions checked applicant criminal convictions.

Drug testing has been challenged as being constitutionally offensive, running counter to the Fourth Amendment that concerns illegal searches and seizures. The right to privacy has been articulated in those challenges. Only two major announcements have come from the United States Supreme Court in *Skinner v. Railway Labor Executives' Association*[14] and *National Treasury Employees Union v. Von Raab*[15] both originating in the public sector. Though both cases involved current employees on the payroll, it seems reasonable that the case announcements could apply to pre-employment consideration. Federal courts have been generally sympathetic toward employers wanting to test employees for illicit drug use. If an employer can show that the nature of work involved justifies efforts to detect impaired employees, and if tests are conducted under a policy narrowly tailored to minimize the intrusion on privacy and to limit the discretion of those giving the tests, the tests will likely withstand judicial scrutiny, even without reason to suspect individual drug use.[16]

Only nine hospitals in this study require applicants to provide a health history and pass a physical exam. Three hospitals require a tuberculosis exam, one re-

TABLE 1

HOSPITAL PRE-EMPLOYMENT SCREENING TOOLS BY POSITION

Hospital employees directly involved with patient care*		Letters of reference	Phone calls to previous employers	Review of employment application form	Information gathered in employment interview	Evaluation of diplomas, licenses, certificates	Discussions with college professors and others	Check of criminal record if accessible	Mandatory drug testing	Physical exam or health record	Polygraph (public institutions)
1. Anesthetists	(63)	45	49	55	59	55	12	13	4	9	0
2. Nurses	(93)	84	89	89	89	89	23	18	12	9	0
3. Nurses aides & orderlies	(92)	84	86	82	82	82	7	18	12	9	0
4. Pharmacists	(90)	84	83	88	88	88	3	19	8	9	0
5. Phlebotomists	(88)	67	78	74	76	74	5	13	6	9	0
6. Technicians (x-ray, laboratory, etc.)	(91)	86	86	77	76	77	5	18	10	9	0
7. Therapists	(91)	73	75	71	70	71	21	19	10	9	0
Hospital employees not directly involved in patient care*											
1. Administrative personnel	(92)	88	90	88	88	88	22	17	10	9	0
2. Custodians	(90)	71	76	90	81	90	0	20	10	9	0
3. Food service personnel	(88)	72	72	86	86	86	3	18	11	9	0
4. Maintenance personnel	(91)	79	80	90	88	90	3	20	11	9	0
5. Medical records personnel	(92)	80	84	90	90	90	6	17	11	9	0
6. Clerical personnel	(93)	72	79	88	88	88	3	14	6	9	0
7. Security personnel	(73)	60	65	60	68	60	4	35	20	9	0

*Numbers in parentheses equal number of responses.

quires rubella screening, and three routinely administer personality tests to applicants for jobs that have direct contact with patients. One respondent routinely checks each applicant's driver's license through the state's division of motor vehicles. Interestingly, the nine hospitals requiring pre-employment physical examinations require it for all positions surveyed.

Considering that the focus of this study was on hospitals, the apparent lack of interest in mandating physical exams and drug tests as part of their pre-employment screening was surprising. The possibility of spreading contagious diseases and the rampant use of drugs in our society perhaps warrant detailed pre-employment exams and tests. Positions such as nurses, nurses' aides and technicians, phlebotomists, and food service workers potentially come into direct physical contact with patients. It would seem that for a carrier of contagious diseases or a drug addict to be hired into any of these positions could result in a heightened risk to hospital patients and therefore to the hospital.

Many state statutes limit acquired immune deficiency syndrome (AIDS) testing or screening in pre-employment policies and procedures. California, Maine, Massachusetts, Iowa, and Vermont prohibit and/or curtail the use of AIDS testing as a condition of hire or continued employment. Florida, Texas, and Wisconsin prohibit AIDS testing unless the employer can demonstrate a legitimate bona fide occupational qualification (BFOQ) for doing so. Federal law in the form of the Rehabilitation Act of 1973[17] is the likely statute that potentially could cover the AIDS victim. A definitive decision by the United States Supreme Court about the status of AIDS as a handicap under the Act has not been made.[18] However, Section 9 of the Civil Rights Restoration Act makes it clear that employers are *not* required to hire or retain an individual who has a currently contagious disease and who poses a direct threat to the health and safety of others or cannot perform essential duties of the job.[19] State statutory legislation, and the fact that federal law having to do with AIDS is still developing, certainly could have influenced participant response to this question.

Respondents' knowledge and sensitivity

Respondents were asked a series of questions that attempted to determine their knowledge of and sensitivity to the doctrine of negligent hiring/retention. Their responses are contained in Table 2.

The majority of respondents exhibit sensitivity toward the matter of negligent hiring as shown in their responses to the survey. Approximately 98 percent either strongly agreed or agreed with the statement that an employer has a duty to hire employees responsibly and to oversee their employment responsibly. Approximately 82 percent either strongly agreed or agreed that an employer who knows that applicants are high-risk individuals is liable for their actions if they are employed. Approximately 87 percent of the survey respondents agreed that an employer has an obligation to inquire into the backgrounds of all job applicants. However, when asked to respond to the statement, "Failure to thoroughly check the backgrounds of job applicant places an employer at risk," the total of respondents who either strongly agreed or agreed fell to approximately 77 percent with the neutral group moving up to 17.5 percent.

In contrast, the response rates to questions inquiring about the use of important pre-employment screening policies and practices were not as consistently high. Approximately 19 percent indicated an interest in asking applicants whether they have tested positive for the AIDS virus. Approximately 65 percent would not inquire. On the issue of drug screening all job applicants, 46 percent agreed while approximately 37 percent disagreed.

In response to the checking of job applicants' criminal conviction records, 49 percent agreed that it should be done. The majority, however, were either neutral (34 percent) or disagreed (20.6 percent). This result is surprising considering that quite a number of court jurisdictions have entertained the introduction of past criminal records into evidence, and such evidence has been used to find employers guilty in some negligent hiring cases. Where an employer is charged with negligently hiring or retaining an incompetent or dangerous person, the individual's character is open to review and evidence may be introduced that focuses on the person's reputation and *prior criminal history* (emphasis added).[20] Some courts expect employers to investigate thoroughly the criminal conviction records of job applicants, particularly for jobs that have a "heightened" duty to the public such as a security guard.[21] Employers filling these types of jobs "are required to exercise a greater amount of care to ascertain whether (applicants)...possessed the specific characteristics such as honesty, that were required by the nature of the employment."[22]

The apparent lack of interest in mandating physical exams and drug tests as part of hospitals' pre-employment screening was surprising.

It should be recognized that for some employers, access to a job applicant's criminal conviction records can be impeded if not foreclosed by some state laws. For example, Minnesota fosters policy promoting the rehabilitation of felons and their reentry into the productive workforce. Other states, including Massachusetts and New York, deliberately restrict employer access to criminal conviction information and encourage ex-convicts to apply for jobs without requiring them to reveal their criminal records. Some states specifically prohibit employer use of conviction information unless it is directly related to the specific employment. Maryland prohibits employers from requiring disclosure of criminal records unless job applicants volunteer such information.[23] Participant response to questions dealing with employers' attempts to use criminal conviction information in making employment decisions could have been influenced by state statutory legislation.

DISCUSSION AND PRACTICAL IMPLICATIONS

United States courts of law have shown and are projected to continue to show a willingness to impose liability on employers who do not take effective mea-

TABLE 2

HOSPITAL PERSONNEL ADMINISTRATORS' ATTITUDES TOWARD SELECTED HUMAN RESOURCE MANAGEMENT ISSUES (97 RESPONDENTS)

Statement	Percentage of responses at each level				
	Strongly agree	Agree	Neutral	Disagree	Strongly disagree
1. "Employment at will" is a thing of the past since employers can no longer fire employees when they desire.	7.2	28.9	20.6	30.9	12.4
2. An employer has a duty to hire employees responsibly and to oversee their employment responsibly.	62.9	35.1	1.0	0	1.0
3. If an employer knows that an applicant is a "high risk" individual, the employer is liable for his/her actions if he/she is employed.	22.7	58.7	12.4	6.2	0
4. Health care facilities have a greater responsibility to hire responsibly than do other types of organizations.	19.6	38.2	14.4	24.7	3.1
5. In some cases, employers are liable for the actions of their employees even if those actions are unrelated to their jobs.	7.2	43.3	11.3	35.1	3.1
6. If an employer learns that an employee poses risks to others after he/she has been hired, the employer has a duty to terminate that employee.	7.2	46.4	24.7	19.6	0
7. Employers are prohibited from obtaining vital information in a pre-employment interview because of discrimination legislation.	5.1	22.7	18.6	37.1	16.5
8. Health care facilities should routinely ask applicants if they have tested positive for the AIDS virus.	5.2	13.4	15.4	39.2	25.8
9. Health care institutions should implement drug screening for all applicants.	15.4	31.0	16.5	24.7	12.4
10. An employer has an obligation to inquire into the background of all applicants.	42.3	44.3	8.3	4.1	0
11. If a person has committed a crime in the past, he/she will likely commit a similar crime in the future.	1.0	6.2	41.2	44.3	7.2
12. An employer has a duty to check the criminal records of job applicants if they are accessible.	6.2	38.1	34.0	20.6	0
13. Failure to thoroughly check the backgrounds of job applicants places an employer at risk.	24.7	52.6	17.5	5.2	0
14. Tortious actions committed by an employee off the job should not affect his/her employment status.	2.1	20.6	30.9	44.3	2.1
15. An applicant who has been convicted of DUI should not be employed by a health care facility.	1.0	3.1	18.5	72.2	3.1
16. If an employer learns that an employee poses risks to others after he/she has been hired, the employer has a duty to terminate that employee.	7.2	46.4	24.7	19.6	0

sures to ensure the hiring and retaining of fit, safe, and competent employees. Where employers hire or retain persons who are unfit, dangerous, and/or incompetent; and, subsequently, those persons do harm to a third party; and, where it can be established that the employee's actions were foreseeable, courts under the doctrine of negligent hiring will hold employers accountable. Results of this study indicate that hospital

HRM understand the doctrine of negligent hiring and its implications and risks.

Hospitals' responsibilities

Due to the sensitive nature of their services, hospitals have a heightened legal duty and responsibility to hire and retain safe and competent employees. As such, employers who have more public contact, such as hospitals, are expected to have an employment screening process designed to make independent inquiry into a

---

Due to the sensitive nature of their services, hospitals have a heightened legal duty and responsibility to hire and retain safe and competent employees.

---

job applicant's background to ascertain the person's fitness, to the extent that there is sufficient basis to conclude the person is fit, safe, and competent.[24] Any employment screening policy or procedure should be consistently followed prior to hiring any person with few if any exceptions.[11] Failure to consistently follow selection policies and procedures potentially leads to inaccurate and ineffective results in employment of hospital workers and opens unnecessary risks to members of the public who could instigate negligent hiring charges against the hospital.

Pre-employment testing

Results of this research indicate hospitals follow basic pre-employment procedures including use of job application, job interview, and reference checking. Participants appeared to put less emphasis on the use of routine pre-employment physical examinations including tests to determine illicit drug usage and evidence of contagious diseases, including AIDS. This was one of the more important findings of this research.

More use of pre-employment physical examinations would seem justified to determine any physical conditions present in the applicant that could impair job performance, be a source of potential harm to the public, or negatively impact the hospital's workers' compensation record in the future. It is recommended that the physical examination include a complete blood count (CBC) and X-ray of chest and spine. The physical examination allows physicians to determine the health of the body system in total including congenital disorders and the presence of infections such as mononucleosis, hepatitis, and tuberculosis. Scoliosis is another important condition to be assessed in a pre-employment physical examination. Presumably the examination could be done at the hospital by staff physicians, thereby making it time efficient and cost effective.

In conjunction with the physical exam, a test to determine illicit drug usage is recommended. The test is justified due to the nature of hospital work and services and the potential for public harm by employee drug abusers. The test procedure should be tightly controlled with minimal discretion by those administering the test. This minimizes potential constitutional "privacy" challenges by the job applicant.

The majority of state and federal courts agree that AIDS victims are protected against adverse employment decisions. Though the United States Supreme Court has not heard a case involving an AIDS victim, in *School Board of Nassau County, Florida, et al. v. Arline*[25] it did decide that tuberculosis, like AIDS, is contagious, and is a handicap as described in the Rehabilitation Act of 1973. As mandated by the Act, federal contractors have to make "reasonable accommodations" to handicapped employees. Hospitals receiving federal funds either directly or through secondary activities are considered federal contractors or subcontractors. As long as job applicants with AIDS are able to perform and could be "accommodated" and pose no threat of transmission, employers covered by the Rehabilitation Act of 1973 and many state and local statutes are prohibited from terminating, refusing to hire, or otherwise discriminating against such persons.[18]

Furthermore, legislation titled "Americans With Disability Act"[26] has passed both houses of Congress and is now in a joint House–Senate conference committee. President Bush has indicated that upon both houses of Congress passing a mutually agreed upon bill, he will sign it. This statute will protect AIDS victims who pose no threat of transmission of the disease and the law will require reasonable accommodations on the part of employers in the employment of handicapped persons including AIDS victims. This statute will also cover more potential employers than The Rehabilitation Act, cited earlier, which is limited to government contractors and recipients of federal funding. Consequently, routine pre-employment AIDS-related inquiries and testing appears inappropriate unless, due to the nature of the job, transmission of the disease is a realistic possibility.

Cautions for hospital employers

Equal employment law and rights to personal privacy cannot be ignored in deference to negligent hiring and retention concerns. Hospital employers need to balance their employment practices and procedures whenever possible in order to minimize liability from either area of employment law. A working knowledge of equal employment opportunity (EEO) mandates, and most particularly of exceptions such as business necessity and BFOQs, is required of any hospital administrator involved in employment.

Hospital employers need to gain a working knowledge of individual rights, including the so-called right to privacy and the potential for defamation allegations relative to checking and providing personal and work references. Public disclosure of embarrassing private facts about a person is an invasion of an individual's interest in acquiring, retaining, and enjoying a good reputation. Violation of this interest is defamation in the form of libel or slander.[27(p.157)] Not having a working knowledge of how to properly handle this two-edged sword could result in peril for the hospital and its management.

It should be emphasized that employment in the hospital industry is characterized by a large percentage of technically trained persons, such as doctors, nurses, radiologists, and therapists. In this industry, labor shortages of medically trained personnel exist and this fact may put pressure on hospitals to hire individuals who are appropriately trained and available without undue concern for individuals' background and character. Although existing labor shortages certainly should not exempt hospitals from liability responsibilities, awareness of the labor-shortage problem and its effect upon delivery of good health care in our society is warranted.

• • •

Clearly, this study has its limitations and additional research is needed. It was exploratory in nature and had scant prior research on which to build. However, landmark court cases that have addressed the negligent hiring/retention issue are discussed herein. Second, this study focused on hospitals as employers. Clearly, it is possible that differences may exist from industry to industry. Furthermore, a study focusing on private business employers could look at the effects of size of business relative to proneness to negligent hiring and retention problems. Another interesting question has to do with an employer who has no human resource management staff and has operating management performing the staffing function. An investigation could focus on how prone such employers are to negligent hiring and retention challenges. The relative importance of selection tools and their impact on reducing negligent hiring/retention risks needs to be addressed.

REFERENCES

1. Fenton, J.W., Jr. "The Negligent Hiring and Retention Doctrine." *Nursing Management* 20 (1989): 28–34.
2. Odewahn, C.A., and Webb, D.L. "Negligent Hiring and Discrimination: An Employer's Dilemma?" *Labor Law Journal* 40 (1989): 705–712.
3. *Fleming v. Bronfin*, 80 A. 2d 915, 917 (D.C. 1951).
4. Collins, T. Comment. "Piercing the Doctrine of Corporate Hospital Liability." *San Diego Law Review* 17 (1980): 380–410.
5. *See, e.g.*, The 1964 Civil Rights Act, Pub. L. 88-352; 78 Stat. 241; 42 U.S.C. §§ 1971, 1975 a-d, 2000a et seq. as amended by the Equal Employment Opportunity Act of 1972, Pub. L. 92-261, 86 Stat. 103; 42 U.S.C. § 2000e et seq. This statute outlaws employment discrimination based on race, sex, religion, color, and national origin absent a narrowly interpreted business necessity. Using neutral employment tools (e.g., check of criminal conviction records) that have disparate impact on employment opportunities for those persons in protected classes (e.g., females, blacks, hispanics) could be in violation of the statute. *See Griggs v. Duke Power*, 401 U.S. 424, 3 EPD π 8137 (1971). However, *see also Antonio v. Wards Cove Packing Co.*, 109 S. Ct. 2115, 50 EDP π 39, 921 (1989) which appears to conflict with *Griggs* interpretations.
6. Cases alleging invasions of privacy have been argued using a constitutional argument employing Fourth Amendment illegal searches and seizures guarantees.
7. Defamation (libel or slander) originating in an employment setting in many cases involves an employer who allegedly gives a prospective employer false oral and/or written statements (e.g., reference checks). For a plaintiff to recover in a defamation suit, he must prove:
    a. A statement was made about him to a third party.
    b. The statement is false and/or given to someone who has no need for such statement.
    c. The statement harms plaintiff's reputation.
    See Keeton, W. *Prosser and Keeton on Torts.* 5th ed. Sect. III. St. Paul, Minn.: West Publishing, 1984.
8. Prosser, W. *Handbook of the Law of Torts*, 4th ed. St. Paul, Minn.: West Publishing, 1971.
9. North, J. "The Responsibility of Employers for the Actions of their Employees: The Negligent Hiring Theory of Liability." *Chi-Kent Law Review* 51 (1970): 719–740.
10. *Garcia v. Duffy*, 491 So. 2d. 435, 438-39 (Fla. App. 1986).

11. Fenton, J.W., Jr. "Negligent Hiring/Retention Adds to Human Resources Woes." *Personnel Journal* 69, no. 4 (1990): 62–73.
12. 125 Ga. App. 1, 186 S.E. 2d 307 (1971), aff'd, 229 Ga. 140, 189 S.E. 2d 412 (1972).
13. 685 P. 2d. 1347 (1984).
14. 109 S. Ct. 1402 (1989).
15. 109 S. Ct. 1384 (1989).
16. Bible, J.D. "Update: Employee Urine Testing and the Fourth Amendment." *Labor Law Journal* 49 (1989): 675–691.
17. 29 U.S.C. § § 701-794, P.L. 93-112, 87 Stat. 357 (1973), approved September 26, 1973. P.L. 93-516, 88 Stat. 1617, December 7, 1974, effective February 6, 1975.
18. Cohen, C.F., and Cohen, M.E. "AIDS in the Workplace: Legal Requirements and Organizational Responses." *Labor Law Journal* 40 (1989): 411–418.
19. Johnston, G.W. "Coping with AIDS: Today's Major Workplace Issue." *Labor Law Journal* 49, (1989): 302–306.
20. *Estate of Arrington v. Fields*, 578 S.W. 2d 173 (Tex. App. 1979).
21. *Williams v. Feather Sound, Inc.*, 386 So. 2d 1238 (Fla. Dist. Ct. App. 1980), petition for review denied, 392 So. 2d 1374 (Fla. 1981).
22. *C.K. Security Systems, Inc. v. Hartford Accident & Indemnity Co.*, 137 Ga. App. 159, 223 S.E. 2d 453 (1976).
23. Gregory, D.L. "Reducing the Risk of Negligence in Hiring." *Employee Relations Law Journal* 14 (1988): 31–40.
24. *Evans v. Morsell*, 284 Md. 160, 395 A. 2d 480 (1978).
25. 107 S. Ct. 1123 (1987).
26. House of Representatives Bill H.R.-2273, U.S. Senate Bill S-933.
27. Sovereign, K. *Personnel Law*. 2d ed. Englewood Cliffs, N.J.: Prentice Hall, 1989.

Index

A

Acquired immune deficiency syndrome (AIDS)
    screening, 230-231, 233
Administrators
    analytical abilities of, 22-23
    career opportunities, 28, 33-34
    as caregivers, 23
    coping with failure, 23-24
    dealing with workplace substance abuse, 188
    gender ratio of, 11-12, 28
    good personal qualities of, 20-21
    high-performing work tendencies, 23-25
    information processing by, 20
    as network builder, 21-22
    perceptions of employee assistance programs surveyed, 216-217
    performance appraisal criteria for, 110
    reducing stress for subordinates, 209
    role stress in, 206-209
    unique stresses on, in health care, 205, 208
    See also Managers

B

Behaviorally anchored rating scales, 110-111, 127-131
Boards of trustees. *See* Governing boards

C

Career development
    barriers to women, 12-13
    degree of planning in, 14
    early stages, for women, 13-14
    and family development, 15-16
    gender issues in, 17
    in Japan, 85-86
    opportunities in health care administration, 28, 33-34
    for women managers, 33-34
    women's expectations, 16-17, 33-34
Comparable worth, 38
    hospital study of, 39-44
Compensation
    changes in benefits surveyed, 30
    comparable worth standard of, 38
    dividing incentive program awards, 152, 155, 156, 159, 172-173
    employee incentive programs, 142-143, 151-152, 155, 172-173, 175
    gender issues in, 12, 34, 37, 162, 166
    health care benefits as, 161-163
    management incentive program studied, 143-147
    practices in Japan, 85, 86

severance, 202-203
studies on comparable worth, 38-44
women managers surveyed, 28, 30-33
Competition in health care, administrator's response to, 24-25
Cost of care
hospital employee health care benefits as percentage of, 162
in productivity formula, 50
programs to reduce, 51-53
and third party reimbursement system, 161
utilization review procedures, 150-151
Culture, workplace
assessing, 60-62, 64
and employee behavior, 60, 63-64, 220-225
in Japan, 84-86
nature of, 59-60
vs. occupational culture, 80

D

Disciplinary procedures, 98
negative implications of, 220-221
for private behavior, 224-226
for role-related misconduct, 222-224

E

Emergency room procedures, 51
Employee assistance programs
effectiveness of, 212-213, 215-217
prevalence in hospitals, 212, 214-215
role of, 211-212
Employment issues
career opportunities in administration, 28, 33-34
cost of health care benefits, 161-163, 168
effects of mergers, 182-186
gender-based wage discrimination, 37-38
legal negligence in hiring, 227-234
See also Job satisfaction; Staff
Equal Employment Opportunity Commission guidelines, 124, 125, 234
Ethical issues, in hospital incentive program, 155-156

F

Focus groups
benefits of, in problem-solving, 48
vs. staff feedback surveys, in problem-solving, 50
Force-field analysis, 8

G

Governing boards
educational development of, 104-105
gender issues in, 16
and hospital staff reduction planning, 199, 202
model for development of, 102
performance assessments of, 103-104
role of, 101
Grievance systems, 97-98

H

Hiring
AIDS screening, 230-231, 233
drug testing in, 188, 191, 230, 233
legal negligence in, defined, 227
manager responsibilities in, 5, 6-7, 64
pre-employment procedures, 228, 231-234
procedures surveyed, 229-231
screening tools, 73, 229-231
use of psychometric instruments in, 73
Hospitals
application of managerial role motivation theory in, 72-74
application of quality circle concepts in, 53, 89-90
corporate culture in, 60-64, 220
cost of employee health care benefits, 162
effects of mergers/acquisitions, 181, 184-186
emergency department procedures, 51
employee health care coverage options, 162-163
and employee incentive programs, 141-143, 171-172
evaluation of jobs in, by gender/wage, 39-44
human resource management in small, 109, 115-120
investor-owned, administrative responsibilities in, 21, 89
legal liabilities, 219, 220, 223, 225, 227-234
management of, in Japan, 87-90
need for behavioral protocols, 219, 223, 225
pay equity, suggestions for achieving, 44-45
performance appraisal study of small rural, 114-120
pharmacy procedures, 52
posting lab reports, improving procedures, 52-53
prelayoff reductions efforts, 198-199
prevalence of employee assistance programs in, 212, 214-215
prevalence of substance abuse in, 194, 212, 215
primary human resource indicators in, 96-99
public, incentive programs in, 143-148, 150, 151-157, 171-172
role of human resource management in, 94-95, 100
special ordering procedures, 51
staff reduction implementation, 202-204
staff reduction planning, 198-202
study of employee health care benefits, 163-168
study of merger, 182-184

utilization review procedures, 150-151
voluntary, administrative responsibilities in, 21
Human resource management
   analyzing personnel data in, 95-96, 98-99
   attitudes toward drug testing surveyed, 189-195, 230, 233
   committee on staff reduction, 199-200
   disciplinary measures, 220-225
   employee assistance programs in, 211, 216-217
   hiring procedures, 227-234
   information needs for, 94-95
   information processing in, 93
   primary indicators in, 96-99
   report formats in, 99-100
   in small rural hospitals, 109, 115-120
   status in hospitals, 94
   See also Employee assistance programs; Hiring; Performance appraisal; Training

I

Incentive programs
   barriers to, 141-142, 171
   data gathering to support, 151
   departmental, 151-157
   employee suggestion programs, 172-176
   ethical issues in, 155-156
   evaluating, 146-148, 152-157, 173-176
   good qualities of, 172
   legal issues, 142-143, 152, 156, 171
   literature review, 142-143, 172-173
   for management, 143-148
   methodology, 143-145, 150-152
   quality of care issues, 155
Information management
   data analysis in, 95-96, 98-99
   of disciplinary actions, 98
   of employee performance indicators, 98
   of grievance indicators, 97-98
   and human resource management, 93-95
   presentation of data, 99-100
   and staffing levels, 198, 200-201
   tracking transfer requests, 97
   in utilization review program, 151
   of withdrawal indicators, 96-97

J

Job satisfaction
   in assessment of corporate culture, 62
   grievance indicators, 97-98
   and performance appraisals, 117
   and role theory, 207
   withdrawal indicators, 96-97

L

Lab reports, timely posting of, 52-53
Layoffs. *See* Staff, reductions
Legal issues
   AIDS screening, 230-231, 233
   in disciplinary procedures, 223, 225
   in drug testing, 188, 192, 193, 230, 233
   employee incentive plans in not-for-profit settings, 142-143, 152, 156, 171
   in employee performance appraisals, 118, 124, 125-126
   gender-based wage discrimination, 38
   in hiring, 227-234
   hospital liability, 219, 225
   personal vs. professional behavior, 220, 222
   physician in quality circle, 87
   in staff reductions, 201

M

Managed care systems, 24
Managers
   attribution of causes of poor performance, 4-5
   benefits of focus-group procedures for, 48
   development program evaluated, 135-140
   educational background surveyed, 28
   empowering, 47-48
   hiring skills, 5, 6-7, 64
   incentive program, in public hospital, 143-148
   increase of women as, 28
   perceptions of employee assistance programs, 212-213
   performance appraisal criteria for, 110, 111
   performance appraisal methods surveyed, 116-118
   problem-solving attitudes surveyed, 76-80
   in problem-solving process, 48-51, 53
   problem-solving skills, 48-50, 53
   and quality circle practices, 53, 87-88, 89-90
   quantifying motivation of, 69-72
   responsibilities during merger/acquisition process, 185
   responsibilities in disciplinary procedures, 221-225
   responsibilities in performance appraisal, 113, 200-201
   responsibilities in staff reduction planning, 199-204
   role in creating culture, 9, 47, 60-61, 63-64, 220-225
   role in quality circle theory, 84
   role motivation theory and, 67-69, 72-74
   self-assessment techniques, 7-8, 120

stress reduction for, 209
support of staff, 5, 7
team-building with, 7-8, 48, 50-51
techniques of Japanese, 87-88
See also Administrators; Supervisors
Maternity benefits surveyed, 30
Mergers/acquisitions
employee perceptions of, 182-184
employee's needs during, 184-186
negotiation process, 182, 184-185
Miner Sentence Completion Scale, 69
Motivation
competency as, 6
employee incentive systems, 141-143, 172
management incentive program, 143
of managers, 67-69, 72-74
and performance appraisal, 113, 125
role motivation theory, 67-68
study of manager's, 69-72

N

Network building, 21-22
Nurses, behaviorally anchored rating scales for, 127, 128-130
Nursing unit
posting of lab reports, 52-53
quality circle concept in, 90
utilization of pharmacy preparations, 52

P

Performance appraisal
behaviorally anchored rating scales in, 110-111, 127-131
by committee, 117-118
compared with other human resource data, 98
criteria for selecting system of, 124-126
critical incident method of, 126
development of, 124
effective systems of, 109-110, 112-113, 118-120
essay method of, 126
forced-choice rating method of, 126
frequency of, 116-117, 118
goals of, 115-116, 123-124
graphic rating scale in, 110, 126
of high-performance teams, 7-8
of hospital governing board, 103-104
job simulation method of, 111
management by objectives method of, 127
paired comparison method of, 110, 127

problems in, 111-112
ranking method of, 110, 127
role in staff reduction planning, 200-201
study of, in small rural hospitals, 114-120
of upper management, 110
Personnel administration. *See* Human resource management
Pharmacy, hospital
in comparable worth study, 40
improving productivity in, 52
Physicians
in financial incentive programs, 147, 150
increasing cost control awareness of, 149
personal vs. professional behavior, and legal liablity, 220-225
in quality circles, 87, 90
role in aggressive utilization review program, 150-151
in staff reduction programs, 201, 203
Preferred provider organizations, 162
Problem-solving
approaches to, 75
focus groups for, 48, 53
gender differences in, 79-80
goals, 48, 53
manager's attitudes surveyed, 76-80
skill-building, 48-50
Productivity
elements of high-performance, 5-6, 23-25
in emergency department, 51
and employee assistance programs, 211, 212
in hospital pharmacy, 52
importance of measuring, 49, 98
in laboratory procedures, 52-53
pressure on managers for increasing, 3, 47
in quality of care ratio, 50, 55
in special ordering procedures, 51
tracking systems, 49-50, 98, 200-201
See also Cost of care; Quality circles concept
Purchasing procedures, standardizing, 51

Q

Quality circles concept
development of, 83-84
implementation of, 53, 89-90
implications for management, 53, 87-89
in Japanese hospitals, 83, 88-90
Quality of care
and employee performance appraisals, 120
monitoring during incentive program, 155

in productivity ratio, 50, 55
unit indicators of, 49-50

R

Role motivation theory, 67-74
Role theory
    role ambiguity in, 206
    role conflict in, 205-206
    in study of health care administrators, 206-209

S

Self-insurance programs, 163
Staff
    attitude, surveys of, 94
    attitudes toward suggestion system surveyed, 173-175
    communication of corporate culture to, 9, 60-61, 63-64, 220-225
    disciplinary actions, 98, 220-226
    drug testing of, 188, 230, 233
    employee suggestion programs, 173-176
    impact of merger on, 182-186
    importance of productivity to, 49
    incentive programs, 141-143, 151-157, 172-173
    lower-level, evaluation methods, 110-111
    management of, in Japan, 84-86
    participation in appraisal instrument design, 124
    participation in problem-solving, vs. focus groups, 50
    reductions
        assessing need for, 198-199
        implementing, 202
        need for flexibility in planning, 199, 201
        planning committee on, 199
        restructuring after, 203-204
        selection criteria, 200-202
    role in quality circle theory, 84
    selection of, 5, 6-7
    support of, 5, 7
    turnover, implications of, 95, 98-99
    See also Employee assistance programs; Hiring; Job satisfaction; Motivation; Performance appraisal
Stress
    on hospital administrators, 205-209
    on nurses, 212
    unique to health care profession, 212, 215
Substance abuse in workplace
    drug testing for, 187, 188-195, 230, 233
    penalties for, 193
    prevalence of, 187-188, 194, 212, 215
Supervisors
    nurse, in comparable worth study, 40
    performance appraisal methods surveyed, 116-118
    problems in performance appraisals by, 111-112
    See also Managers
Surveys/studies
    on comparable worth, 38-44
    effect of hospital merger, 182-186
    evaluating employee assistance programs, 214-217
    hiring procedures, 229-231
    of hospital culture, 61-62
    of hospital employee's health care benefits, 163-168
    incentive programs in public hospital, 143-148, 150-157, 173-176
    managers' attitudes toward drug testing, 189-195, 230, 233
    manager's attitudes toward problem-solving, 76-80
    performance appraisal systems in small rural hospitals, 114-120
    quantifying manager motivation, 69-72
    on role stress in health care administrators, 206
    of women health care managers, 13-17, 27-34

T

Team-building
    for problem solving, 50-51
    steps for managers, 7-8, 48
Training
    economic concepts in, 133-134
    example of cost-benefit evaluation of, 135-140
    for hospital board members, 104-105
    managers in problem-solving, 48-50, 53
    methods of evaluating cost of, 134-135
    in performance appraisal, 111
Transfer requests, interpreting, 97

U

Utilization review
    aggressive, 150-151
    incentive program based on, 144, 151-152

V

Vertical integration of health services, 24

W

Women's issues
   analysis of hospital jobs, by gender/wage, 39-44
   in career development, 12-17
   effect of family on career strategy, 15-16
   gender differences in problem-solving, 79-80
   representation in health care professions, 11-12, 28, 29
   representation in professional organization's leadership, 11, 29-30
   survey of managers, 27-34
   wage discrimination, 37

About the Editor

Montague Brown is a consultant and Director of Strategic Management Services, Inc. a national management consulting firm, with offices in Shawnee Mission, Kansas and Washington, D.C. Brown is Of Counsel with the law firm of Calligaro and Mutryn, Washington, D.C. and is also Editor of *Health Care Management Review*.

Dr. Brown's practice focuses on strategic issues and policies including mergers, vision and strategy, active aging, and building comprehensive vertically integrated systems of health service.

He has served as a hospital trustee and is currently on the boards of the Johnson County Foundation on Aging, the Consumer Health Information and Research Institute, and the Forum for Health Care Planning. He also serves as founding director of the Florida Foundation on Active Aging.

Dr. Brown's work in health care includes service with the New Jersey Hospital Association, Director of the Program and Associate Professor in the W.K. Kellogg School of Management, Northwestern University, and Professor, Department of Health Administration, Duke University. Dr. Brown also serves as adjunct professor, University of Kansas. He has held numerous voluntary positions over the years and has written many books, articles, and reports on issues dealing with the profession of health administration and ways of improving organizational performance.

Dr. Brown holds an AB and a MBA from the University of Chicago, a Doctor of Public Health (Dr PH) and a Juris Doctor (JD) from the University of North Carolina. He has lectured at dozens of universities and participated in international workshops in Canada, Austria, Germany, and England. He continues to lecture, write and participate actively in ongoing research and education programs.